Aristoula Georgiadou
Plutarch's *Pelopidas*

Beiträge zur Altertumskunde

Herausgegeben von
Michael Erler, Ernst Heitsch, Ludwig Koenen,
Reinhold Merkelbach, Clemens Zintzen

Band 105

B. G. Teubner Stuttgart und Leipzig

Plutarch's *Pelopidas*

A Historical and Philological Commentary

Aristoula Georgiadou

B. G. Teubner Stuttgart und Leipzig 1997

Die Deutsche Bibliothek – CIP-Einheitsaufnahme

Georgiadou, Aristoula:
Plutarch's Pelopidas: a historical and philological commentary / Aristoula Georgiadou. – Stuttgart; Leipzig: Teubner, 1997
(Beiträge zur Altertumskunde; Bd. 105)
ISBN 3-519-07654-3

Printed in Germany
Druck und Bindung: Röck, Weinsberg

Preface

This Commentary on the *Life of Pelopidas* originates with my doctoral dissertation of which it is a completely revised and expanded version. My aim has been to present and assess the information available on this *Life* and, where possible, to resolve its outstanding problems. In some cases, however, where I feel that the evidence does not allow for certainty, I follow Wittgenstein's precept "Wovon man nicht sprechen kann, darüber muss man schweigen".

I wish to thank John Buckler of the University of Illinois for his many insightful suggestions and corrections which he provided with unfailing generosity; David Larmour of Texas Tech University, who offered numerous helpful comments and stylistic advice; Gerald Browne and David Sansone of the University of Illinois for their valuable advice, especially with regard to philological matters. I am also grateful to Ludwig Koenen, Christoph Konrad and an anonymous reader, all of whom made many essential corrections and saved me from several infelicities. It is also a pleasure to acknowledge my debt to Angeliki Tzanetou, who always gave most generously of her time and intellect, and to Carol Buckler, who has been a source of thoughtful and provoking conversation over the years. Work on the manuscript was greatly facilitated by the technical assistance of Devin Casenhiser, Mary Ellen Fryer and Walter Spencer.

I would like to express my gratitude to Pennsylvania State University and the Fondation Hardt for financial support, and to the University of Illinois for granting access to its facilities and its remarkable Classics Library, where I was ably assisted by Mardell O'Brien and Bruce Swann.

To Anabella I. Nefele who has become a source of immeasurable happiness. My sweetest debt goes to Stratis Gallopoulos of the University of Patras for all the grants and fellowships he has awarded me throughout the years. It is to him that this book is dedicated.

. . . ἀτὰρ σύ μοί ἐσσι πατὴρ καὶ πότνια μήτηρ
ἠδὲ κασίγνητος, σὺ δέ μοι θαλερὸς παρακοίτης.

Contents

Introduction
1. Plutarch
[a] Life

Plutarch was born c.45 AD in Chaeronea, a small town in Boiotia in Central Greece. The plain near the town was associated with two major battles: the defeat of the Thebans and the Athenians by Philip II in 338 BC, evoked by the biographer in the *Pelopidas* in his digression on the Sacred Band (18.7; cf. *Alex.* 9.2-4, *Dem.* 19-20.2, *Thes.* 27.8, *Cam.* 19.8), and Sulla's victory over the Pontic king Mithridates in 86 BC (*Su.* 15-19). The affluence of Plutarch's distinguished family gave him the opportunity to attain a high degree of learning by pursuing his studies at Athens, which was still a prominent cultural center in the Graeco-Roman world. There he must have attended classes in rhetoric, although he mentions nothing about his rhetorical training, probably because of the traditional contempt the Platonists showed to this discipline.[1] He became the student of the Egyptian Academic philosopher Ammonios c.66/7 AD, and no doubt Plutarch owes to him his allegiance to Platonism and, at least in part, his own profound interest in religious matters.[2] Plutarch travelled widely in Greece, Egypt, Italy, and Rome on

[1] On Plutarch and rhetoric, see R. Jeuckens, *Plutarch von Chaeronea und die Rhetorik* (Strassburg 1907), F. Krauss, *Die Rhetorischen Schriften Plutarchs und ihre Stellung im plutarchischen Schriftenkorpus*, Diss. Munich 1912, Ziegler, 1964: cols. 291-301, J.R. Hamilton, 1969: xxi-xxiii, G.W. Bowersock, *Greek Sophists in the Roman Empire* (Oxford 1969) 110-112, C.P. Jones, 1971: 14-16, 67-71, Wardman, 1974: 221-244, G.W.M. Harrison, "Rhetoric, Writing and Plutarch", *Ancient Society* 18 (1987) 271-279, Stadter, 1987: 251-269, esp. 251-252, Flacelière and Irigoin, 1987: ccvi-ccx.

[2] On Plutarch's philosophy and religion, see R.M. Jones, 1916 [1980], Latzarus, 1920 [1987], Soury, 1942, Flacelière, 1959: 197-215, Barrow, 1969: 72-118, Ziegler, 1964: cols. 301-308, Babut, 1969, id. 1994: 77-102, Russell, 1973: 63-83, Dillon, 1977: 184-230, Brenk, 1977, id. 1987: 248-349, Donini, 1986b: 203-226, U. Bianchi, "Plutarch und der Dualismus", *ANRW* 2.36.1 (1987) 350-365, Froidefond, 1987: 184-233, Gallo, 1988, Valgiglio, 1988, Hershbell, 1992a: 3336-3352, id. 1992b: 3353-3383, Valdés, 1994, Opsomer, 1996: 165-194. On Ammonios, see C.P. Jones, 1966b: 205-213, id. 1971: 16-19, Dillon, 1977: 189-192, Donini, 1986a: 97-110, Brenk, 1987: 257-260, Aalders and De Blois, 1992: 3384-3404.

political missions but also as a lecturer,[3] and yet, notwithstanding the constraints that the limited library resources in Chaeronea imposed on his work (*Dem.* 2.2, *De E ap. Delph.* 384E), he remained faithful to his native town. He became an honorary Athenian citizen and was enrolled in the tribe Leontis (*Quaest. conv.* 628A); he also received Roman citizenship probably through his friend L. Mestrius Florus. Among the various offices which he held (J.R. Hamilton, 1969: xv-xvii, C.P. Jones, 1971: 25-28), especially significant was his long tenure as priest at the temple of Apollo at Delphi, an office which appears to have had a considerable impact on his literary activity (cf. *De E Delph., De Pyth. orac., De def. orac.*).[4] In fact, both "Delphi and Chaeronea were to be", in C.P. Jones' words "the twin poles of Plutarch's adult life" (1971: 4; for the various stages of Plutarch's career, *ib.* 13-38).

[b] The Biographies

The bulk of Plutarch's literary output was most likely produced in the later period of his career, after 96 AD, during the reigns of the emperors Nerva, Trajan and Hadrian.[5] Previously his travels would have allowed him the opportunity to collect abundant material for his diverse writings. Later, far less burdened with public duties and travelling, he had more freedom to devote his leisure time to the composition of essays and dialogues (*Moralia*)[6]—which cover a wide range of topics from philosophy to religion, and from ethics to rhetoric and mathematics—and the *Parallel Lives.*[7] As a prominent citizen of the Roman empire enjoying a wide circle of influential Greek and Roman friends (C.P. Jones, 1971: 39-64,

---

3 On Plutarch's travels, see Barrow, 1969: 36-50, C.P. Jones, 1971: 15-16, 20-27, Buckler, 1992: 4799-4829.

4 On Plutarch's priesthood at Delphi, see Soury, 1941: 50-69, Flacelière, 1943: 72-111, C.P. Jones, 1971: 26, Brenk, 1987: 330-336.

5 On the dating of Plutarch's works, see Stoltz, 1929, Ziegler, 1964: cols. 71-82, 262-268, C.P. Jones, 1966a [1995] 61-74, id. 1971: 28-38, 135-137, JR. Hamilton, 1969: xxxiv-xxxvii, Pelling, 1979 [1995] 80-81, Brenk, 1987: 255-256.

6 For a survey of his essays, see Russell, 1973: 63-99, C.P. Jones, 1971: 67-71, 110-121, Flacelière and Irigoin, 1987: vii-cccii.

7 For general introductions to Plutarch's *Parallel Lives*, see Wilamowitz, 1926 [1967, 1995] 247-279, Flacelière, 1957: xix-xxvi, Ziegler, 1964: cols. 257-277, Russell, 1966 [1995] 139-154, id. 1973: 101-116, Gossage, 1967: 45-77, Barrow, 1969: 51-65, C.P. Jones, 1971: 81-109, Wardman, 1974, Schneeweiss, 1985, Barigazzi, 1994: 287-301, and most recently, Scardigli, 1995: 1-46.

Ziegler, 1964: cols. 30-60, Puech, 1992: 4831-4893), he was in a particularly favorable position to witness and express through his literary work the merging of the Greek and Roman cultures. The arrangement, for instance, of most of these *Lives* in pairs,—the biography of a Greek statesman or general followed by that of his Roman counterpart—, together with the concluding formal comparison (*synkrisis*) of the two protagonists' various qualities that usually accompanies each pair, typifies the fusion of cultures.[8] His essays are also affected by this amalgamatory tendency (cf. *Praec. ger. reip.*, *Aet. Rom. et Gr.*).

Plutarch's extant biographical works include twenty-two sets of *Parallel Lives*—if we count the set of the biographies of *Tiberius and Gaius Gracchus/Agis IV and Cleomenes III* as one pair—, two single *Lives* of *Aratos* and *Artaxerxes II*, and the *Lives of Galba* and *Otho*, which belong, in all likelihood, to a series of *Lives of the Caesars*. The *Lives of the Ten Orators*, a valuable source of information on the Attic orators from Antiphon to Deinarchos, is generally considered spurious. A list of some 227 titles of Plutarch's works—not all of which are genuine—is provided by the so-called 'Lamprias Catalogue'.[9] This inventory is included in some of the medieval manuscripts of Plutarch's writings and immediately invites comparison with his extant works. Clearly time and the vicissitudes of fortune appear to have favored the survival of his biographies and were less kind to his treatises. Sixteen out of sixty-six *Lives* are lost, while only seventy-eight treatises survive out of one hundred and eighty-six.

Plutarch's imperial biographies running from Augustus to Vitellius (Nos. 26, 27, 29-33 in the 'Lamprias Catalogue') appear—if we can judge on the basis of those of Galba and Otho, which alone exist—to have formed a series of interdependent annalistic narrations; the two *Lives* in question, despite their sketchy character, present some interesting similarities with the *Parallel Lives*, so far as their overall structure and the attitude of the biographer are concerned: the programmatic statement at the beginning of the *Galba* with the characteristic disclaimer of narrating events in a formal historical fashion, the synkritistic technique,

---

[8] For a discussion of Plutarch's moral and political purposes in the *Lives*, see C.P. Jones, 1971: 103-109, Valgiglio, 1992: 4026-4028, 4046-4051.

[9] On the 'Lamprias Catalogue', see Ziegler, 1964: cols. 60-66, Barrow, 1967: 193-194, Sandbach, 1969: 3-7, Irigoin, 1987: ccxxviii-ccxxix, ccciii-cccxviii.

the subject-matter and their pervading moralizing tone clearly reflect some of the typical features of the *Parallel Lives*.[10] The composition of the *Lives of the Caesars* is assigned to the period before Domitian's death in 96 AD (Jones, 1966a: 71, id. 1971: 72-73) or to the brief reign of Nerva (96-98AD; Syme, 1980: 104-128, esp. 108-110, Geiger, 1981: 86-87).

Plutarch owes his reputation and his popularity chiefly to his *Parallel Lives*. The idea of drawing parallels certainly does not originate with Plutarch but is tightly linked with his rhetorical training and wide-ranging reading. It is a distinct structural component of Valerius Maximus' collection of *exempla*, which Plutarch must have known and used.[11] In addition, the concept of confronting and comparing outstanding public figures in Greece with those in Rome, as well as the general framework and arrangement of the biographies, may owe a great deal to Nepos' series in his *de viris illustribus*. In fact, Plutarch's acquaintance with his writings is confirmed by the use he made of this work in some of the Roman biographies (*Marc*. 30.5, *synkr. Pel./Marc*. 31(1).8, *Luc*. 43.2, *TiGr*. 21.3; on his use of Nepos, see Scardigli, 1995: 21). Nepos' book on Foreign Generals, which consists of twenty-two *Lives*, indicates at the end that it was to be followed by a book on Roman Generals (*Hann*. 13.4). It is likely that Nepos' comparisons of Greek and Roman generals were on the basis of groups, and not of individuals as favored by Plutarch, and that the former's book on the Greek subjects would precede its Roman counterpart.[12] Plutarch's originality seems to lie, as Geiger says (1988: 249-250), in the scale of his biographies—book-length pairs of *Lives* as opposed to Nepos' brief sketches—which allowed more space to character-development. We should note that the method of setting Greeks and Romans side by side was practised also by

[10] For an introduction to the *Lives of the Caesars*, see C.P. Jones, 1971: 72-80, Syme, 1980: 104-128; for a study of features common to this series and the *Parallel Lives*, see Georgiadou, 1988: 349-356; on the proems of the *Caesars*, see Stadter, 1988: 294-295.

[11] See Gossage, 1967: 60, Russell, 1973: 109. Cicero's works, most of which Plutarch appears to have known, could have inspired him for some of his comparisons; see further Scardigli, 1995: 20 and n. 140.

[12] See Gossage, 1967: 75n. 48, Geiger, 1981 [1995] 96. For a survey of Plutarch's indebtedness to the Peripatos for his comparative technique and to Roman models for his contrasts, see Scardigli, 1995: 7-12.

a near contemporary of Nepos, the polymath Varro, in his *Hebdomades* or *Imagines*. This was an illustrated collection of 700 portraits of prominent men of Greece and Rome each followed by an appended epigram. Plutarch shows familiarity with Varro's work (Helmbold and O'Neil, 1959: 74) and it is not unlikely that the polymath's layout of portraits in the *Imagines* may have played some part in the literary format of Plutarch's *Parallel Lives*.

Of the 24 Greek heroes of the *Parallel Lives*—including the lost *Epameinondas*—, ten are furnished by the history of Athens—Theseus, Solon, Themistokles, Aristides, Kimon, Perikles, Nikias, Alkibiades, Demosthenes, Phokion—, five are Spartans—Lykourgos, Lysander, Agesilaos, Agis IV and Kleomenes III—, two are Thebans—Epameinondas and Pelopidas—, and, as for the two liberators of Syracuse, Timoleon and Dion, the former is Corinthian and the latter Syracusan; the remaining heroes of the series are derived, with the exception of the Macedonian King Alexander, from the Hellenistic period, and include the liberators of Syracuse Dion and Timoleon, the Diadochi Eumenes of Cardia, Demetrios Poliorketes, and Pyrrhos of Epiros, and, finally, 'the last of the Greeks', Philopoemen of Megalopolis.

The bulk of the subjects of Plutarch's Greek *Lives* of the Classical period are also included in Nepos' list of heroes, and it is therefore tempting to see here an indication of Nepos' impact on Plutarch's choice of at least some of his Greek characters.[13] Plutarch's Roman subjects—all 24 possibly including Scipio Aemilianus (see [**1c**]) are exclusively selected from the early history of Rome down to the end of the Republic.[14] A few fragments which have been securely assigned to Nepos' *Lives* of Marcellus, Lucullus and Scipio Aemilianus, and were most likely included in his book on Roman generals, suggest that Nepos' influence on Plutarch may have extended beyond the latter's selection of the Greek subjects, since Plutarch himself also composed biographies of these individuals (Geiger, 1981 [1995] 92-99, id. 1985: 104-106).

---

[13] For recent general Introductions and bibliographical surveys of Plutarch's Greek *Lives*, see B. Bucher-Isler, 1972, Scardigli, 1986: 7-41, ead. 1995: 2-18, 32-46, Podlecki and Duane, 1992: 4053-4127.

[14] For recent surveys of Plutarch's Roman *Lives*, see Scardigli, 1986: 42-59, ead. 1995: 18-24, 32-46, Titchener, 1992: 4128-4153.

Not all the *Lives* appear to have been part of Plutarch's original plan, but were added to the series progressively. On his own admission, he expanded the series because of the delight he derived from it (*Aem.* 1.1). Analyzing the frequency of Plutarch's references to subjects from the Classical and Hellenistic Age in the *Moralia*—those from the classical period far exceeding the Hellenistic—Geiger has proposed that the biographies of Greeks from the latter period were probably added at a later stage.[15]

[c] The Lost *Epameinondas/Scipio* Pair

A most regrettable loss in the entire corpus of Plutarch is the *Epameinondas/Scipio Pair*, which appears to have been the opening set of biographies in the series of the *Parallel Lives*—No. 7 in the 'Lamprias Catalogue'. Because of its leading position among the *Lives* it might have contained a programmatic statement applying to the entire series; as such, it would give us a deeper insight into Plutarch's biographical scheme and the motivations and expectations that lay behind it which we must now piece together from the scattered comments which we find in the opening chapters of a number of *Lives* (*Alex.* 1.2, *Kim.* 2.4-5, *Tim.* 1.5, *Per.* 1-2.4, *Nik.* 1.5, *Demetr.* 1.3-6). Moreover, given Plutarch's numerous references to Epameinondas in the extant *Lives* and the *Moralia*, where he makes no effort to hide his unbounded admiration for the Theban's combination of political activity and interest in philosophy, one might expect to trace in the *Epameinondas*, more than in the *Pelopidas*, the degree to which the biographer's local patriotism and admiration for his Boiotian compatriot might have interfered with the selection and shaping of the available source-material; we could have, therefore, obtained some control of Plutarch's choice of material and his use of sources in the composition of the Theban biographies. And, last but not least, the *Epameinondas* could have shed more light on the attitudes and viewpoints of his principal authorities whose extensive writings on the period in question have unfortunately not survived. In its absence, our understanding and assessment of the above features is confined almost exclusively to the extant *Pelopidas*. Of course, to assume that material lost to us could have contained all the information we desire and that the bio-

---

[15] On Nepos' possible impact on Plutarch in the selection of his characters, see Geiger, 1981 [1995] 92, 95-99, id. 1985: 104-108, id. 1988: 245-256, Valgiglio, 1992: 4028-4030.

grapher would have given us a deeper insight into his scheme than what he offers at *Alex.* 1.2 may be an illusion.

Various attempts to reconstruct the *Epameinondas* through other ancient sources which supposedly drew on Plutarch's lost biography, have yielded interesting, yet somewhat speculative results.[16] Wilamowitz, followed by several other scholars, was the first to advance the view that Pausanias 9.13.1-15.6 represents an epitome of the lost *Life*. This has been contested by Tuplin who stresses the difficulties in establishing a direct connection between the two writers.[17]

Finally, the identification of Epameinondas' Roman counterpart, Scipio, has been a matter of an ongoing dispute, and has divided scholarly opinion between P. Cornelius Scipio Africanus Maior (236-184 BC) and P. Cornelius Scipio Aemilianus Numantinus (185/4-129 BC), who is known simply as Scipio Minor. The 'Lamprias Catalogue' lists a *Life of Scipio Africanus* (No. 28) separately from the *Epameinondas/Scipio* Pair (No. 7). Peper (1912: 129-131) and Ziegler (1964: cols. 258-259) favored the selection of the elder Africanus as Epameinondas' Roman counterpart in the lost Pair; Ziegler based his arguments on quotations from certain of the *Lives* and on two incidents, which he considered to be appropriate points of contact between them, the battles of Leuktra and Zama, and the trials in which both heroes were involved (cf. also Sandbach, *Loeb Moralia*, vol. 15, 1969: 74-78, Scardigli, 1986: 20). Peper and Ziegler are countered by Wilamowitz (1926: 260) and Herbert (1957: 83-88; cf. Stadter, 1989: xxviii n. 12), who, comparing the credentials of the two Scipios and also those of Epameinondas and Scipio Aemilianus arrived at the conclusion that Aemilianus' 'scholar-statesman' type of personality would be the more likely choice for the Roman hero in the first pair of the *Parallel Lives*. Geiger (1981: 87n. 6) points out that the *Life* known in the 'Lamprias Catalogue' (No. 28) as *Scipio Africanus*—i.e, not the parallel to Epameinondas—must have been written prior to the inception of the *Parallel Lives*, "since it is inconceivable

16 For systematic reconstructions of Epameinondas' career, see L. Pomtow, *Das Leben des Epaminondas, sein Charakter und seine Politik* (Berlin 1870), Vischer, 1877, Swoboda, 1900, Peper, 1912, Carrata Thomes, 1952, Fortina, 1958, Shrimpton, 1970, Bengtson, 1983.

17 Tuplin, 1984: 346-358; for a detailed bibliography on this issue, see *ib.* 346n. 1. On Pausanias' dependence on local Boiotian sources in this section, see Fuscagni, 1975: 45-46 and n. 53, Sordi, 1989: 125-126.

that no room would have been found in the *Parallel Lives* for both Scipios". The question thus is not which Scipio could best be paired with the great Theban, but which of the two could have triggered the philosopher-biographer's interest independently, and before the *Parallel Lives*. The same arguments adduced by the supporters of Scipio Aemilianus—the kind of 'scholar-statesman' Plutarch so much appreciates in the person of Epameinondas—are then better applied to the case for an individual *Life*; the Scipio paired with Epameinondas must therefore, in all probability, have been Africanus, not Aemilianus.

[d] Plutarch and Thebes[18]

It is only natural to assume that Plutarch's admiration for the architects of the Theban hegemony, Pelopidas and Epameinondas, would also extend to the Thebans in general. Certain restricting factors are, however, at play here. As has been observed, the loss of the *Epameinondas* greatly limits the material from which to form a coherent picture of Plutarch's views about the Theban people. In the *Pelopidas*, his occasional remarks about the Thebans do not, in any way, betray any excessive bias towards them comparable to Xenophon's transparent pro-Spartan fervor. Plutarch applauds them for their military training and prowess (15.4-7, 17.13, 18.7, 19.3-5), the confidence they inspired in their allies in times of danger (24.5-8), their sense of justice (26.5), their hatred for tyranny and the panhellenic spirit of their external policy (31.6). Interestingly enough, his laudatory comments on the Thebans' valor are not made in isolation from the role played by either Pelopidas, or Epameinondas, or both. In fact, they are meant to be understood merely as the result of the two Thebans' military and political virtues.

But Plutarch is not uncritical of the Theban people. He speaks disapprovingly of their unpleasant reception of Pelopidas and Epameinondas, after the two men's first expedition to the Peloponnese, and of their envy, which was increasing as the two grew in reputation (25.1). He also refers, in a rather disparaging manner (29.1), to their failed mission to liberate Pelopidas from the hands of the tyrant Alexander. Moreover, he would have known Ephoros' criticism of the Thebans' lack of culture and disregard for learning which the historian considered as the main

---

[18] I thank Prof. J. Buckler for having generously shared with me his extensive notes on the views of Plutarch, Xenophon and Diodoros regarding Thebes from 382 to 362; a short version of these notes is included in Appendix 2, 1980a: 263-272.

causes of the Thebans' loss of hegemony in Greece (Diod. 15.39.2, 88.4, Str. 9.2.2). Only Epameinondas, according to Ephoros, stood out among the Thebans as a leader who combined both intellectual and military virtues and Plutarch's lavish remarks about the Theban's philosophical pursuits seem to bolster this view. He clearly resented the fact that the Theban people suffered under ill-repute and were criticized for their meager achievement in the intellectual sphere,[19] and that Pelopidas was not exempt from such a criticism; rather than tarnish the latter's image, however, he tactfully refrains from expounding his views on this issue and shifts the emphasis to the other virtues and qualities of the Thebans. Only once does he draw attention to their superstitious belief in demons who take delight in human sacrificies, thus allowing the reader to detect their lack of intellectual sophistication and rationalization, but even then he does so obliquely and does not elaborate on the matter (21.2-3).

2. The *Life of Pelopidas*

[a] Scope and Methods

The *Pelopidas*, parallel to the *Marcellus*, is Plutarch's only extant Theban biography after the loss of the *Life of Epameinondas* [**1c**]. This *Life*, despite the limited scope and purpose expected from a work of biographical nature, constitutes one of the three major literary sources for the history of the Theban hegemony; the other two are Xenophon's *Hellenika* (6.3.1-7.5.27) and Diodoros' *Bibliotheke* (15.50.4-15.88.4), on the scope and worth of which see Buckler, 1980a: 263-272.

Plutarch's *Pelopidas* is also particularly valuable for the historian because it preserves the most extensive account of the career of one of the principal architects of the Theban hegemony.[20] It fills some of the gaps left by Xenophon in his *Hellenika*, whose extreme bias in favor of Sparta and unsympathetic attitude towards Thebes pervade his entire account of the events in this period. In his reluctance to acknowledge the two Thebans' achievements, and especially those of Pelopidas, he blatantly

[19] For a full discussion of the literary tradition on the proverbial Boiotian μισολογία, see Roberts (1895); for Plutarch's polemic against the opprobrious epithets applied to the Thebans for their lack of intellectual sophistication, see *De gen.* 575D, *De Herod. malign.* 864 D.

[20] Wilamowitz (1926 [1995] 60n. 22) rather unfairly attributes Pelopidas' inclusion in the corpus of Plutarch's *Lives* merely to his Boiotian origin and claims that the Theban does not really fit the company. For earlier discussions of the *Life of Pelopidas*, see Vater, 1842, Queck, 1876, Bersanetti, 1949: 43-101.

suppresses events that were humiliating to the Spartans or readily omits victories of the Thebans.[21] He omits all mention of Pelopidas and his past military exploits until 367, when the Theban was sent as an ambassador to the Persian court (7.1.33-37). It is clearly no coincidence that he picked this incident to introduce Pelopidas, since despite the latter's diplomatic success at Sousa the Thebans did not succeed, Xenophon remarks, in convincing the other Greeks to accept the peace or gain the leadership in Greece (7.1.40).

Diodoros' description of the events in central and northern Greece from 371 to 362 (Bk. XV) is the most complete account we possess from this period. He lacks Xenophon's biases and is eager to allot praise both to Spartan and to Theban generals. Owing to the wide scope of his work he is forced to treat most events in a sketchy manner conducive to chronological confusion. In his account of the Theban hegemony he concentrates largely on the achievements of Epameinondas, whom he lauds above all other men. Despite the lack of depth in his work, Diodoros is especially useful because he supplements Xenophon's *Hellenika* and gives due credit to the achievements of Pelopidas and Epameinondas.

By his own admission, Plutarch does not seek to provide a comprehensive account of various historical periods (*Alex.* 1.2), but is mainly concerned with the character of men who affected the course of events in their own lifetime. Consequently, not all important events of the period in question are included in the *Life*, while certain incidents which are more closely connected with Epameinondas than with Pelopidas will doubtless have been treated in the lost *Epameinondas* (cf. *Ages.* 28.6). In addition, a number of episodes in Theban history which took place in the last decade of Agesilaos' reign in Sparta and brought about the decline of the Spartan hegemony in Greece, such as the Battle of Leuktra, Epameinondas' first invasion of the Peloponnese and the refounding of Messene, are dealt with from a different angle in the *Agesilaos* (28-29.1, 31-32, 34.1-2) and therefore complement the accounts given in the *Pelopidas*.

---

[21] On Xenophon's silences and distortions, see Cawkwell, 1972: 255-257, id. 1973: 56-57, Westlake, 1975: 23-40, Buckler, 1980a: 263-268, Fernández, 1984: 41-70, Proietti, 1987, Lévy, 1990: 125-157, Riedinger, 1991: 41-60, Dillery, 1995: 195-230; for a more moderate approach, see Higgins, 1977: 174-175, Gray, 1989: 69, Wickersham, 1994: 86-90.

To determine and illustrate Pelopidas' character Plutarch sketches the notable deeds in which he faced danger and adversity and achieved prosperity and glory. He also frequently uses anecdotes and sayings for the same moralizing purpose. He is willing even to suppress certain incidents or aspects of Pelopidas' character that would suggest a different evaluation of his ethos [**2d**]. In brief, he appears at times to be shaping his narrative so as to throw into relief only those aspects of his hero's career which will best reconfirm the virtues outlined in the proemium or in the formal *synkrisis*. Failing to look for larger historical causes and motives in Pelopidas' acts, he simply sees and interprets them as a result of his virtue. And yet, despite his tendency to present Pelopidas in as favorable a light as possible, Plutarch provides a valuable supplement and corrective to Xenophon's account and, as Buckler points out (1980a: 272), gives "the fullest account of the careers of Epameinondas and Pelopidas".

Local patriotism[22] and forthright admiration for Epameinondas' education and, in particular, his philosophical training must have prompted Plutarch to give him a leading position in his biographical series. Pelopidas, on the other hand, lacked his friend's proper education and philosophical inclinations (4.1), but is greatly commended by Plutarch for his harmonious cooperation with Epameinondas throughout his career (4.2-4). Plutarch is fully aware of the precarious position of the Greek magistrates in his own time, who, subject to the authority of the Roman governors, are often embroiled in unnecessary quarrelling; in the *Praecepta gerendae reipublicae*, a political treatise composed roughly during the period of the composition of the *Parallel Lives* (C.P. Jones, 1966a [1995] 72), he gives practical advice to future politicians. Although he draws his Greek examples chiefly from the politicians of the 5th and 4th centuries BC,[23] his suggestions are made with a view to his own time (cf. 805A-B, 814C-815A, D, 824E-825A; see also Pelling, 1995: 205-220, esp. 215-217). He strongly disapproves of the disgrace-

[22] A number of non-extant biographies of mythological and literary subjects who are Boiotian in origin, such as Herakles, Hesiod, Pindar and Krates, Nos. 34, 35, 36, 37 in the 'Lamprias Catalogue', respectively, and the extant treatise *De Herodoti malignitate* attest to Plutarch's desire to bring Boiotia to the fore; on Plutarch's local chauvinism, see Aalders, 1982: 14-15.

[23] On Plutarch's tendency to use models of the past rather than of his own period, see Wardman, 1974: 20-21.

ful rivalry among the Greeks of the ruling class and urges them to strive for friendship and concord, which he regards as the most elevated form of the art of politics (824C-D).[24] In the *Pelopidas*, the theme of concord or envy among politicians has a central place in the Introduction (ch. 4); it informs the entire *Life* and is also used to stress continuity with the *Marcellus* [**2d**].[25] The political terminology is the same used in the *Praecepta*, and Plutarch's examples are, likewise, not of contemporary political leaders, but of statesmen from classical Athens who were given to enmity and envy. He juxtaposes them to the two Thebans, who emerge as the ideal model of collaborative leadership (4.3). The *Pelopidas* is as close to a political treatise as a Plutarchean biography can be. Though it does not contain any explicit political messages or guidance for the preservation of unity and concord in the ruling class whose spokesman he is, the entire career of Pelopidas, inextricably bound up for the most part with that of Epameinondas', is one of the best examples illustrating Plutarch's political lessons in the *Praecepta*.

Plutarch has very little to say about Pelopidas' early life. The lack of personal details about his family, his childhood and upbringing, private life, or physical appearance in the introductory chapters,[26] in conjunction with the unusual number of *chreiai* in the proemium (1-2.8) and the significant amount of space devoted to digressions (16.5-8, 18-19, 34) and to unnecessary amplification (1-2.9, 8.7-8, 21-22, 28.5-10, 29.6-11, 30.9-12, 35.6-12), may indicate that Plutarch did not make use of biographical sources, but had to rely on historians who would have little or nothing to say about Pelopidas' early life. We cannot be certain that Plutarch knew of a *Life* of Pelopidas—other than that of Nepos—that

---

[24] On Plutarch's concern for the political issues of his time, see C.P. Jones, 1971: 110-121, Aalders, 1982: 48-60, Caiazza, 1993: 7-23. He is criticized for failing to weigh his judgements of characters against the moral norms and expectations of the societies in which they lived: Bucher-Isler, 1972: 73-74, A. Wallace-Hadrill, *Suetonius: The Scholar and his Caesars* (New Haven, London 1983) 108; for a more moderate opinion, see Pelling, 1995: 205-220, esp. 211-213.

[25] Other themes providing a link between the two *Lives* are: the tyranny (Alexander of Pherai, Dionysios of Sicily), the political envy (*Pel.* 25.1, *Marc.* 22.1), the trials (*Pel.* 25.3-5, *Marc.* 23, 27), superstitious beliefs and omens (*Pel.* 21, 31.4, *Marc.* 3.6-7, 5.5-7, 6.11-12, 20.9, 28.3), people buried alive (*Pel.* 29.6, *Marc.* 3.6).

[26] Cf. for instance the *Lives* of Fabius Maximus, Camillus, Coriolanus, Crassus, Antony, as well as Pelling's comments on the occasional absence of biographical sources (1979: 85, id. 1988b: 30).

had also been used by the Roman biographer, whom he names as a source in the *Marcellus* (*Marc.* 30.5, *synkr. Pel./Marc.* (31(1).8), since Nepos himself points to the historical and not biographical nature of his sources in his *Pelopidas* **[2f]**. It has been argued, however, probably with too much strictness, that no political biographies were composed before Nepos, but only *Lives* of philosophers, poets and other intellectuals,[27] and probably Pelopidas, whose career was, to a great extent, military, would not qualify for such a biography. There is no evidence of such a work in the period between Nepos and Plutarch. We only know of a certain Xenophon of Athens who is credited by Diogenes Laertios (2.59 [*FGrH* 111T1]) with a biography of Epameinondas and Pelopidas, but nothing is known about the content or dating of this work other than its description as a βίος. The absence of the biographical flavor from the *Pelopidas* and the occasional loose linking of episodes make it read like a historical narrative whose author is at pains to tailor it to the needs of a biography.[28]

Plutarch, like Nepos, had apparently a difficult time collecting material pertaining exclusively to Pelopidas, since his career was interwined with that of Epameinondas. Pelopidas is presented as acting on his own, without being overshadowed by the genius of his Theban friend, only on four occasions: in the preparation of the coup against the Theban oligarchs which led eventually to the liberation of the Kadmeia (5-13), at the battle of Tegyra (16-17), in the expeditions to Thessaly against the tyrant Alexander of Pherai (26-28, 31-32) and in the embassy to Sousa (30). In all four incidents Plutarch goes out of his way to stress either the personal nature of Pelopidas' success or Epameinondas' dissociation from his friend's exploits (5.4, 13.7, 16.1, 26.1, 30.1-2, 13). The abundance of passages of a moralizing character in the *Pelopidas* is re-

---

[27] See Geiger, 1985: 30-116, criticized by Moles, 1989: 229-233, and Scardigli, 1995: 17-18.

[28] Plutarch's Greek *Lives* have been criticized as occupying "a fluid and intermediate position in the history of biography", because they "approximate more closely to the work of a historian than to the more straightforward biographical forms": Scardigli, 1995: 17; see further, Valdés, "Approximación a la *Vida de Cimon* de Plutarco", in *Unidad y Pluralidad en el mundo antiguo* (Madrid 1981) 317-324, esp. 320, A. Dihle, *Die Entstehung der historischen Biographie*, Sitzungsberichte der Heidelberger Akademie der Wissenschaften, Philosophisch-Historische Klasse, 1986, vol. 3 (Heidelberg 1987) 19, A. La Penna, *Plutarco: vite parallele, Cesare* (Milan 1987) 28.

markable, and although Plutarch generally tends to incorporate material in the biographies which lends itself to moralistic judgments, his reliance on such tendentious material in the Theban biography seems to have been dictated mostly by the inadequacy of his sources.

[b] Chronology

Plutarch is hardly ever in the habit of referring explicitly to the order of the biographies and the dates at which he composed and published them (cf. *Su.* 21.8). As a result, the reconstruction of their relative chronology is based on cross-references,[29]—although they appear occasionally to contradict each other—, and on other indirect indications in the *Lives*. Once the general period of the composition of the *Lives* has been established [**1b**], it remains to determine the approximative sequence of publication. The *Pelopidas/Marcellus* is, of course, not the only Pair which has endured various permutations within the general scheme of the *Parallel Lives*. Mewaldt (1907: 564-578, esp. 575), in his sequential arrangement of Plutarch's first four books, places it second immediately after the missing Pair *Epameinondas/Scipio*, to which he assigns the first place in the biographical series, with Solon/Publicola and Kimon/Lucullus coming third and fourth, respectively. His proposed order differs radically from Michaelis', who maintains that the *Pelopidas/Marcellus* constituted pair seven.[30] Theander (1958: 18-19), being in agreement with Michaelis and Stoltz (1929: 46-47), argues that the *Pelopidas/Marcellus,* which he assigns to ninth place, was written about the same time as the *Perikles/Fabius Maximus*, on the grounds that the two Roman generals were coeval. Theander's argument has been countered by Van der Valk (1982: 302n. 8) and Ziegler (1964: col. 265), who firmly support the chronological proximity of the two Theban *Lives* on the grounds of their related subject-matter. C.P. Jones, whose suggested order of the *Parallel Lives* has won a fairly wide acceptance, proposes a few modifications on Mewaldt's chronological arrangement and

---

[29] On the reliability of the cross-references as a criterion for establishing the sequence of the *Lives*, see Mewaldt, 1907; Stoltz, 1929, disagrees, as do C.P. Jones, 1966a [1995] 66-68, J.R. Hamilton, 1969: xxxiv-xxxvii and Pelling, 1979 [1995] 80-82; on the use of citations as a means of determining the sequence of the *Lives* and the limitations of this methodology, see Frazier, 1988: 297-309.

[30] Michaelis, 1875: 20-47: (1) Epameinondas/Scipio, (2) Sertorius/Eumenes, (3) Kimon/Lucullus, (4) Lysander/Sulla, (5) Demosthenes/Sulla, (6) Agis-Kleomenes/Gracchi, (7) Pelopidas/Marcellus.

offers the second, third and fourth places to *Kimon/Lucullus*, *Pelopidas/Marcellus*, and either *Sertorius/Eumenes* or *Philopoemen/ Flamininus*; he leaves the specific order of each pair undetermined.[31] Using the double reference to the completed *Life of Marcellus* in the *Fabius Maximus* (19.2, 22.8), C.P. Jones demonstrates that the composition of the *Pelopidas/Marcellus* was undoubtedly prior to that of the *Perikles/Fabius Maximus*. Finally, Van der Valk, in his defense of the proximity of the two Theban *Lives* draws attention to Plutarch's methodology in writing the biographies (1982: 302); he claims that the biographer was clearly more interested in the two Thebans than in Marcellus or Fabius Maximus, and so the order of the first two pairs, at least, must have been dictated by his personal taste; he suggests, therefore, that Plutarch followed "a definite line of thought" while composing his *Lives*, according to which he could pass more naturally from the writing of the *Life of Epameinondas* directly to that of the other great Theban, Pelopidas. It is reasonable to suppose that the two Theban *Lives* were composed within a short time of each other, since they would involve research and source material of a very similar type. The assumption that they were both written at the same time has, of course, no real bearing on their respective dates of publication.

[c] Sources for the *Pelopidas*[32]

Kallisthenes and Ephoros

Plutarch is noticeably reluctant to identify his sources for the *Pelopidas*,[33] while he is clearly more generous in that respect in the Theban's Roman counterpart, the *Marcellus*, where he cites by name seven different authorities. Only once do we get a glimpse of the kind of material which he appears to have consulted in the composition of the *Life*: in 17.4 he mentions the historians Ephoros, Kallisthenes and

---

[31] C.P. Jones, 1966a [1995] 68; see also J.R. Hamilton, 1969: xxxiv-xxxvii, and Stadter, 1989: xxvii-xxix, who agree, in general, with Jones' scheme. For a brief survey of the relative chronology of the *Lives*, see Podlecki and Duane, 1992: 4064-4065.

[32] For a more extensive discussion of Plutarch's possible use of Kallisthenes and Ephoros in the *Pelopidas*, see Georgiadou, in *Boeotia Antiqua*.

[33] This practice is not uncommon; cf. the *Coriolanus* and the *Sertorius*; on this issue, see C.P. Jones, 1971: 84-85, P. Desideri, "I documenti di Plutarco", *ANRW* 2.33.6 (1992) 4538-4539.

Polybios for their diverging views on the size of the Spartan *mora*.[34] Of the three named authorities only the first two are believed to have dealt systematically with the period of the Theban hegemony. In fact, Ephoros and Kallisthenes were very likely the most authoritative concerning Thebes in the late fourth century,[35] and it is reasonable to suppose that either historian, or both, could have provided the basis of Plutarch's narrative in the *Pelopidas*, and also probably in the lost *Epameinondas*. Xenophon was among Plutarch's sources for another fourth century *Life*, the *Agesilaos*, but there is no indication that the biographer drew on him for the *Pelopidas*, since the historian systematically avoided associating the Theban hero with any of the major events of the Theban hegemony [**2a**]. Nothing is known about the work of the Boiotian historians Dionysodoros and Anaxis referred to once by Diodoros as having concluded their narrative of Greek history in 361/0 (15.95.4).

Ephoros of Kyme, a fourth century historian, wrote a *Universal History* in 29 books starting with the Return of the Heracleidae and ending with the siege of Perinthos in 341. It is generally agreed that Books XX-XXV covered the period of the collapse of the Spartan hegemony, which started with the Spartan seizure of Mantinea in 385, and included the gradual increase of the power of Thebes, the Theban invasions of the Peloponnese and ended with Epameinondas' death at Mantinea in 362.[36] The fact that Ephoros dedicated one fifth of his entire work to record events which span a limited period of twenty-three years should be of no surprise to us, since he was dealing at that point with events of his own lifetime. Apart from the extant fragments, much of the information concerning Ephoros' *Universal History* comes from the epitome of it made by Diodoros. The latter, wrote in the first century BC and composed his *Bibliotheke*, a world history, in forty books. He compiled Book XV of his work almost entirely from Ephoros' account of the Theban

---

[34] On his tendency to cite sources when he introduces variant views, see Ziegler, 1964: col. 274.

[35] Shrimpton argues (1970: 28-46, id. 1971: 316-317) that a Boiotian point of view first emerged and gained ground with the publication of Ephoros' *Histories* and Kallisthenes' *Hellenika*; see also Alonso-Núnez (1990: 173-192) for the development of the notion of universal historiography in the fourth century BC.

[36] On Ephoros, see *FGrH* 70, Schwartz, *RE* 6.1, 1907: cols. 1-16, R. Laqueur, "Ephoros", *Hermes* 46 (1911) 161-206 and 321-354, Barber, 1935: 36-37, Wickersham, 1994: 119-177.

supremacy,[37] thus covering in one book, by way of abridgement, the years from the King's Peace in 387/6 to 361/0 (Barber, 1935: 34-38).

Kallisthenes of Olynthos, historian and philosopher (*Alex.* 52.3) of the fourth century BC, wrote his *Hellenika* in ten books starting with the Peace of Antalkidas in 387/6 and ending with Philomelos' seizure of the temple at Delphi in 356. Polybios finds a place for Kallisthenes in the canon of the λογιώτατοι τῶν ἀρχαίων συγγραφέων together with Xenophon, Ephoros and Plato (6.45).[38]

Plutarch's familiarity with the works of Ephoros and Kallisthenes is well attested, and to judge from the way in which he occasionally compares the one with the other, it is obvious that he had read them both.[39] Throughout the *Lives* and the *Moralia* he cites Ephoros at least fifteen times as his source[40] and Kallisthenes at least fourteen times, if we exclude all the other instances in which he refers to them indirectly. The frequency of his references to them does not, of course, prove that he consulted either of these two historians directly in the *Life*, nor does it

[37] On Diodoros' extensive borrowing from Ephoros for Books XI-XVI, see C.A. Volquardsen, *Untersuchungen über die Quellen der griechischen und sizilischen Geschichten bei Diodor, Buch XI bis XVI* (Kiel 1868), Hammond, 1937: 84-85, Herbert, 1954: 37-38, 41-42, Drews, 1962: 390, C.I. Reid, *Diodorus and his Sources*, Ph.D. diss. Harvard University 1969, Sordi, 1969: xii, Buckler, 1980a: 268, Andrewes, 1985: 189, Sacks, 1990: 13, Munn, 1993: 133; for a more reserved approach, see Frost, 1980: 32, M. Casevitz, *Diodore de Sicile: Bibliothèque Historique, Livre XII* (Paris 1972) xiii-xv.

[38] On Kallisthenes, see *FGrH* 124, Schwartz, 1900: 106-130, Kroll, *RE* 10, 1917: cols. 1674-1726, A. Schaefer, 1930: 21-33, Pearson, 1960: 22-49, Shrimpton, 1970: 35-46, Pédech, 1984: 15-69. How widely known Kallisthenes was is illustrated by a fragment published by G. Manganaro, "Una biblioteca storica nel ginnasio di Tauromenion", *La Parola del Passato* 29 (1974) 389-409, esp. 391 and 394.

[39] I see no reason to doubt that a widely-travelled and widely-read person like Plutarch should not have read these authors and others in the original himself. Plutarch's ample use of Ephoros in *Mul. virt.* has been demonstrated by Stadter: 1965: 37-38, 59-68, 127-128; cf. also Theander, 1951: 62-63.

[40] As an indication of Plutarch's familiarity with Ephoros, note his joke about a friend who had read some Ephoros and was always holding forth about Leuktra, and so got the nickname "Epameinondas" (*De garr.* 514C [*FGrH* 70F213]). Clearly reading Ephoros meant learning a lot about Epameinondas, and no doubt about Pelopidas as well.

imply that he preferred Ephoros to Kallisthenes.[41] A parallel examination of the *Pelopidas* and the corresponding sections in Diodoros' Bk. XV reveals significant similarities, but also some noteworthy differences between the two narratives. It would be tempting to attribute the existing coincidences to Plutarch's systematic consultation of Ephoros, were it not for certain discrepancies, mainly of an argumentative nature, which, as Westlake observes (1939: 17), are "too frequent and too wide to admit of this explanation".

The task of determining Plutarch's sources in the *Pelopidas* becomes more complicated both because of the loss of the *Epameinondas* [**1c**] and because Ephoros may well have used Kallisthenes especially for the last three decades of his history.[42] This suggestion, now commonly accepted by scholars, raises considerable problems for a satisfactory assessment of either historian's work on its own merits. In view of the complexity of the process of transmission from the original source to Diodoros, and in the absence of any extensive fragments from Kallisthenes' *Hellenika*, or from Ephoros' history other than Diodoros' abridgement of it, it is difficult, if not impossible, to reconstruct with any certainty their attitude towards Thebes and its protagonists during the period of history in question,[43] or to trace distinguishing features of their historical narrative in the *Life*;[44] likewise, to decide which of the compressed sections in Diodoros'

---

[41] On Plutarch's use of sources, see J.R. Hamilton, 1969: xliii-xlix, Pelling, 1988b: 31-36, Stadter, 1989: xliv-li, Desideri, 1992: 4536-4567, Buckler, 1995: 43-58.

[42] Eusebius quotes a statement by Porphyry in which Ephoros is accused of copying from Kallisthenes, Daimachos and Anaximenes (Porph. ap. Euseb. *Praep. Ev*. 10.3, 464 B [*FGrH* 70T17]); on Ephoros' plagiarism, see Barber, 1935: 131-133, Momigliano, 1935a: 110-111, Westlake, 1939: 18, Carrata Thomes, 1952: 10-11, Herbert, 1954: 107. Very little is known of the work of Daimachos and Anaximenes; cf. Diod. 15.89.3 [*FGrH* 72T14], Aristeid. 26.51 (vol. II, 105 Keil) [*FGrH* 72F21, *FGrH* 65F1-4].

[43] On Ephoros' attitude towards the Thebans, see Accame, 1936: 343-355, Wickersham, 1994: 158-160.

[44] There is, in fact, more evidence of Ephoros' attitude towards the Theban hegemony and its protagonists, especially towards Epameinondas, than of Kallisthenes'; see Momigliano, 1935a: 114-117, id., 1935b: 192-199, Accame, 1936: 343-344. Besides Diodoros' epitome, Ephoros' views are partly reflected in Nepos' *Epameinondas* (Bradley, 1968: 138-140, 142, 146-147) and *Pelopidas* (id. 151-152), in Strabo 9.2.2, in Justin 6.7-8; on Kallisthenes' pro-Theban attitude, see Kroll, *RE* 10, 1917: cols. 1696-7; cf. Pédech (1984: 28-30), who objects to this view.

Bk XV is genuinely Ephorean, and which has been contaminated by borrowings from Kallisthenes or may even reflect Diodoros' own views remains to a large extent an object of speculation.[45] If Shrimpton (1970: 36-39) is right in maintaining that Kallisthenes centered his account of the Theban hegemony around the figure of Pelopidas, and Ephoros around Epameinondas, we could assign Diodoros' specific references to Pelopidas' exploits to the Ephoros/Kallisthenes tradition, and regard the rest of the account of the Theban hegemony as genuinely deriving from Ephoros. Yet, although this hypothesis can adequately explain certain narrative inconsistencies in Diodoros' account, it cannot be used as the only criterion for discerning Kallisthenes' traces in Ephoros' work. To avoid further complication and confusion, Diodoros' narrative of the events in Greece in Bk. XV will be generally assumed for our purposes to represent the Ephorean tradition, unless striking discrepancies between passages in his account strongly argue for a change of source.[46] Moreover, besides the errors which usually occur in incorporating a detailed exposition, we should allow for some creativity in Diodoros' abridgement, and not regard his work as a simplistic copy of Ephoros' history.[47] Similarly, it would be wrong to view Plutarch's *Pelopidas* as a servile imitation and reproduction of any of his alleged sources. As Scardigli points out (1995: 26 andn. 182), he was "more than a mere collector of secondary material".

In spite of the lack of conclusive evidence, scholars generally agree that Kallisthenes, and not Ephoros,[48] must have been Plutarch's principal

---

[45] On Diodoros' sources, see Reid, 1969.

[46] For instance, the detailed presentation and emphasis in the section containing the eulogy of Pelopidas (15.81.1-4) are quite different from the rest of the narrative and can argue for a change of source. Unfortunately, Diodoros shows no awareness of Ephoros' borrowings from Kallisthenes, nor does he provide variant views of the two historians.

[47] On Diodoros' use of his sources, see Hammond, 1937: 79-91, Drews, 1962: 383-392, id. 1963: 244-255, Andrewes, 1985: 189-197, Gray, 1987: 72-89, Wickersham, 1994: 165-173.

[48] Ephoros as Plutarch's main or additional source has been favored by: Queck (1876: 22-25; he finds traces of Ephoros in chs. 26-35), A.H.L. Heeren (*De fontibus et auctoritate vitarum parallelarum Plutarchi commentationes quattuor* [Göttingen 1820] 50; he suggests Ephoros, who in turn draws from Anaxis and Dionysodoros), G.R. Sievers (*Geschichte Griechenlands vom Ende des peloponnesischen Krieges bis zur Schlacht bei Mantineia* [Kiel 1840] 171-172n. 20), Haug (1854: 58-60; he remains undecided among Ephoros, Kallisthenes, Anaxis and Dionysodoros), Christ

source.[49] Although the identification of Kallisthenes with the historian followed by the biographer cannot be proven with any certitude, the meager fragments of Kallisthenes which pertain to the period of history in question and a relatively consistent variance in the reports of Ephoros/Diodoros and Plutarch do appear to weigh more heavily in favor of Kallisthenes.[50] Nevertheless, given the limitations of such an assumption, Plutarch's dependence on this historian should not be overestimated (cf. Buck, 1994: 67), nor should the *Life* be used as the model in order to reconstruct Kallisthenes' own account of the events of the Theban hegemony.

A number of episodes in the *Pelopidas*, discussed at a greater length in the appropriate sections of the Commentary (see also Georgiadou, in *Boeotia Antiqua)*, have been shown to bear the undeniable imprint of Kallisthenes. Without detracting from the merit of certain arguments advanced in support of this view and given the biographer's creativity and personal judgement in the depiction of his protagonists,[51] we should point out that the weight of the evidence strongly favors a more sceptical approach to Plutarch's admittedly extensive indebtedness to the historian.

(1901: 67-74; he considers Xenophon as Plutarch's principal source, and Ephoros as his secondary), Westlake (1939: 17-18; he acknowledges certain striking similarities between Plutarch and Ephoros/Diodoros [Bk. XV] and is inclined to attribute them to their common source, Kallisthenes), Accame (1936: 351; he sees an Ephorean tradition in chs. 6-13), and Flacelière and Chambry (1966: 127 propose Ephoros as a main source for the entire *Life*).

[49] Kallisthenes as Plutarch's main or only source has been proposed by: Haug (1854: 59), Queck (1876: 28-30; for chs. 5-25), Stern (1884: 249, id. 1887: 51-67; he assumes that almost the entire *Life* is drawn from Kallisthenes, who, in turn, used the Boiotian historians Anaxis and Dionysodoros), Schwartz, (1900: 106-130), Bock (1910: 62), Peper (1912: 33-39; Kallisthenes used in conjunction with Ephoros), A. Schaefer (1930: 21-33), Momigliano (1934: 196), Westlake (1939: 18-21), Bersanetti (1949: 47n. 3), Herbert (1954: 103-110; he does not deny the occasional use of Ephoros), Ziegler (1964: col. 205), Shrimpton (1970: 42; Kallisthenes is supplemented by local sources and traditions), Fuscagni (1975: 31-55), Pédech (1984: 32-34), Prandi (1985: 70-74) and Sordi (1989: 123).

[50] Note also Plutarch's use of Kallisthenes as a source with respect to Epameinondas' second invasion of the Peloponnese in *Ages*. 34.4.

[51] It is argued, for instance, that the unfavorable representation of Nikias should not be attributed so much to Plutarch's sources as to his personal evaluation of the Athenian's character; cf. Nicolaides, 1988: 319-320, L. Piccirilli, "Nicia in Filisto e in Timeo", *RFIC* 118 (1990) 389-390.

a) The portrayal of Sphodrias (*Pel.* 14.3 [*FGrH* 124F9], Diod. 15.29.5-7)

Among the meager extant fragments of Kallisthenes' *Hellenika* fragment 9 is the only one which allows us to establish a concrete connection between the historian and the biographer. Kallisthenes' characterization of Sphodrias betrays a demonstrable affinity with Plutarch's description of the same character. The historian's comment on Sphodrias' character is probably associated with the latter's raid on the Peiraeus, and it may be reasonably assumed that Plutarch depended on him for the entire episode (14.2-6)[52] and not on Ephoros.

b) The Oracle of Apollo at Tegyra (*Pel.* 16.5 [*FGrH* 124F11]) and the Spartan *mora* (*Pel.* 17.4 [*FGrH* 124F18], Diod. 15.37.1-2, 81.2)[53]

Fragment 11 of Kallisthenes mentions Tegyra as the birthplace of Apollo and also refers to the oracle of the god in this location. It is probable that Kallisthenes' discussion of the mythological connections of the place was occasioned by his account of the battle of Tegyra. Since both Plutarch and Kallisthenes associate Tegyra with the birth of Apollo, it has been conjectured that the biographer's digression on the oracle of Apollo Tegyraeus in *Pelopidas* which precedes the account of the battle of Tegyra derives from Bk. III of Kallisthenes' *Hellenika*. Nevertheless, as a priest at the temple of Delphi Plutarch doubtless had first-hand information about Apollo's oracle at Tegyra—where he had probably been (Buckler, 1992: 4804-4805)—and had therefore no reason to resort to Kallisthenes' *Hellenika*; cf. Theander (1951: 29-30) and A. Schaefer's reservations (1930: 21; see also note on 16.5-8) about Plutarch's dependence on Kallisthenes here. Furthermore, Plutarch's unique reference to the authority of Kallisthenes with respect to the strength of the Spartan *mora* (17.4) has been used as additional evidence for the biographer's reliance on the historian for the entire episode (chs. 16-17). Yet, there is no special reason to suppose that Plutarch drew exclusively from Kallisthenes, since he also includes the variant opinions of Ephoros and Polybios on the *mora*; nor is there any hint in Plutarch's account from which to infer his indebtedness to Kallisthenes rather than to Ephoros.

---

[52] Queck, 1876: 28, Stern, 1887: 60-61, Bock, 1910: 61, Fuscagni, 1975: 36, Prandi, 1985: 37-39, 72, Buckler, 1980b: 75-93, esp. 75-76, Pédech, 1984: 34-35.

[53] My thanks go to Prof. P. Harvey for his valuable comments on this passage.

On the other hand, a parallel examination of Diodoros' remarks on Pelopidas' exploit (15.37.1-2, 81.2) and those of Plutarch (16.1, 17.11) shows that the latter may have made use of Ephoros' rhetoric and language in his eulogy of Pelopidas. The possibility, therefore, that Plutarch resorted to Ephoros for the Tegyra episode should be seriously considered (see also note on 17.4). Furthermore, Polybios should not be discounted entirely as a possible source for this incident, since he may have dealt with the battle of Tegyra by way of digression.

c) The portents at Leuktra (*Ages.* 28.6, *De Pyth. orac.* 397E-F) and the battle of Leuktra (*Pel.* 23, Diod. 15.54-56, Xen. *Hell.* 6.4.8-15)

Kallisthenes has generally been acknowledged as Plutarch's source for the description of the battle of Leuktra[54] mainly on the basis of the exclusion of Ephoros and Xenophon[55] as the other two possible sources known to us. This assumption has also found partial support in a number of portents which, according to Plutarch, occurred before the battle (*Ages.* 28.6, *De Pyth. orac.* 397E-F). It has been demonstrated that some of these omens go back to the authority of Kallisthenes, but Plutarch makes no mention of any portent in the *Pelopidas*. Although his familiarity with Kallisthenes' list of omens has been established, it does not constitute in itself conclusive evidence of his reliance on this historian for the entire episode at Leuktra.

d) The panhellenic ideology of Kallisthenes' *Hellenika* and Plutarch's *Pelopidas*: the liberation of Messene (*Pel.* 24.9, 30.7, *Ages.* 34.1, Diod. 15.66.1) and Pelopidas' campaigns in Thessaly and Macedonia (*Pel.* 26-27, 31-32, Diod. 15.61.2-5, 66.1, 67.3-4, 71.2-7, 75.1, 80)

Scholars have sought to demonstrate that Kallisthenes in his *Hellenika* combined a panhellenic ideology with the glorification of Theban hegemony, and that Philip functioned in the historian's work as the continuator of the panhellenic aims of Pelopidas and Epameinondas;[56] they

[54] Bock, 1910: 61-62, Fuscagni, 1975: 37, Buckler, 1980b: 75-93, esp. 76; for Plutarch's use of Kallisthenes in accounts of other battles, see *FGrH* 124F15 and 124 F37.

[55] On Plutarch's independence of Xenophon with respect to Leuktra, see Buckler, 1980b: 75-93, esp. 76, 86-89.

[56] Momigliano, 1934: 196, Sordi, 1958: 193-202, Fuscagni, 1975: 31-55, Prandi, 1985: 70-74. For a discussion of Panhellenism as a political concept serving

have, therefore, drawn attention to Plutarch's special emphasis on the liberation and refounding of Messene and the political significance of Pelopidas' campaigns to Thessaly and Macedonia. Plutarch attaches, indeed, a panhellenic tone to all these episodes and may have found in Kallisthenes the proper source of inspiration in order to invest the two Thebans with panhellenic motives for their exploits. This hypothesis, however, is far from furnishing any firm grounds on which to base Plutarch's dependence on Kallisthenes for Pelopidas' campaigns to Thessaly and Macedonia. Moreover, to attribute all the omissions, emphases and biases in the *Pelopidas* to Kallisthenes' political agenda in the *Hellenika* does, at best, an injustice to Plutarch's individual literary methods, moralistic purposes and the complexity of his biographical technique [**2d**, **e**].

Plutarch and Diodoros, on the other hand, agree very closely when they are dealing with Pelopidas' campaigns to Thessaly and Macedonia. Nevertheless, certain discrepancies between their accounts, which have readily been explained away by the two writers' different ideologies, have prompted some scholars to dismiss the presence of an Ephorean tradition in Plutarch's narrative.

On the whole, it would be cautious to say that, given our very limited knowledge of Kallisthenes' treatment of the Theban hegemony and the scarcity of the extant fragments from his *Hellenika*, we cannot evaluate to what degree Plutarch's historical portions of the *Life* were inspired by or modelled after the *Hellenika*; nor, for that matter, can we assess how far Plutarch went in adapting and interpreting his material in the interests of his Boiotian compatriot and at the expense of his Roman counterpart [**2d**]. Furthermore, there is no convincing reason to believe that Plutarch used Kallisthenes so extensively that he excluded Ephoros, nor should Ephoros' account, as it is preserved to us by Diodoros, be seen as deriving solely from Kallisthenes. As Herbert notes (1954: 110) in his discussion of Ephoros as a potential source in the *Pelopidas* "it would be rash to eliminate entirely from consideration one [i.e. Ephoros] who devoted a considerable part of his work, probably four complete books, to the Theban hegemony".

---

to advocate and promote the interests of the polis in foreign policy, as well as in internal politics, see S. Perlman, "Panhellenism, the Polis and Imperialism", *Historia* 25 (1976) 1-30.

Finally, to return to the initial statement about Plutarch's tantalizing silence concerning his sources, there remains the possibility that despite the different role that Kallisthenes and Ephoros are assumed to have assigned to the two protagonists of the Theban hegemony, they may, in fact, have exhibited only insubstantial differences in their accounts of the period in question. Consequently, the biographer, being familiar with both works, but relying to a great extent on his memory for the composition of the *Life*, might have found it difficult to say whose version he presented and have chosen instead to refrain from naming either of the two. If this hypothesis holds true, we may have here an indication of Ephoros' extensive reliance on Kallisthenes for the events of this period.

Theopompos and Xenophon

Theopompos is usually dismissed as Plutarch's primary source for the *Pelopidas*, since he is believed to have avoided the period of the Theban hegemony; his *Hellenika* finishes with the battle of Knidos in 394, "having approached the times of Leuktra", as Polybios puts it, and his *Philippika* begins with Philip's accession to the throne in 360 (Polyb. 8.11.3 [*FGrH* 115T19.3]). He appears, however, to have dealt, by way of digression,[57] with some incidents of the Theban hegemony:[58] Plutarch names him as a source in his account of Epameinondas' first invasion of the Peloponnese in 370/369 (*Ages.* 31.4 [*FGrH* 115F322], *Ages.* 32.14-33.1 [*FGrH* 115F323]).[59] He is also known to have treated at some length the story of Alexander of Pherai (Polyb. 38.6.2 [*FGrH* T29]) and there is adequate evidence in support of this claim; in *Non posse* 1093C (*FGrH* 115F337), for instance, Plutarch delights in Theopompos' account of Thebe (cf. also *FGrH* 115F352). Plutarch's extensive description of Alexander's brutality (27.6, 29.5-10), the conversations between Pelopidas and Thebe (28.3-10), and the tyrant's

---

57 See Stern, 1887: 51, Shrimpton, 1970: 26, id. 1991: 40-41, 51, Flower, 1994: 30.

58 Flower suggests (1994: 30 and n. 22) that Theopompos may have even narrated in digressions in the *Philippika* most of the events which took place between 394 and 359.

59 On this incident, see Shrimpton, 1991: 50-51; on Plutarch's use of Theopompos, see Theander, 1951: 58, Stadter, 1989: lxxii-lxxiii.

murder at the hands of Thebe's brothers (35.4-12),[60] all of which abound in dramatic detail, may have derived from a digression on Alexander in Theopompos' *Philippika*.[61] Fragment 409[62]—author uncertain—echoes Plutarch's sharp criticism of Alexander's iniquity when the latter imprisonned Pelopidas (*Pel.* 27.7). The authenticity of this fragment would be strengthened if it could be demonstrated that Plutarch's moral condemnation of the tyrant was not his own, but derived from Theopompos (Westlake, 1939: 15n. 1). Furthermore, the biographer's low opinion of Philip's moral virtues at 26.8 may have been inspired by Theopompos (see also note on 26.7-8). Although he started his *Philippika* with 360, and Philip's sojourn at Thebes was in the year of Pelopidas' first intervention in Thessaly (369), he must have had various opportunities in his work to include information about the period during which Philip II was a hostage in Thebes.[63]

It has been recently suggested by Sordi[64] that Plutarch's and Nepos' accounts of the occupation of the Kadmeia by the Spartan Phoebidas and its liberation derive independently from the same source, which is not Kallisthenes or Ephoros,[65] but Theopompos. This claim is based mainly on certain features of the description of this incident common to both biographers, such as their emphasis on the valor and audacity of the Theban conspirators (Plut. *Pel.* 13.5, Nep. *Pel.* 3.1) and the element of paradox and fortune in the entire enterprise (Plut. 13.4, Nep. 2.3-4),

---

60 Alexander's murder is also recounted by Xenophon (*Hell.* 6.4.35-37), but bears very few similarities with Plutarch's account; cf. Westlake (1935: 159), who claims that Xenophon's account is almost identical with Plutarch's.

61 On Theopompos' lengthy digressions, see Dion. Hal. *Ad Pomp.* 6 [*FGrH* 115T20], Theon, *Prog.* 4 [*FGrH* 115T30], Phot. *Bibl.* 176 [*FGrH* 115T31]; Plutarch's possible indebtedness to Theopompos in the *Pelopidas* is discussed by Westlake, 1935: 159, id. 1939: 14-15.

62 See Suid. s.vv. παρ' οὐδὲν θέμενος: "having trampled on the treaty and counted his oaths as nothing, Alexander imprisoned Pelopidas and put him under guard", transl. Shrimpton, 1991: 274.

63 For further discussion, see note on 26.7-8. On Theopompos' treatment of Philip, see Connor, 1967: 33-154, Flower, 1994: 98-115.

64 See Sordi, 1995: 415-423, whom I wish to thank for showing me a draft of this article before its publication.

65 On Plutarch's use of Xenophon as a *Hauptquelle* for chs. 5-13, see Christ, 1901: 67-70, refuted by Bock, 1910: 62-68, Schäfer, 1930: 13 and Hani, 1980: 41-42. Accame (1936: 348-351) maintains that Plutarch used here a romantic elaboration of Ephoros' account written by some Hellenistic author, but also consulted Xenophon.

which are not to be found in other major historical sources. Given Theopompos' keen interest in tales of the marvellous, it has been maintained that the account upon which both Plutarch and Nepos drew would have probably belonged to a digression in Bk. VIII of Theopompos' *Philippika* part of which was dedicated to the Θαυμάσια (*FGrH* 115F64-76). Sordi's suggestion, although intriguing, is unconvincing. Plutarch, despite his familiarity with Theopompos' work, is more likely to have used major sources like Ephoros or Kallisthenes to describe this significant chapter in Theban history than the Chian historian. Moreover, Sordi suggests that Theopompos, known and used independently by both biographers, would also have incorporated parts of Xenophon's account of the same incident (*Hell.* 5.2.25-36, 4.1-12). Plutarch's use of Xenophon as his principal source for the *Pelopidas* has found minimal support (cf. Christ, 1901: 67-74, and above s.v. "Kallisthenes and Ephoros").

Polybios

Plutarch introduces the *Pelopidas* with a surprisingly long list of *exempla* on the multifaceted notion of bravery, its excesses and effects. The names of Cato the Elder, Antigonos, Iphikrates, Kallikratidas and Timotheos are listed in connection with the theme of bravery in battle, soon to be followed by the two subjects, Pelopidas and Marcellus, whose inconsiderate rashness cost them their lives (chs. 1-2). Such accumulation of instances does not appear to be Plutarch's own invention, but will have formed the basis of whole compilations such as the one by Valerius Maximus. Polybios (8.35) also discusses individuals who lost their lives or put themselves in danger, such as Ti. Sempronius Gracchus (cos. 215, 213; pro cos. 212), the Spartan king Archidamos, the Theban Pelopidas, and Cn. Cornelius Scipio Asina (cos. 260, 254). In a manner similar to Plutarch's he expresses his approval or disapproval of each type of death on the basis of the particular circumstances that existed at the time of the battle. He distinguishes between the unavoidable type of death, as typified by Tiberius (cf. Plut. *Pel.* 2.7), and that caused by recklessness and lack of consideration, exemplified by the other three leaders (cf. *Pel.* 2.8, 11, *synkr. Pel./Marc.* 33(3).1 and 8). Polybios disapproves entirely of the latter kind of death (8.36.1) and elaborates on Pelopidas' rashness (8.35.6-8). In the same vein, Plutarch denounces the two heroes' death, but tends to be less critical of Pelopidas than of Marcellus [**2d**]. Polybios' list of examples of temerity and his particular interest in Pelopidas may indeed have suggested to Plutarch the idea of prefacing the *Pelopidas*, but also the Pair as a whole, with similar examples;[66] or perhaps we should assume that both Polybios and Plutarch drew for their respective lists on a common source, perhaps Kallisthenes (Sordi, 1958: 196-198). On Plutarch's possible use of Polybios for the Tegyra episode, see [**2c**] sect. b, s.v. "The Oracle of Apollo at Tegyra and the Spartan *mora*".

Phanias and Hegesander

The account of Pelopidas' embassy to Sousa is enriched with anecdotal material which is not likely to have derived from Plutarch's principal source(s). The anecdote about the Athenian Timagoras (30.9-11) Athenaios associates with Timagoras [the Cretan?] or Entimos from Gortyna, and he cites the Peripatetic Phanias[67] (Phaenias) as his source (*Deipn.* 48d-e), while the anecdote about the Athenian Epikrates (30.12) he attributes to Hegesander (251b).

Philistos

Plutarch enhances his discussion of the posthumous honors paid to Pelopidas with examples of funeral rites offered to other statesmen; he cites the Syracusan historian Philistos as his source for the extravagant funeral of Dionysios the Elder (34.1) and leaves unnamed his source for Alexander's expression of immoderate grief at the death of Hephaestion (34.2; cf. *Alex.* 72.3-8).

Autopsy and Oral Tradition

In the composition of his writings Plutarch did not rely solely on the information furnished by the historical, philosophical, poetic, rhetorical and other sources, but supplemented them with material gathered from the inspection of inscriptions, statues, paintings, sanctuaries and battlefields.[68] It is surprising that he makes no mention of the statue of Pe-

---

[66] Plutarch's familiarity with Polybios' work is well attested; see Helmbold and O'Neil, 1959: 63-64; cf. also *Pel.* 17.4.

[67] On Plutarch's use of Phanias as a source for *Themistokles*, see bibliography in Scardigli, 1995: 8; see further, Helmbold and O'Neil, 1959: 54.

[68] On Plutarch's methods of autopsy, see Theander, 1951: 2-32, id. 1959: 99-131, Stadter, 1989: lxxxiii-lxxxv, Buckler, 1992: 4788-4830.

lopidas dedicated at Delphi by the κοινόν of the Thessalians.[69] Even if the statue were destroyed by the Phocians during the Sacred War, he must have known about it, at least through Nepos, who refers to the "bronze statues" offered to Pelopidas after his death (*Pel.* 5.4). Moreover, he has nothing to say about the Theban treasury at Delphi (Paus. 10.11.5) dedicated most likely from the spoils taken at Leuktra (see J.-P. Michaud, *Le Trésor de Thèbes*, in *Fouilles de Delphes*, vol. 2, part 1, fasc. 8 [Paris 1973] esp. 1-6, N.Δ. Παπαχατζῆς, *Βοιωτικὰ καὶ Φωκικά*, vol. 5 [Athens 1991] 336); nor about the Theban τρόπαιον made up of the arms of the fallen Spartans on the field of Leuktra (Diod. 15.81.2), —probably to be identified with that mentioned in an epigram inscribed on the base of a monument dedicated to Zeus at Thebes (*IG* VII 2462; Tod, 1962, No.130 [Harding 1985, No.46]),[70] but to be distinguished from the pre-battle trophy recorded by Pausanias, 4.32.5-6)[71]—and replaced later on with a bronze one, apparently the first such battlefield memorial to a Greek victory over Greeks (Cic. *De invent.* 2.23; cf. Diod. 13.24.5-6, Plut. *Aet. Rom. et Graec.* 273C-D). Likewise, he does not mention anything about the Lion of Chaeronea, which marked the common grave of the Theban Sacred Band, although he has a lengthy digression on the valor and death of the Sacred Band in this location (*Pel.* 18). On the other hand, to judge from the amount of topographical information he gives about the oracle of Apollo of Tegyra (16.5-6), it is very likely that Plutarch inspected the site, as well as the battlefields of Tegyra, Leuktra and Chaeronea, in order to test the accuracy of his literary sources.[72]

---

[69] On the inscription that came from the statue, see Wilhelm, 1941: 35-45, Bousquet, 1939: 125-132, id. 1963: 206-208, Sjôqvist, 1973: 19-33, esp. 10-11.

[70] A reference to the Theban trophy at Leuktra is made in *Non posse* 1098B, but is associated only with Epameinondas.

[71] On the unmentioned τρόπαιον of Leuktra, see Pritchett, 1974: 253, Buckler, 1992: 4801-4802n. 43, Tuplin, 1987: 98n. 87. For a discussion of the epitaph in *IG* VII 2462, see E. Beister, "Ein thebanisches Tropaion bereits vor Beginn der Schlacht bei Leuktra", *Chiron* 3 (1973) 65-84, countered by Tuplin, 1987: 72-107, esp. 94-107.

[72] See González, 1991b: 89-100, Buckler, 1992: 4801-4802 and n. 43, 4804-4805, id. 1995: 43-58.

[d] Comparison: Pelopidas and Marcellus

The study of the *Parallel Lives*, and their source criticism in particular, were dominated in the past by the tacit belief that Plutarch employed a single process of composition within a set biographical framework with restrictive boundaries. Erbse (1956: 398-424, esp. 418; see also Constanza, 1955: 127-156) was the first critic to question the value of such an assumption in his analysis of the *synkriseis*, with which Plutarch ends most of his *Lives*. He demonstrated that these conclusive epilogues are not merely rhetorical devices intended to epitomize the similarities and differences between the subjects studied, but, instead, they are the *konstitutive Prinzipien* of the pairs, and offer valuable information on the biographer's compositional methods in the main narrative of his biographies; furthermore, that the comparative technique is an integral component of his biographical narrative, which carries the tension of his strenuous selectiveness and shaping of the source-material; and that the key themes in a *Life* are often determined by its pair. Recent studies on a number of *Parallel Lives* have, indeed, shown the presence of several guiding principles in Plutarch's biographical methodology: his omission, selection and presentation of the source-material appear to have been influenced, to a certain extent at least, by his desire either to maintain a balance within a pair of *Lives* or to demonstrate the moral and/or military superiority of a hero in a *Life* to his counterpart in the paired *Life*.[73] This is also the case with the delineation of the personalities and careers of the subjects in the *Pelopidas/Marcellus* Pair. The comparative technique therein does not simply underlie the entire narrative, nor is it confined to the two protagonists of the Pair, but extends to other leading characters also, such as Epameinondas and Fabius Maximus, whose presence determines the inclusion or exclusion of material and certain emphases or omissions in these two paired *Lives* [**2e**]. In evaluating the *Pelopidas*, therefore, it is helpful to scrutinize the characterization of the hero and the portrayal of his actions in the parallel *Life of Marcellus*.[74]

---

[73] Stadter, 1975 [1995], Pelling, 1986, id. 1988b: 31-36, Swain, 1988, Larmour, 1988, id. 1992, Bosworth, 1992a, Valgiglio, 1992: 4022-4026.

[74] For a detailed discussion of Plutarch's choice of the protagonists in this pair and his occasional manipulation of his source-material therein, see Georgiadou, 1992a: 4222-4257.

While it is possible, as has been observed earlier **[1b]**, that the choice of Marcellus may have been suggested by Nepos' book on Roman Generals, the idea of pairing Pelopidas with the Roman general belongs, most likely, to Plutarch himself. In the proemium he leads us to believe that the two protagonists were similar in their leading characteristics: "they both were great men, but both fell by their own rashness... they did honor to their respective countries by their conduct in most significant campaigns... but were prodigal of their lives and carelessly threw them away when there was the greatest need of men and commanders such as they" (*Pel.* 2.9-11). Although the above list of common traits could just as easily have been used for some of the other heroes, Plutarch obviously regards them as providing an adequate substratum of characteristics on which to build and compare the two *Lives*; of all these aspects of their ethos, however, it is the negative ones that seem to have served, at first sight at least, as the starting point for the joining of the two heroes, namely their reckless attitude in war and their untimely deaths. Surprisingly enough, even these undeniable common features collapse into the larger pool of their differences; for one would expect that since both Pelopidas and Marcellus are depicted as equally unable to check their passion in battle they would be reproached and criticized in relatively similar terms; but this is not so. Marcellus is criticized rather more fiercely for his rashness than Pelopidas (*synkr. Pel./Marc.* 33(3).3). The two men's reckless bravery, which initially constituted their most obvious similarity, degenerates in the end into an unfair dissimilarity on not especially convincing grounds (*Pel.* 32.9, *Marc.* 28.6, *synkr.* 33(3).3). One naturally wonders about the purpose of choosing Marcellus to form a pair with Pelopidas when it is evident that Plutarch's conclusions in the *synkrisis* are in conflict with the objective he promulgates in the proem (*Pel.* 2.9-12). The choice of Marcellus seems to be rather forced, made out of the necessity of providing a counterpart to Pelopidas rather than out of some convincing similarities of their characters. It is reasonable then to suggest that the Greek hero was selected first, and that the Roman was suggested by him; the latter's obsession with Hannibal no doubt furnished a comforting semblance of Pelopidas' hatred for the tyrant Alexander, enough at least to form a convenient starting point for recounting and parallelling the two *Lives*.

Throughout the Theban *Life* Plutarch shows a pronounced tendency to illuminate Pelopidas' virtues and award praise to his acts, while he is clearly less generous in that respect with Marcellus in the Roman biography. It is not a mere coincidence that he consistently chooses to highlight only those aspects of the Theban's personality which will best reconfirm his positive qualities outlined in the Introduction (chs. 3-4) or in the *synkrisis*; likewise, he tends to tone down or suppress other aspects of his ethos, in order to suggest Marcellus' moral inferiority to the Theban hero. The trial of Pelopidas and Epameinondas (*Pel.* 25.3-4), for instance, is a good example of the biographer's skills in refining a tradition which is unfavorable to Pelopidas; although he was aware of two different traditions related to this episode (cf. *De laud. ips.* 540D-E) he consciously selected that one which would not undermine Pelopidas' ethos vis-à-vis both Epameinondas and Marcellus (*Marc.* 27.6-7). Equally worth noting is Pelopidas' inefficient policy in Thessaly which ended up with his imprisonment by Alexander of Pherai (*Pel.* 27). Plutarch must have known how harshly this act had been criticized by Kallisthenes to whom Polybios may owe his remarks on Pelopidas' foolhardy conduct (8.35.6-8), but refrains, at all costs, from passing a negative remark on the Theban's naiveté in trusting the tyrant.

In his straightforward characterization of Marcellus Plutarch is usually not inclined to award blame for his acts; it is chiefly through subtle and unsuspected contrastive comparisons with other characters in the Roman *Life* that the biographer displays all his mastery in letting the negative aspects of Marcellus' ethos emerge. For instance, Archimedes' fondness for a life spent in freedom from war (*Marc.* 14.15) creates a sharp, but not an obvious, contrast with Marcellus' bellicose nature (1.2, 24.1-2, *Fab.* 19.2, *Reg. et imp. apoph.* 195D; on Plutarch's portrayal of Archimedes and Marcellus, see Georgiadou, 1992b: 162-163). Hannibal's modesty in face of his success, or his piety, are juxtaposed implicitly to Marcellus' mad desire for battle (29.3) or his neglectfulness of the divine will (29.9-10); Plutarch's emphasis on Fabius' sagacity and excessive care in his military tactics make Marcellus' boldness and aggressive spirit appear more destructive than they would have, had Marcellus been characterized separately (9.4-7).

The necessity of reading the two *Lives* as a single piece consisting of two parts is apparent. A combined study of the two biographies together with the formal *synkrisis* promises a fuller assessment and understanding of Plutarch's character-delineation techniques in the ***Pelopidas*** and a more thorough knowledge of his use of the source-material in the interests of the Theban hero and at the expense of Marcellus.

[e] Pelopidas and Epameinondas: a Pair within a *Life* (*Pel.* 3-4, *synkr. Pel./Marc.* 31(1).3, 32(2).2)

After an expansive prefatory section on the theme of recklessness being wrongly substituted for courage (1-2.8) comes the expected, albeit too brief, summary of the common features of the careers of Pelopidas and Marcellus (2.9-11). Plutarch then proceeds to the typical opening topic of his biographies, which consists of a description of his hero's lineage, character and education. Faced apparently with an unfortunate gap in his sources on Pelopidas' private life and early activities [**2a**] the biographer is forced to supplement the compressed portrait of his subject by way of internal contrastive comparisons with other characters, especially Epameinondas, Pelopidas' friend and collaborator, about whom the available range of sources was far less restricted, as one can infer from Nepos' *Life of Epameinondas*.

The idea of treating the two Thebans as a pair in the Introduction (chs. 3-4) and in the *synkrisis* (31(1).3, 32(2).2) can be partly understood in view of their leading political and military roles during the period of the Theban hegemony. Yet, apart from their friendship and their common aspirations, namely their so-called "divine desire to see their country become most glorious and powerful in their day by their own efforts" (4.3-4), Pelopidas and Epameinondas had too little in common to justify such an elaborate comparison. The themes of friendship and patriotism are, of course, of considerable moralizing and structural importance in the Theban *Life*: not only do they shed light on the biographer's characterization techniques and means of achieving emphasis [**2a**], but they also inform the entire narrative and provide further links with the paired *Marcellus*. A closer look, however, at the parallel characterization of the two Thebans can give us an insight into the less obvious reasons which seem to have led Plutarch to define Pelopidas almost exclusively through Epameinondas and to attach so much importance to their amicable relationship (cf. *De amic. mult.* 93E, where Plutarch includes the two

Thebans in his list of ideal friends) and their tendency to share in undertakings equally.

In describing the two Thebans' different educational backgrounds, Plutarch points overtly to Pelopidas' lack of enthusiasm in intellectual pursuits and Epameinondas' dedication to the study of philosophy (4.1). The contrast is, however, distinctly less focused and illuminating than we might have expected and the biographer does not use it to explore and explain other aspects of Pelopidas' character,—such as his inability to control his anger—, as he does in some other *Lives*.[75] In fact we feel frustrated that Plutarch does not examine the relationship of the two men in order to uncover any traces of Epameinondas' influence on Pelopidas' character and policies.[76] Instead, he limits his criticism of their conduct to occasional and isolated remarks (25.4, 26.8, 29.5, 32.9). It is apparently left to the reader to recognize the causal connection between the two men's education and character, and to judge for himself the effects it had on their careers. The reader is, of course, familiar with Plutarch's emphasis on the ability to control a particular pathos, which he sees as a process of taming and humanizing the natural instincts through education and philosophy, or with the concept that πρᾳότης is the product of self-discipline and training and that it is a thing learned and is produced by *logos* and *paideia*.[77] It suffices to read the *De genio*, a work akin to the *Pelopidas* in its historical subject-matter, to grasp Plutarch's profound interest in exploring the effect which the study and practice of philosophy has on appeasing human passions.

---

[75] See *Them.* 2.7, *Arat.* 10.5, *Tim.* 6.1, *Cor.* 1.3-6, 15.4-5, *Mar.* 2.2-4; for a study of Plutarch's interest in the question of education and its effects on the individuals' passions, see Swain, 1987: 96-196; id. 1989: 63, 65, 1990 [1995] 126-145, Pelling, 1988a: 260-261, id. 1989: 199-232. Swain observes (1987: 103, id. 1990 [1995] 129, 134) that Plutarch does not regularly explore the effects of the Greek heroes' παιδεία to any great extent, while he is more scrutinizing with the Romans.

[76] Note, for instance, how differently Plutarch contrasts Chabrias and Phokion (*Phok.* 6.2-3): he describes the former as being "controlled by his passions in battle, which eventually led to his death", while the latter as "cautious and energetic"; he also elaborates on the beneficial influence of Phokion on Chabrias (6.1, 6.3-4).

[77] On Plutarch's discussion of the effects of philosophical training, see *De rect. rat.* 48D, *Quom. virt. sent.* 80A, *Praec. ger. reip.* 802E, *Plat. quaest.* 999E-F.

Although in the *Life* Plutarch never exploits Pelopidas' lack of proper education to explain his failings,[78]—which he does implicitly in the *De genio* (577 A)—, his stray remarks on Epameinondas' self-control suggest that certain emotional over-reactions or misjudgments in the former's career (25.4, 27.6-7, 28.3-4) are indirectly attributable to deficient *logos* and *paideia*. One therefore wonders why Plutarch compares Pelopidas' pursuits with those of Epameinondas at the risk of undermining the former's image. I would suggest that Plutarch's portrayal of the two Thebans as a pair should be seen as an artful device aiming at extenuating, not accentuating, any failings in Pelopidas' character. The description of his ethos is made in such a way that the leading traits of his personality become harmoniously fused with those of Epameinondas. As a further evidence of this blending, the biographer expresses his unreserved admiration for the two men's ideal friendship and collaboration (4.2-8). It is certainly no coincidence that his extensive comments on their amicable comradeship immediately follow his statement about their different pursuits. By so doing, he removes the tension felt initially with the incongruity of the two Thebans' activities, and prepares the ground for a theme so central to the *Life*, namely their joint undertakings. Pelopidas' physical activities then emerge as blending comfortably with the esoteric philosophical tastes of Epameinondas. Through their complementary qualities, and by virtue of their friendship and common aspirations the two characters come to embody the "philosopher-statesman" (ὁ πολιτικὸς φιλόσοφος), a combination which Plutarch greatly admired and encouraged.[79]

Plutarch would have known the privileged position occupied by Epameinondas in Ephoros' list of distinguished generals, a position

[78] Note, however, how differently Plutarch treats Agesilaos' proclivity to anger against the Thebans, which he regards as one of the major causes of Sparta's defeat (*synkr. Ages./Pomp.* 81 (1).7).

[79] Cf. for instance Plutarch's particular interest in public figures who are versed in philosophy, like Cicero, Brutus, Dion, Cato the Elder; on this theme, see *Max. cum princ. philos. esse disser.* 776C-777A, 777C-E, 778B, *Ad princ. inerud.* 779F; also J.R. Hamilton, 1969: xxxvii-xxxix, Wardman, 1974: 211-220, A. Pérez Jiménez, "El ideal de buen rey segun Plutarco", in J.M. Candau et al., eds., *La imagen de la realeza en la antigüedad* (Madrid 1988) 89-113, Sansone, 1989: 12-15, De Blois, 1992: 4601-4611, Barigazzi, "Scritti politici: *Maxime cum principibus philosopho esse disserendum*", and "Scritti politici: *Ad principem indoctum*", in *Studi su Plutarco*, Università degli Studi di Firenze, Dipartimento di Scienze dell'Antichità (Florence 1994) 235-261 and 263-286.

which he owed to his intellectual and military virtues (Diod. 15.39.2, 88.3-4). But the biographer is far from adopting Ephoros' competitive ranking of the two Thebans. He offers instead an alternative model of leadership—similar to the one he promotes in the *Praecepta* [**2a**]—which comes about through a combination of the positive aspects of the personality and qualities of people in power.[80] This model works favorably for Epameinondas also, who was severely reproached by his compatriots for refusing to get involved actively in the overthrow of the Theban oligarchy despite his excellent education. Plutarch avoids blaming him for his inactivity at the time and excuses it on purely philosophical grounds (*Pel.* 5.4, *De gen.* 576D-E).[81]

Moreover, the theme of the fusion of contrasting qualities constitutes an interesting point of contact with the *Marcellus*. One cannot fail to notice how differently Plutarch treats the pair Marcellus/Fabius Maximus in the Roman *Life*, at least as regards the delineation of their opposite qualities. They are depicted as having different natures (*Marc.* 9.4-7)—the bold and aggressive spirit of Marcellus is contrasted with Fabius' caution—but while Marcellus' boldness and activity were thought to suffice in an attack, these same qualities were considered by the Romans, Plutarch notes, to be inadequate in defence. The drawing of the contrast between their natures illuminates an aspect of Marcellus' character, which can indeed be harmful if it is not combined equally with Fabius' opposite qualities (cf. *Marc.* 24.2). The type of complementarity proposed in the Roman *Life* is of a different sort; it points crudely to the imperfection and deficiency in either Roman's qualities, while at the same time it invites a comparison with the two Thebans, who are described as equal in their virtues (*Pel.* 4.1). The moral to be extracted is obvious: it is the unconditional friendship and uncompetitive collaboration of the two Thebans, which are absent in the Roman pair, that warrant success in political leadership.

---

[80] On Plutarch's emphasis on the complementarity of virtues, see Wardman, 1974: 27-32, 57-60.

[81] Nepos also draws attention to Epameinondas' refusal to participate in the expulsion of the Spartan garrison from Thebes in 379 (*Epam.* 10.3), but does not elaborate on it; for a discussion of this incident, see Georgiadou, 1995: 187-200, and note on 5.4.

Plutarch does not look into the political reasons for Pelopidas' and Epameinondas' cooperation. He has no intention of suggesting the political expediency which made their cooperation in policies even more desirable. As Wardman notes (1974: 29) "the position of Thebes at this time was such that her politicians had every reason for combining in order to make their country emerge as a leading city-state".

Plutarch's treatment of Pelopidas' qualities in conjunction with those of Epameinondas provides yet another link with the *Marcellus*. Had Pelopidas been treated in isolation from his friend, he would not have emerged as an adequate model for imitation, nor would he have fared well in his comparison with Marcellus. Plutarch tells us that Marcellus was "an enthusiast for Hellenic learning and culture... but was prevented by his occupations [i.e. as a soldier] from reaching the levels of practice and learning that he desired" (1.3). Marcellus then emerges as combining in one person both Pelopidas' dedication to the state (*Pel.* 3.7) and Epameinondas' intellectual pursuits, even though the Roman's exposure to Hellenic culture is explicitly stated to have been limited. Marcellus' moral attributes of πρᾳότης, φιλανθρωπία and other related qualities are brought out in the *Life* on various occasions (1.3, 10.6, 19.6, 23.3-10, 28.1), and at 1.3, his σωφροσύνη and φιλανθρωπία are connected implicitly with his Hellenic culture (Swain, 1990 [1995] 131-132, 140-142). As in the case of Pelopidas, Plutarch makes no attempt to blame insufficient education for Marcellus' obsessive passion to fight a decisive battle with his adversary (28.4). He simply qualifies this failing as "a youthful ambition that ill became his old age and a career replete with military honors" (28.6, *synkr.* 33(3).6). The association is again left for the reader to make.

Finally, Plutarch's stress on the two men's well motivated collaboration and sharing of their accomplishments appears to have been prompted by other considerations as well. He tells us that the two Thebans "used to treat each other's successes as their own" (4.4). Are we expected to deduce from this statement that there is no need, or that it is even pointless, to apply strict measures in order to determine each hero's exact contribution to any of their joint operations? Throughout the *Life* Plutarch never fails to bestow equal praise on both men in their joint operations (12.7, 23.6, 24, *synkr.* 31(1).3). While his laudatory comments on the two Thebans' exploits can be seen as reflecting the biographer's own views, one can hardly fail to notice a valuable hint in the *synkrisis* (32(2).2)[82] at Pelopidas' questionable leading role at the battle of Leuktra and his first invasion of the Peloponnese, which are emphatically stated elsewhere in the *Life* to have been performed jointly with Epameinondas.[83] Plutarch's uncertainty about Pelopidas' share in these exploits seems to result, I suggest, from diverging information furnished by his sources. Shrimpton has argued quite convincingly about the existence of different traditions in the fourth century; in them Epameinondas was in the spotlight and his name had become synonymous with every successful Theban venture,[84] while the other Theban leaders, like Pelopidas, stood on the periphery (1970: 11-46). His suggestion that Ephoros, "was the fountainhead of the laudatory Epameinondas tradition" (*ib.*, 30-34, esp. 34), while Kallisthenes may have been responsible for Pelopidas' heroization (*ib.*, 45) tends to strengthen such a hypothesis about Plutarch's perplexity vis-à-vis the two traditions. Being sensitive to their different emphases and to the damaging effect the laudatory Epameinondas tradition might have on the image of Pelopidas, Plutarch carefully avoids tackling the issue by leaving his sources unnamed and uncriticized; he then proceeds to reconcile their conflicting information by laying equal stress throughout the *Life* on the two men's common exploits; at the same time, he seems to have felt the scholarly obligation to point to an alternative version of these events in which Epameinondas, and not Pelopidas, was the protagonist.[85]

[f] Nepos' *Pelopidas*

Nepos' sketchy *Life of Pelopidas* is the only other extant biography of the Theban hero, and it looks more like a supplement to his *Life of Epa-*

---

[82] For Plutarch's different interpretation of incidents in the narrative and in the *synkrisis*, see Pelling, 1986: 88-91.

[83] Elsewhere in his works Plutarch connects only Epameinondas with these two achievements: *De glor. Athen.* 346B, *De garr.* 514C, *De laud. ips.* 540D-E, 542C, *An sen. resp.* 786D, *Non posse* 1098A; cf. also Diod. 15.81.2, 87.6, Str. 9.2.39; Deinarchos (*contra Dem.* 72-73) extols both Thebans.

[84] Ephoros' account of the battle at Leuktra, as is preserved to us by Diodoros, makes Epameinondas, and not Pelopidas, responsible for the Theban victory (15.55.1-2); cf. Plut. *De garr.* 514C.

[85] Plutarch commonly associates the Theban victory at Leuktra with Epameinondas: *Cor.* 4.6, *De tuenda sanit.* 136D, *Reg. et imp. apoph.* 193A, *An sen. resp. ger.* 786D, *Non posse* 1098A.

meinondas than a *Life* standing on its own merit. Nepos' explanatory introduction in the Preface to the *Pelopidas* (1.1) is of particular interest for throwing light on the wider distinction between biography—political, not intellectual—and history (for an extensive study of these two genres, see Geiger, 1985: 21-65). It has been argued that it was the very novelty of the differentiation between the two genres, a distinction expounded later on more elaborately by Plutarch in the *Alexander* (1.2), that prompted Nepos, but also Plutarch, to define it precisely (id., 38). Nepos makes it clear from the outset that Pelopidas "is better known to historians than to the general public" (1.1); this declaration has been used as a pointer to Nepos' sources in the *Pelopidas*: in order to compose this *Life* he did not refer to a work of a biographical nature, not because it was unavailable to him, but simply because no such work existed [**2a**]. As a consequence, Nepos had to confine himself to a different literary genre, that of history.

Nepos introduces Pelopidas' biography with a continuous narrative of the hero's exploits, unlike Epameinondas,' whose career is preceded by a lengthy characterization and categorization of his virtues (*Epam.* 1-6). After a summary account of the occupation of the Kadmeia by the Spartans and the establishment of a Spartan garrison in the Theban citadel Nepos proceeds to describe, in greater detail, the coup of the Theban exiles, giving full credit to Pelopidas for his role in it, and concludes it with the liberation of the Kadmeia. This episode, which runs from chapters 1.2 to 3.3, occupies a disproportionately large amount of space in the biographical narrative—two-thirds of the total *Life*—with respect to the other sections, and is comparable to the lengthy account of the same episode in Plutarch's *Pelopidas* (5-13).

Before embarking on the Theban's subsequent enterprises Nepos makes it clear that almost all of these were undertaken jointly with Epameinondas, and that although Pelopidas played one of the two major roles at Thebes, he was, the biographer points out, secondary to Epameinondas (4.1). The theme of the two Thebans' joint undertakings plays a decisive part in the characterization of Pelopidas in the Greek Theban *Life* [**2d**], but Plutarch clearly uses it in order to redress the balance between the two heroes' achievements, and not to undermine Pelopidas' role.

Adhering to a strict chronological sequence, Nepos offers merely an outline of Pelopidas' other military operations, very similar to Diodoros' summary of the hero's accomplishments at 15.81. He omits Pelopidas' role at Tegyra and proceeds to his part at the battle of Leuktra (4.2), his service in the first invasion of the Peloponnese (4.3), his mission to Sousa and his activities in Thessaly from 368 until his death in 364, and concludes with a description of the honors bestowed by the Thessalians upon Pelopidas after his death, and upon his children (5).

Nepos, like Plutarch, is silent about the type of historical sources he consulted for the *Pelopidas*. Considering, however, the brevity of the Latin biography, one would hardly ascribe the similarities which exist between the two *Lives* to Plutarch's dependence on Nepos' work. There remains the possibility that both writers drew from the same tradition, but whereas Plutarch appears to have supplemented his principal narrative source(s) with a relatively large amount of extraneous material drawn from other minor sources, collections of anecdota and local tradition [**2c**], Nepos favored abridgment. Ephoros and Kallisthenes are proposed as the most likely authorities used for the Latin biography.[86]

[g] The *Pelopidas* and the *De genio*

The Spartan seizure of the Kadmeia (382) and its liberation by the Theban democrats (379) are twice treated by Plutarch: in the *Pelopidas* these events occupy a sizeable portion of the biography, and an equally ample amount of space is devoted to the narration of the same events in the *De genio*.[87] While the length of the account in the *Life* can easily be justified by its importance not only for illuminating Pelopidas' first major achievement which led to the liberation of Thebes from Spartan rule, but also for understanding the subsequent rise of Thebes as a leading power in Greece, there are no apparent reasons which can explain satisfactorily another detailed treatment of this memorable chapter of Theban history.[88]

---

[86] On Nepos' sources in the *Pelopidas*, see Bradley, 1968: 150-166, 278-282, Westlake, 1939: 11. Sordi (1995: 415-422) raises doubts about Kallisthenes or Ephoros as Nepos' sources.

[87] *Pel.* 5.2-13 (one-fourth of the *Life*); *De gen.* 575F-577D, 586B-588B, 594D-598F (one-third of the treatise).

[88] Since the *De genio* contains no references to any events of Plutarch's own time, it is impossible to assign any absolute date of composition or publication to it.

Unlike the other philosophical works of Plutarch, the *De genio* represents the amalgamation and synthesis of its author's varied skills as historian, philosopher, storyteller and dramatist; it uses a historical setting for a series of philosophical debates and purports to be an eye-witness description of the secret operations which resulted in the liberation of Thebes and also of the conversations that took among the conspirators during that period. The treatise is an experimentation with a rather uncommon form of historical representation, which combines historical narrative and philosophical debates on issues not obviously connected with the dramatic setting of the work.[89] The two types of narrative are not integrated with each other in any ostensible way, but freely alternate without impairing the organic way in which the facts of history are related to the debates of philosophy.[90] The inspirational model for this compositional form may have been Plato's *Phaedo*, in which a series of philosophical debates is presented in a quasi-historical context, viz. Sokrates' last days before his execution.[91] In both works the historical narrative begins and ends the philosophical dialogues, but, whereas the structural unity of the *De genio* and the causal relationship between its two types of narrative are not instantly recognizable, in Plato's dialogue the links are clear between the account of Sokrates' imminent death and the discussions on the nature of the soul and the philosopher's attitude to death.[92] In other words, the central theme of the philosophical debates in the Platonic dialogue is generated from the particular circumstances that

[89] Cf. also the admixture of history and fiction in the *Septem sapientium convivium* and the *Amatorius*; see Ziegler, 1964: col. 204.

[90] The *De genio* has often been the object of unfair negative criticism for its supposed lack of originality; cf. Hirzel, vol. 2, 1895: 151; for the recognition of the masterful admixture of its diverse narrative components, see for instance Russell, 1973: 37, Borthwick, 1976: 149.

[91] For extensive discussions of the structural similarities between Plato's *Phaedo* and the *De genio*, see Hirzel, vol. 2, 1895: 148-151, Christ, 1901: 58-61, Kahle, 1912: 17-19, 78-85, Lattanzi, 1933: 15-17, Corlu, 1970: 82, Riley, 1977: 257-273.

[92] For a discussion of the allegorical content of the *De genio*, see Lattanzi, 1933, Corlu, 1970: 45-92, Riley, 1977, Aloni, 1982: 11-38, Babut, 1984 [1994], id. 1988 [1994], Barigazzi, 1988, Georgiadou, 1995: 187-200, Hardie, 1996: 123-136; for further bibliography, see Barthelmess, 1986: 76-77. For a recent interpretation of the *De Genio* as reflective of a new ideology and critique of the Theban past seen through the spectacles of Plutarch's contemporary intellectual culture, see Brenk, 1998; also Alcock, 1997: 103-115. For an analysis of the notion of time as a structural element in the *De genio*, see Brenk, "Time as Structure in Plutarch's *The Daimonion of Sokrates*", *Studia Hellenistica* 32 (1996) 29-51.

inform it, while Plutarch disguises it amidst discussions on topics such as the decipherment of a mysterious inscription found in the tomb of Alkmene, an interpretation of the geometrical problem given to the Delians by their oracle, and the elaborate presentation and exploration of the nature and modus operandi of the Socratic *daimonion.*

Though still an important figure among the Theban conspirators and liberators of the Kadmeia, Pelopidas has lost the protagonistic role assigned to him in the *Life,* and the success of the operation is, therefore, equally distributed among the Theban conspirators. The shift of emphasis and the existence of certain discrepancies between the two accounts[93] are to be explained by the different scope of each work. The *De genio* is a quasi-historical work, and is clearly not meant to function as a supplementary source from which to extract historical information not available in the briefer account in the *Life*. The bold, if not reckless, plot prepared by the Theban conspirators and its execution offer the opportunity for a study of "the entire panorama of humanity seen from the angle of its relation with the divinity" (Babut, 1984 [1994] 68-69). Three differerent groups of people are delineated:[94] the first encompasses individuals such as Sokrates, Epameinondas and Theanor, who by virtue of their philosophical training are undisturbed by passions and are, for this reason, morally and spiritually superior to all other human beings; the second includes people like the tyrannical rulers of Thebes, who are totally uninspired, odious and morally blind (578B-D); and the third, which lies between the first two extremes, is coextensive with the Theban liberators, all of whom are moved by passion but still perform acts of valor. Their course of action is divinely inspired and guided, although they are portrayed as not being able to perceive directly the demonic guidance, but only through omens and signs (cf. Alkmene's tomb and the Delian oracle). Unlike Sokrates, who is represented as having achieved the ideal combination of philosopher-citizen, Epameinondas is sharply criticized by his countrymen for refusing to participate in the conspiracy on philosophical grounds, despite his superior education. While the gap between theory and practice is, as Riley observes (1977: 268), entirely resolved in Sokrates, who was one of the very few born inspired, it is only partially

---

[93] On the narrative differences between the *Pelopidas* and the *De genio*, see Corlu, 1970: 22-31.

[94] On the tripartite division of humanity, see Babut, 1994: 51-76, esp. 69-70.

bridged in Epameinondas, who tries to repress, through habit and training, his innate desires.[95]

In the *De genio* Plutarch is not particularly interested in the actual outcome of the conspirators' operations, as he is in the *Life* (13.4-7), but is, as Riley points out (ib. 272), more concerned with "the mechanism of choice, not the verification of choice". His schematic division of humanity into three types exemplified by those Thebans who were directly or indirectly involved with the conspiracy against the tyrants seems to be aimed at exploring the old dichotomy between the active and the contemplative life, or—to use the terms which form the twofold structure of the treatise—πράξεις καὶ λόγοι (*De gen.* 575D). Though he typically belongs to the first group described earlier, Epameinondas stands half way between the ordinary Thebans and the philosophers. While he is depicted as refusing to become an active conspirator, he does not totally neglect his civic duties (*De gen.* 598D, *Pel.* 7.5). In the real world of action, furnished by the historical background of the treatise, Epameinondas is the only one who comes close to resolving the tension between detachment from politics and full assumption of civic responsibilities. At a later stage of his life he will be seen to achieve the ideal blending of philosophical training and political activity [**2e**].

There is no consensus as to the relative chronology of the composition of the *Pelopidas* and the *De genio*. Hirzel (1895, vol. 2: 155-156) claims that the *Life* was later than the treatise on the grounds that the former, with its historical character did not need another more poetic work to illustrate it; A.O. Prickard takes the same view (*The Return of the Theban Exiles 379-378 B.C.* [Oxford 1926] 9). Christ (1901: 85-89), agreeing with Hirzel's proposed sequence of the two works, regards the *Life* as an excerpt of the treatise and attributes all the additional information of the treatise that appears neither in the biography nor in Xenophon's account of the same events to Plutarch's imagination. The reverse view, which advocates that the *Pelopidas* was written earlier than the dialogue, is supported among others, by Kahle (1912: 73), Lattanzi (1933: 116), Ziegler (1964: col. 205) and Aloni (1982: 24-25), while Corlu (1970: 22-27) and Hani (1980: 58-59) have reasonably concluded

---

[95] For a discussion of the Socratic example, see Hershbell, 1988, Georgiadou, 1996: 113-122.

that the chronological order of the two works cannot possibly be determined.

[h] Text

The text used is H. Gärtner's revised Teubner edition (1994, vol. 2, fasc. 2) of Ziegler's text (1968, vol. 2, fasc. 2) to which the reader is referred for an account of the manuscript tradition. I list below a number of readings in which I differ from Ziegler/Gärtner; I discuss these in the Commentary in the appropriate sections, where I also cite the manuscript evidence in greater detail:

Georgiadou	**Ziegler/Gärtner**
1.2: ⟨- - - ἐν μὲν ταῖς στρατείαις⟩ ἰταμός, (Georgiadou)	*** ἰταμός,
1.7: Λακεδαιμόνιοι δ' ἀμφότερα καὶ ζῆν ἡδέως καὶ θνῄκειν ἀρετῇ παρεῖχον (codd.)	Λακεδαιμονίοις δὲ καὶ ζῆν ἡδέως καὶ θνῄσκειν ἀμφότερα ἀρετὴ παρεῖχεν (Reiske)
3.4: ἀόκνῳ καὶ κατὰ στρατείας ἀδόλῳ (codd.; κατὰ om. P)	ἀόκνῳ [καὶ] κατὰ <τὰς> στρατείας [ἀδόλῳ] (Schaefer; Reiske; Ziegler)
7.1: τοῖς λέγειν δυναμένοις (codd.; *PHeid.* 209, col. II, 7-8; Gärtner)	τοὺς λέγειν δυναμένους (Ziegler)
9.2: <καὶ διὰ τοῦτο> (*De gen.* 594D; Georgiadou); also possible is ⟨διὸ⟩ (cf. δι' ὃ Reiske)	<ᾧ> (Ziegler)
9.9: παρασχεῖν ἑαυτὸν ⟨δεῖν⟩ ἀνυπόπτως (Sintenis; δὴ codd. except K)	παρασχεῖν ἑαυτὸν ἀνυπόπτως (K; Ziegler)
11.8: συνεπλέκετο (KCB, Sintenis)	συνεπέπλεκτο (reliqui codd.)
14.4: † ἰδίᾳ ἔμπορόν τινα † (codd.)	[ἰδίᾳ] (Sintenis) <Δι>έμπορόν τινα (C. Keil; ἰδίᾳ Διέμπορον Kröhling)
14.6: συνταράξας οὐ φαῦλον οὐδὲ ῥᾴδιον ... πόλεμον (Xylander)	συνταράξας ⟨πόλεμον⟩ οὐ φαῦλον οὐδὲ ῥᾴδιον... πολεμεῖν (Ziegler)
16.5: ὑπὲρ τὰ ἕλη (Wilamowitz)	ὑπὸ τὰ ἕλη (codd.)
17.8: τὴν δεδομένην ὁ Πελοπίδας ⟨***⟩ ἡγεῖτο (Georgiadou)	τὴν δεδομένην ὁ Πελοπίδας ⟨χαίρειν ἐάσας⟩ ἡγεῖτο (⟨παριδὼν⟩ Flacelière)
22.4: λόγον... ⟨δια⟩διδόντες (Coraes)	λόγον... διδόντες (codd.)
28.8: καὶ γὰρ ἐγὼ σέ (codd.)	καὶ γὰρ ⟨καὶ⟩ ἐγώ⟩ σε (Ziegler)
30.6: ᾧ γε (Ziegler, not in text)	ᾧ (Coraes; Kronenberg; ὅτι codd.)
32.3: ἀνατεινόντων (Bryan)	ἀντιτεινόντων (codd., Ziegler)

34.5: καὶ τύχῃ (codd.) μεταβάλλεσθαι ⟨μὴ⟩ ἀπολιπών (A[m])	καὶ ⟨τῇ⟩ (Ziegler) τύχῃ μεταβάλλεσθαι ⟨μὴ⟩ ἀπολιπών (A[m])
35.2: ἡγουμένου (Lindskog) Μαληκίδου καὶ Διογίτονος (*IG* VII 2408; Flacelière)	ἡγουμένων Μαληκίδου καὶ Διογείτονος (codd.)
35.5: ἐντὸς τῶν ὅπλων καὶ τῶν φυλάκων (Coraes)	ἐντὸς τῶν ὅπλων καὶ τῶν φυλακῶν (Schaefer; φυγάδων codd.)

Chs. 1-2: Proem

Recklessness in battle and its potentially disastrous consequences constitute the theme of P.'s elaborate Introduction to the *Lives of Pelopidas and Marcellus.* The proem, starting out with a *chreia* (1.1), invites the reader to inquire into a significant theme of the ethical life, that of the multifaceted notion of bravery and its excesses, and also to explore the various circumstances in which it is expressed. The *chreia* points with its typically succinct formulation to a misconception held about immoderate bravery, namely that it is wrong to suppose instantly that readiness to face death is identifiable with courage. P. then provides more *chreiai* (1.2-2.6), in order to support the veracity of the introductory saying with confirmatory testimony. Inspired probably by Aristotelian ethic, he demonstrates the excesses of courage, which range from recklessness—due to despair, one's upbringing and education, ignorance, or self-confidence—to cowardice deriving from fear of death. Each saying illustrates a different nuance and variation of recklessness, and serves as a model for imitation or avoidance. P. concludes the list of his *chreiai* with a maxim (2.7-8), in which the special circumstances that occasionally necessitate a general's exposure to danger are described. Finally he moves to the comparison of the two protagonists and focuses on their similarities, the most prominent of which is their thoughtless daring in battle, which led to their death (2.9-12). Thus the link between the types of courage and the two heroes' recklessness is established.

P.'s examples of courage and cowardice in the *chreiai* are reminiscent of the types described by Aristotle in the *Eudemian* and *Nicomachean Ethics* (*EE* 1228a27-1230a35-36, *EN* 1115a6-1117b22): the type of rashness exhibited by Antigonos' soldier (1.2-4) is seen as an act of despair, an indifference to life, caused by his poor health and impaired body (*EE* 1229b33-1230a4, *EN* 1116a13-16). The Spartan example of honorable life and death (1.5-7) stands out in the proem as a positive model (*EE* 1230a29-34, *EN* 1116a18-22). The daring of Antigonos (Gonatas or Monophthalmos, see note on 2.4), which is applauded by P. (2.4-5), derives from his experience and knowledge of ways of encountering the formidable (*EE* 1229a14-16, 1230a4-10, *EN* 1116b3-5). The opposite example is provided by Timotheos (2.6) and corresponds to the Aristotelian type of bravery which stems from lack of experience and knowledge and is usually exhibited by children and madmen (*EE* 1229a17-18, *EN* 1117a22-23). There is one type of bravery which P. wholly recommends to the military leader at 2.7, that in which the whole issue of the fight is determined by exposing oneself to danger; there he should fight and risk his person (*EE* 1229a1-5, 1230a27-34, *EN* 1115b17-24).

The type of recklessness exhibited by Pelopidas and Marcellus (*synkr.* 33(3).3-6; cf. also the similar case of Chabrias, *Phok.* 6.2-3, and Cyrus the Younger, *De cohib. ir.* 458F) does not fall within the specific types outlined in the proem—for none of the listed demonstrations of rashness is attributable to some irrational and uncontrollable passion, as happened in their case—but it does correspond to the type of ἀνδρεία proposed by Aristotle, namely that stemming from πάθος ἀλόγιστον (*EE* 1229a21-23, *EN* 1116b24-1117a9).

The theme of bravery, or lack of it, and its various manifestations runs through the narrative of both *Lives*, but is especially stressed in the *Pelopidas*. P. puts on the stage several imitable or cautionary examples of brave figures, such as Thrasyboulos (7.2, 13.4), Charon (8.4, 9.13), Hippostheneidas (8.5), the Theban liberators (10.5), Sphodrias and his soldiers (14.3, 6), the Theban soldiers (15.2, 17.13), the Spartan soldiers (17.11-12), the Sacred Band (18.3-7), Pelopidas (19.4, 31.4-5, 32.2), Epameinondas (4.7, 29.4-5), Thebe (35), and occasionally explores the special circumstances from which this or that manifestation of courage or cowardice originates. For a general study of P.'s proems, see Stadter, 1988, Rosenmeyer, 1992; for a discussion of his possible indebtedness to Polybios for his examples of recklessness in battle, see **[2c]** s.v. "Polybios"; for bibliography on P.'s familiarity with and use of Aristotle and the Peripatetic tradition, see Scardigli, 1995: 6-12. Sandbach (1982: 207-232) claims that P. had a restricted first-hand knowledge of the Aristotelian Corpus; for a contrary view, see P. Merlan, *From Platonism to Neoplatonism* (The Hague 1960) 219, Verbeke, 1960: 236-247, G.J.G. Aalders, "Political Thought in Plutarch's *Convivium Septem Sapientium*," *Mnemosyne* ser. 4, 30 (1977) 28-39, esp. 36, id. 1982: 63-65, Aalders and De Blois, 1992: 3384-3404, esp. 3397-3398. On P.'s personal compilation of *chreiai* and the ample use he makes of them in his work, see Hock and O'Neil, 1986: 7-8, 38.

1.1 Κάτων ὁ πρεσβύτερος: Marcus Porcius Cato the Elder (234-149 BC) was born in Tusculum. He became quaestor in 204, consul in 195, and censor in 184; see Astin, 1978, Sansone, 1989.

πρός τινας... νομίζειν: this anecdote does not figure among the sayings listed under Cato's name in the *Reg. et imp. apoph.* 198B-199C, or in those included in P.'s biography (*Cat.Ma.* 8-9). A very close version of this anecdote is ascribed by Stobaios to Aratos (4.12.25). Cicero credits the old Cato with a compilation of apophthegmata (*De Off.* 1.104, *De Or.* 2.271). P. observes, probably on the basis of his own reading, that Cato's writings are embellished with Greek ideas and anecdotes, and that many literal translations (from Greek) have found a place among his sayings and maxims (*Cat.Ma.* 2.6). He also identifies two of Cato's sayings translated from the Greek (*ib.* 8.4-5, 24.7-8). We do not know, however, whether P. refers to Cato's own collection, or to some other compilation of sayings independent of Cato's, which may have included some anecdotes of Greek origin. Nevertheless, if P. is right in suggesting that Cato occasionally passed off Greek sayings as his own, we may perhaps assign the anecdote in question, as well, to a Greek source. The view, however, that Cato was exposed to Greek education only late in life should also be given due consideration (Astin, 1978: 148-149, 157-181, 186-188, Sansone, 1989: 206).

The message conveyed by Cato's saying can conceivably be attributed to Aratos himself on the basis of his personality: Polybios tells us that Aratos' courage and endurance in devising stratagems and ruses against the enemy and in executing them are contrasted with his timidity and cowardice when he undertook field operations (4.8.3-8), while P. draws attention to his puzzling attitude in the face of dangers (*Arat.* 10. 2-3; 29.7-8); the two writers' combined portrayal of the Sicyonian statesman may be interpreted as exhibiting his reluctance to expose himself recklessly to any danger.

τὴν ἀρετήν: used here in the narrow sense of ἀνδρεία (Arist. *EN* 1114b26-1115a6); ἀνδρεία is singled out as one of the several other ἀρεταί exhibited by Pelopidas and Marcellus (cf. *synkr.* 33(3).8). The emphasis on the notion of ἀνδρεία in ἀρετή can be detected in *Pel.* 2. 5,

9. 8, 13. 3, 19.3-4, *Marc*. 29.12, *synkr*. 33(3).4. Elsewhere it refers more generally to the quality of the hero as a political leader (*Pel*. 4.4, 7.2, 26.8). For a discussion of political ἀρετή in P., see Wardman, 1974: 31-32, 44-45, 49-104; on the concept of ἀρετή generally in P.'s *Lives*, see J.J. Rose, *The Concept of 'arete' in Plutarch's 'Parallel Lives*', Ph.D. diss. Rutgers University, 1988.

διαφέρειν... νομίζειν: contempt for death deriving from the desire for glory, freedom, or fear of suffering is a common theme in the Lakonian Sayings (210F, 216C, 219F; cf. *De gen*. 586D, Xen. *Ages*. 11.8-9). P. also talks about man's deliberative faculty which is often expressed in contrary opinions regarding what is expedient; the advantages of both a glorious death and a pleasant life are at issue in *De virt. mor.* 447E (Eur. fr. 854, Nauck; Stob. 3.7.8):

τὸ μὲν σφαγῆναι δεινὸν εὔκλειαν δ' ἔχει,
τὸ μὴ θανεῖν δὲ δειλὸν ἡδονὴ δ' ἔνι.

Cf. also Hom. *Il*. 9.408-420, where Achilles expresses his doubts about a hero's death and toys with the preference that life could be dearer to him than honor; for a compilation of *sententiae* in which life and death are compared, see Stob. 4.53.

1.2-4 γοῦν: used here to introduce a statement which is evidence for the preceding statement, called for this reason "part proof"; see Denniston, 451.

παρ' Ἀντιγόνῳ: it is hard, if not impossible, to determine which Antigonos is referred to here. At 2.4 P. assigns another anecdote specifically to Ἀντίγονος ὁ γέρων, i.e. Gonatas. Although Tarn suggests that "a number of anecdotes merely labelled Antigonos must refer to Gonatas, as well as those expressly assigned to him" (1969: 251n. 107), I see no reason why P. should designate the same Antigonos differently within the Introduction. The possibility that the appellation γέρων mentioned at 2.4 was meant to point to the unspecified Antigonos at 1.2 is remote. There is a considerable number of anecdotes assigned to Antigonos the first, the so-called Monophthalmos (*Reg. et imp. apoph*. 182A-183A). The gist of the anecdote at 1.2-4 is that military prowess, or at least its semblance, is not an invariable trait of one's character, but is determined by the special circumstances that allow it to emerge; cf. Theophrastos, who discusses whether character changes or not according to the symptoms and ailments of the body: fr. 146, Wimmer. In the case of Antigonos' soldier, his venturesome attitude in battle is explained as due to his poor health and impaired body (cf. Soph. *Phil.* fr. 636 Nauck^2). In the hope of escaping his suffering permanently, the soldier chose to expose himself to danger; but when he was cured of his disease, he was no longer daring and impetuous, so that even Antigonos scolded him and expressed his surprise at the change. P. correlates physical defects with impetuosity in battle and singles out Philip II, Antigonos the First, Hannibal and Sertorius, who were all one-eyed, as the most valiant generals (*Sert.* 1.8). Unlike the soldier, Antigonos' impairment led him to great bravery in order to demonstrate that physical deficiency was not to become an impediment to his career, nor a reason for despair; cf. also Agesilaos' lameness, which made him more ambitious, *Ages.* 2.3, Androkleidas' brave attitude toward war despite his crippled leg, *Apoph. Lac.* 217C, or Caesar's military service as a cure for his feeble health, and not an excuse for soft living, *Caes.* 17.3. Although the attribution of this anecdote to Antigonos the First is not subject to proof, P.'s comment on Antigonos' negative reaction to the change in the soldier's attitude can be better explained in the context of the former's positive attitude to his defect. On Antigonos the First, see Briant, 1973, R.A. Billows, *Antigonos the One-Eyed and the Creation of the Hellenistic State* (Berkeley, Los Angeles 1990).

στρατευόμενος * ἰταμός:** P. introduces the reader to an apparently well-known *exemplum*, unattested elsewhere. Ziegler's suggested restoration of the assumed lacuna (καὶ ἐν μὲν τοῖς ἀγῶσι φαινόμενος or simply ἄνθρωπος—the latter is mentioned in 1.4—presupposes a homoioteleuton. If it is necessary to posit a lacuna—for ἰταμός is often absolute in P.—then the supplement <ἀνὴρ ἐν μὲν ταῖς στρατείαις>, though still speculative, is perhaps more likely, owing to P.'s tendency to vary his vocabulary; a parechesis with στρατευόμενος/στρατείαις, which is a common doublet in P. (cf. *Pel.* 4.5, 16.2, *Alk.* 7.3, *Cor.* 2.3, *Cat.Ma.* 1.8), may have been responsible for the loss of the μέν-clause; for similar types of contrast, see *Pel.* 14.3, *Nik.* 2.4-6.

φιλοκίνδυνος... ῥαγδαῖος: for the element of impetuosity (ῥαγδαῖος) which accompanies the daring person (φιλοκίνδυνος), see *Tim.* 3.7, *Alk.* 18. 3, *Arist.* 17.2. Note also P.'s variety of synonymous expressions in his conscious effort to avoid repetition: ἀλογίστως παράβολον καὶ τολμηρὸν (1.1), ἰταμός (1.2).

ἐκείνων τῶν κακῶν: the unspecified disease mentioned earlier at 1.2.

1.5 Συβαρίτης ἀνὴρ... δίαιταν: the Spartan heroic ideal of bravery is questioned by the Sybarite. Similar, but more detailed, versions of this anecdote are found in Athenaios (138d, 518e) and Diodoros (8.18.2). Although it has been observed that Athenaios occasionally lifted material from P. without acknowledgement (Sansone, 1988: 311), the details provided in his versions of the Sybarite anecdote cannot account for such a borrowing in this instance.

Despite what the man of Sybaris perceived as the real reason for the Spartans' indifference to death, it is widely attested that the virtue which the Spartan state desired above all others to inculcate in its soldiers was ἀνδρεία; this was tightly linked with the ideal of preferring death to dishonor (cf. the ἀνύβριστος τελευτή in *Pel.* 9.13); the dishonorable status of the cowards and the scorn and reproaches heaped upon them are widely attested: cf. Tyrt. frr. 7, 8 (Diehl), Xen. *Reip. Lac.* 9.1-2, *Hell.* 4.5.10, 6.4.16, Plut. *Reg. et imp. apoph.* 190B, *Apoph. Lac.* 210 F, 222F, 225D, *Inst. Lac.* 238A, *Mul. virt.* 240F, 241A-C, F.

πόνους... δίαιταν: the rigid training was regarded more than anything else as the code of honor for the Spartans; it made them, while still boys, accept pain and hardship without flinching. In Plato's *Laws* there is agreement that the bravery of the Spartan citizens is demonstrated by endurance (καρτερήσεις) of pain and discomfort, as well as, of pleasures (633B-C; cf. Plut. *Apoph. Lac.* 210A). The καρτερίας ἀγών, for instance, which, as Chrimes notes (1949: 136), was probably a comprehensive examination for the selection of the most courageous men, closed a young man's ephebic training and marked the beginning of his military career. On the simplicity of Spartan habits and their rigorous physical training, see Plut. *Lyk.* 10-12, 16-18, 22, *Ages.* 1.2, *Apoph. Lac.* 237A, *Inst. Lac.* 239D; cf. Thuk. 2.39, Xen. *Reip. Lac.* 2-5, *Anab.* 4.6.14-15, Isokr. *Panath.* 217, Ael. *V.H.* 7.13; on penalties for cowardice, Plut. *Ages.* 30.2-4, Xen. *Reip. Lac.* 9.1-6. On the Spartan agōgē, see H. Jeanmaire, *Couroi et Courètes*, Travaux et Mémoires de l'Université de Lille, ser. 21 (Lille 1939 [1975]) 463-588, Marrou,

1946: 216-230, Chrimes, 1949: 118-136, Michell, 1952: 165-204, A.H.M. Jones, *Sparta* (Oxford 1967) 34-39, Forrest, 1968: 51-53, A. Brelich, *Paides e Parthenoi*, vol. 1 (Rome 1969) 113-228, Cartledge, 1977: 11-27, esp. 16, id. "The Politics of Spartan Pederasty", *PCPhS* (NS) 27 (1981) 17-36, Lazenby, 1985: 3-4, 25, Kennell, 1995. For a recent interpretation of the state of humility and baseness imposed by the rigors of the Spartan agōgē, see Vernant, 1991: 220-243. On P.'s views about the philosophical aspect of the Spartan agōgē, see G. Schneeweiss, "History and Philosophy in Plutarch", in G.W. Bowersock, W. Burkert, M.C.J. Putnam, eds., *Arktouros, Hellenic Studies presented to Bernard M.W. Knox* (Berlin, New York 1979) 376-382.

1.6 Συβαρίταις... μαλακίας: the Sybarites (S. Italy) were known for their pursuits of self-indulgence and luxury: Arist. *EE* 1216a17-19, *Mirab.* 96 (Apelt), Athen. 138d, 518c-521a, 528b, Diod. 8.18.1-2; cf. also Suda s.v. Συβαρῖται: γάστριες ἦσαν καὶ τρυφηταί. Owing to their mode of life συβαρίζειν was identified with τρυφᾶν; see also Zenob. 5.19, 87 (*Paroem. Gr.*, 1, Leutsch/Schneidewin).

Athenaios, drawing on Timaios, connects their license with their subsequent destruction by Croton (*FGH* 566F50 [Athen. 519b-520c]; cf. also Str. 6.1.13); he also comments on their luxuriousness (518c) quoting lines from the *Δαιταλεῖς* of Aristophanes (Kassel–Austin, fr. 225); elsewhere, Athenaios mentions the works Περὶ Ἡδονῆς of Chamaeleon of Pontos (273c), Herakleides of Pontos (512a, 525f-526a, and Theophrastos (511c), as well as Aristotle's Θαυμάσια, as his sources for some of the anecdotes about the Sybarite mode of life. Such works could have been the source for P.'s anecdote here.

The mention of the Sybarite luxury and softness, conditions which instantly imply cowardice and attachment to life, creates a strong antithesis with the simplicity of the Spartan diet and their training to withstand pleasures and hardships; for a similar contrast between the Spartans and the Ionians, cf. Plut. *Alkib.* 23.5, or between the Spartans and the Persian prisoners of war in Ephesos, *ib.* 9.8, Xen. *Hell.* 3.4.19, *Ages.* 1.28. On τρυφή in Greek literature, see A. Passerini, "La τρυφή nella storiografia ellenistica", *SIFC* 11 (1934) 35-56.

ἐκτετηκόσιν: "wasted away": used metaphorically; cf. Eur. *O*. 134-135, Pl. *Resp*. 411b. The ἐκ is intensive.

τὴν πρὸς τὸ καλὸν ὁρμὴν καὶ φιλοτιμίαν: accusative of respect referring to the Sybarites; the prepositional constructions διὰ τὴν (KPQL[2]) or πρὸς τὴν (C) introduce a jarring hyperbaton.

μαλακίας: softness, hence cowardice, as opposed to Spartan bravery.

εἰκότως... τὸν θάνατον: with respect to Spartan fearlessness in war and the reasons that lie behind it, we find in the *Apoph. Lac*. a view contrary to that of the Sybarite, expressed by the Spartan Anaxandridas: it is not out of hatred of life that the Spartans are so bold, but out of respect for it (αἰδεῖσθαι περὶ βίου, 217A); on the notion of αἰδώς, see C.E. von Erffa, "ΑΙΔΩΣ und verwandte Begriffe in ihrer Entwicklung von Homer bis Demokrit", *Philologus Suppl*. 30.2 (1937) 1-203. Elsewhere in the same collection of sayings a certain Zeuxidamos claims that war is better than peace only for those men for whom death is better than life (221C). Fear of blame and ill-repute also occasions boldness in battle (Plut. *Kleom*. 9.4; see note above on 1.5). Xenophon expresses a view which echoes remotely the Sybarite's saying; after he has described in detail the threat of disgrace and dishonor that hangs over the head of a cowardly soldier, he concludes that "death seems preferable to a life so dishonored and ignominious", *Reip. Lac*. 9.6; cf. *ib*. 9.1.2, where he makes it clear that "to escape from death goes rather with valor than with cowardice". Death as a welcome release from a miserable life is advocated by Antigone (Soph. *Antig*. 463-464) and Sokrates (Xen. *Apol*. 9 and 33; cf. Pl. *Phaed*. 68d), but is strongly discouraged by Perikles in his funeral oration (Thuk. 2.43.5; see Gomme, 1956, vol. 2: 139) and is firmly disapproved of by Aristotle (*EE* 1230a27). What Perikles proposes, instead, is the blending of great daring with reflection upon the ventures to be undertaken (2.40.3; cf. the ἔμφρων ἀνδρεία recommended by Epameinondas, Plut. *Apoph. Lac*. 214C). The idea that death should be accepted and not sought is seen in an anecdote concerning Aristodemos, the most valiant Spartan warrior at Plataia, who was deprived of honors by his fellow citizens, because he aspired to death (Hdt. 9.71) having been accused in the past of cowardice (7.230-231); to seek one's own death is considered as a selfish act in *Agis/Kleom*. 52.10; on the characteristics of a 'fine death', see Loraux, 1986: 99-100.

1.7 Λακεδαιμονίοις... παρεῖχεν: the codd. have Λακεδαιμόνιοι δ' ἀμφότερα καὶ ζῆν ἡδέως καὶ θνῄκειν ἀρετῇ παρεῖχον ("the Lacedaemonians displayed the ability both to live pleasurably and to die virtuously"); this seems to be a better reading than Reiske's emendation, despite the shift of subject (Συβαρίταις μέν... Λακεδαιμόνιοι δὲ), which is not uncommon in P. (*Pel.* 3.6-7, 13.3, 31.1-2, 33.10); cf. an analogous construction in *Phok.* 16.7.

P. counters the Sybarite's objection to the noble motivation of Spartan bravery and proceeds to reinstate the heroic status of their ἀνδρεία by granting them the memory of song. He celebrates their ideal of an honorable death, because it is inextricably bound with an equally honorable life; cf. also *Inst. Lac.* 238A: τῶν γεννικῶς ζησάντων καὶ ὑπὲρ τῆς Σπάρτης ἀποθανόντων. For the model of ἀρετή in both life and death, see Thuk. 2.44.1, Soph. *Aj.* 479-480, Xen. *Ages.* 10.3, Plut. *synkr. Pel./Marc.* 33(3).4; cf. *De aud. poet.* 24D (Eur. fr. 994, Nauck). On death in battle in Greek and Latin literature, see M. Sordi ed., *"Dulce et decorum est pro patria mori"*, in *La morte in combattimento nell'antichità, Contributi dall'istituto di storia antica*, vol. 16 (Milan 1990).

ζῆν... θνῄσκειν: appositives to ἀμφότερα (Kühner-Gerth, 4:4), hence without the article; for arguments against ἀμφότερα, see Ziegler, 1934: 234.

οὐ τὸ ζῆν θέμενοι... ἐκτελέσαι: the same couplet—completed with οἳ θάνον at the beginning—is found in *Consol. ad Apoll.* 110C, and P. there specifies its Lakonian origin. Goettling suggested that the elegiac couplet is clearly Lacedaemonian and that it must have belonged to a Spartan tomb (1851: 320); he assigns these verses to Tyrtaios' *Εὐνομία* (Diehl, 1949, 7-18). Stobaios (4.44.83) attributes the same couplet, preceded by the words καὶ ἐπιγράφουσι Λακεδαιμόνιοι, to the *Περὶ 'Απαθείας* of Teles (Hense[2], 1969, 59, 4-5). Bergk attributes the lines to Simonides, and proposes that it was inscribed on the Spartan tomb at Leuktra (1882, vol. 3: 516). C. Buresch ("Consolationum a Graecis Romanisque scriptarum historia critica", *Leipziger Studien zur Klassischen Philologie* 9 [1886] 37) assumes that either P. derived the couplet from Teles, or that they both drew from the same source. P.

omits here the first foot of the dactylic hexameter, since it is not related to the context. A similar idea is expressed by Menander (fr. 592, Kock):

⟨οὐκ ἔστι τὸ ζῆν αὐτὸ καθ' ἑαυτὸ γλυκύ,⟩ (suppl. Cobet)
οὐδ' αὖ τὸ θνῄσκειν δεινόν, ἀλλ' ἐν τῷ καλῶς
ἑκάτερον αὐτῶν τὴν διάγνωσιν φέρει.

καλῶς: the adverb may also refer to the care the Spartans took before the battle to die in "a fair and fitting manner"; cf. Hdt. 7.208-209, Xen. *Reip. Lac.* 13.8, Plut. *Lyk.* 22.1-2. For an analysis of the theme of καλὸς θάνατος in the Athenian funeral oration in the classical city, see Loraux, 1986: 98-118; for a study of the same topic in the Spartan agōgē, see ead. 1977: 105-120; on the Homeric ideal of 'beautiful' and 'glorious death', see Vernant, 1991: 50-74, esp. 65-67.

1.8 οὔτε γὰρ φυγὴ... τοῦ ζῆν: P. reformulates the compressed message conveyed by the preceding couplet, in order to ensure its broader applicability, and not confine it to the Spartan ethic. He makes a conscious effort to avoid any type of repetition by way of periphrasis: φυγὴ θανάτου (τὸ ζῆν), οὔτε... μεμπτόν (καλόν), μὴ αἰσχρῶς (καλῶς), and ὑπομονὴ (sc. θανάτου) parallels τὸ θνῄσκειν. P. seems to have in mind the Aristotelian idea about courage as the observance of the mean, which the philosopher expresses in similar negative terms (*EN* 1116a13): τὸ δ' ἀποθνῄσκειν... οὐκ ἀνδρείου, ἀλλὰ μᾶλλον δειλοῦ· μαλακία γὰρ τὸ φεύγειν τὰ ἐπίπονα, καὶ οὐχ ὅτι καλὸν ὑπομένει (i.e. ὁ ἀποθνῄσκων), ἀλλὰ φεύγων κακόν. ἔστι μὲν οὖν ἡ ἀνδρεία τοιοῦτόν τι. For similar views, cf. Eur. *Hek.* 377-8 (Stob. 4.53.20), Philemon, fr. 203, Kock (Stob. 4.53.8). The observance of the mean is, however, beyond Ajax' reach (Soph. *Aj.* 473-480): καλῶς ζῆν is denied to him; therefore the next best thing is to die honorably (καλῶς τεθνηκέναι); cf. also Lys. 2.69, Pl. *Menex.* 246, *Crit.* 48B.

γάρ: causal and explanatory. P.'s further explanation forms a confirmation of the preceding views about bravery and cowardice.

1.9 ὅθεν Ὅμηρος... ἀγῶνας: the themes of bravery and cowardice cannot be adequately explored without reference to the epic tradition; cf. the newly discovered poem of Simonides for Plataia, which ties the actions of the Spartans at Plataia to the heroic age discussed in *Arethusa* 29.2 (1996). P. is, of course, not interested in the Homeric concept of martial ἀρετή and the logic of heroic honor, according to which the renown of a hero is guaranteed by his glorious death which alone will ensure his imperishable glory; on the Homeric ideal of life and death, see J. Griffin, *Homer on Life and Death* (Oxford 1980). Even though the celebration of the epic 'beautiful death' retained its importance during the classical period (cf. the Athenian funeral orations, Loraux, 1986), death is not seen exclusively as the only source of all ἀρετή, and poets attribute no less valor to the living than to the dead warriors (Pind. *I.* 7.23-30, Tyrt. 9.35-42, Diehl). The historical context of the city-state is far removed from the Homeric world, and P. has to confront the study of ἀνδρεία in terms of the political and national interests of Pelopidas' times. Loraux draws attention to the transition, in the early fourth century, from a civic ethic to a private one, which she attributes to the ever-increasing importance acquired by exceptional individuals on whose initiative and decision depended the fate of the entire army (*ib.* 111); she also notes that the private stelae of the fourth century lay emphasis on individual wisdom consisting of moderation and domestic virtues; the epitaphs declare that virtue is practiced throughout life and that ἀρετή is always accompanied by σωφροσύνη (*ib.* 109). The traditional view of military bravery identified with the 'beautiful death' is, therefore, no longer an adequate definition of ἀρετή, since the significance of the latter is no longer coextensive with that of ἀνδρεία (*ib.* 110). Without rejecting the time-honored model of ἀνδρεία, P. adapts it to the new political exigencies: the security and safety of the commander of the army have become of foremost importance, much more so than in the Homeric value system.

1.10 οἱ δὲ τῶν Ἑλλήνων νομοθέται: P. probably refers to Solon's legislation of the punishment prescribed for the ἀστράτευτος, the λιπὼς τὴν τάξιν and the δειλός, described at length by Aeschines (*In Ctes.* 175-176). P. makes no mention of such provisions in the Athenian legislator's biography.

τὸν ῥίψασπιν... προέμενον: this legislative injunction is recorded more extensively and with minor differences in the *Apoph. Lac.* 220A in the form of a saying attributed to Demaratos: the Spartan king justifies the scorn reserved for those who have lost (ἀποβαλόντας; cf. ῥίψασπις at 1.10) their shields but not their helmets or their breastplates (cf. the ξίφος and λόγχη at 1.10), on the ground that while the former guarantee the common good of the whole formation, the latter are intended for the soldier's individual safety. Since in a phalanx drawn up in close order the bearer's shield protected also his left-hand neighbor, one can guess how the loss or casting away of a shield could put at stake the safety of the line in a phalanx. The shield is thrown away in connection with turning and running, being the heaviest piece of equipment easily divested. It seems, however, that the Spartans punished the loss of a shield regardless of circumstances, even if, for instance, it slipped from the bearer's grip in the thick of battle. For the disgrace involved in the loss of the shield and the value attached to it, cf. Archil. 6, Diehl, Plut. *Inst. Lac.* 239B, Anakr. 5, Diehl, Ar. *Nu.* 353 (cf. Dover's note in *Aristophanes' 'Clouds'* [Oxford 1989] 147-148), Plat. *Leg.* 944c), Aeschin. *In Tim.* 29, *In Ctes.* 175-176, Dion. Hal. *Ant. Rom.* 9.53.4, Diod. 15.87.6, Polyb. 6.37.11, Plut. *Reg. et imp. apoph.* 194A, *Lacaen. apoph.* 241F, Sext. Emp. *Pyrrh.* 3.216, Plaut. *Trin.* 1034, Cic. *De orat.* 2.294, Hor. *O.* 2.7.9-10, Tac. *Germ.* 6.6; on the military significance of the shield, see Anderson, 1970: 14-17, Pritchett, 1985: 31-33; for a discussion of the famous passage about Archilochos' throwing away his shield and the act of ῥιψασπία, see T. Schwertfeger, "Der Schild des Archilochos", *Chiron* 12 (1982) 253-280.

2.1 εἰ γάρ... στρατηγός: a similar version is preserved in Polyainos' collection of stratagems (3.9.22); while the Macedonian rhetorician's dependence on P.'s *Mulierum virtutes* for his collection of examples of generalship has been convincingly shown by Stadter (1965: 13-29), it is difficult to demonstrate a similar type of dependence in this particular instance. P.'s collection of Iphikrates' six sayings in *Reg. et imp, apoph.* 187A-C—if indeed this work is genuine—has absolutely nothing in common with that of Polyainos, which includes sixty-three anecdotes (3.9). The verbal similarity between the two accounts may be easily explained if we suppose that both authors closely followed the same source. Different versions of the same anecdote are found in P.'s. *De fort.* 99E, *Reg. et imp. apoph.* 187B (cf. Stob. 4.13.52) and *An virt. doc. poss.* 440B. For bodily similes used in a military context, see Plut. *Pel.* 19.4, *Per.* 12.6, *Ages.* 33.3, *Marc.* 9.7, *Philop.* 2.5-6, 7.6, Stob. 4.13.50.

'Ιφικράτης: Athenian general (415-353 BC), son of Timotheos of Rhamnous; he was one of the most distinguished generals according to Diodoros (15.88.2, 16.85.7); he was sent by the Athenians in 370 to rescue Sparta during Epameinondas' invasion of the Peloponnese (Xen. *Hell.* 6.549-52); see Diod. Bks 14-16 passim, Xenoph. *Hell.* 4-6 passim, Dion. Hal. *Lys.* 12, Plut. *Reg. et imp. apoph.* 187A-C, Polyain. *Strateg.* 3.9, Nepos, *Iphicrates*, Ampel. *Lib. Mem.* 15.16. See further, Rehdantz (1844), Kirchner, *PA*, 7737, Kahrstedt and Thalheim, *RE* s.v. "Iphikrates", 9, 1916: cols. 2019-2022, Davies, *APF*, 7737, Mosley, 1974: 65-68, Pritchett, 1974: 62-72, Kallett, 1983: 239-252, Develin, 1449.

οὐχ αὐτοῦ... τοὐναντίον: for a similar idea, see Xen. *Mem.* 3.2.1, Eur. *Phoen.* 599. P. explains and elaborates on Iphikrates' apophthegm laying stress on the significance of the role of the commander of an army for the safety of the army (cf. Stob. 4.13.50). Almost all of the previous anecdotes (*Pel.* 1.2-10) centered on ordinary, anonymous soldiers and their attitude to death. The new set of *exempla* is strictly confined to well-known commanders, since the general is, in Aristotle's terms (*Metaph.* 1075a), the army's greatest asset. Iphikrates' comparison of the military units to specific parts of the body—the most important of which is the head representing the general—is heartily approved of by P. The wisdom of the saying appears to be in agreement with an anecdote preserved by Polyainos in which Iphikrates is praised for his military sagacity (3.9.28). In the *Life*, Pelopidas is seen to emphasize before the battle at Leuktra the importance of the commander for the safety of his army (20.2), although he does not abide by this principle at Kynoskephalai (32.9). The theme of the loss of a considerable military force incurred by the loss of the commander of the army recurs often in P.; cf. *Ages.* 11.5, *Arist.* 14.7-8, *Fab.* 14.20, *Marc.* 9.2-4.

2.2-6 P. relates three risk-involving incidents. His level of approval or disapproval is shown by the climactic use of the qualifying adverbs οὐκ εὖ (2.2), βέλτιον (2.4) and καλῶς (2.6). P. rounds off the set of his examples with Timotheos' forthright statement in the presence of boastful Chares about the latter's recklessness as a commander at the siege of Samos, thus reconfirming and reinforcing the validity of his own argument about the consequences of the commander's safety upon the army (2.1).

2.2 ὁ Καλλικρατίδας: a Spartan admiral, who commanded the Spartan fleet at the battle of Arginousai in 406 BC against the Athenians under Thrasyboulos, and was killed in that battle (cf. *Pel.* 2.3, *Lys.* 7.1); on Kallikratidas, see Plut. *Apoph. Lac.* 222C-F, Xen. *Hell.* 1.6.1-36, Diod. 13.76-79 and 97-99, Plut. *Lys.* 5.5-7.1. See further G. Busolt, *Die Lakedaimonier und ihre Bundesgenossen* (Leipzig 1878) 25-26, Praechter, *RE* s.v. "Kallikratidas", 10, 1917: cols. 1641-1643, Poralla, 1985, No. 71, G. Ronnet, "La figure de Callicratidas et la composition des 'Helléniques'", *RPh* 55 (1981) 111-121, Roisman 1987: 21-33, Proietti, 1987: 10-27.

καίπερ ὢν τἆλλα μέγας: P.'s introduction of Kallikratidas' saying with a concessive clause immediately predisposes the reader for a negative evaluation of his response to the seer and the impending disaster, namely his defeat and death at the battle off the Arginousai islands.

οὐκ εὖ... τὰ Σπάρτας: in the *Apoph. Lac.* 222E-F P. reports two distinct exchanges between Kallikratidas and Hermon the helmsman (not the seer, as P. reports in the *Life*). The first exchange, as in Xenophon (*Hell.* 1.6.32-33), revolves around the enemy's overwhelming superiority in numbers, but then—in a departure from Xenophon—has Kallikratidas respond that to retreat would be a disgrace for Sparta. The second, as in *Pelopidas*, has the omens presage death for the commander and victory for the fleet—although this is an instance of a major failure of divination—with Kallikratidas retorting οὐ παρ' ἕνα τὰ Σπάρτας. P. chose the second version for his biography, but omitted the detail of the omens' predicting victory. He hints at Kallikratidas' recklessness—he does not criticize him overtly, but by implication—in the preceding sentence, when he makes a general observation about the commander, who in taking undue risks and being overbold, would seem to neglect not himself, but everyone. Moreover, he does not mention the reason for the Spartan's death. Kallikratidas, unlike Pelopidas or Marcellus, in reality did not rashly expose himself to personal danger, but fell overboard as his flagship rammed an enemy (*Hell.* 1.6.33). His fault, as P. says, consisted in dismissing a sign portending his own death as a purely personal matter: for the general is not just one man, but the head of the army; his life, therefore, is more important than the common soldier's, and if his reckless loss of it causes defeat, he has failed in his duty. It should be noted that by implication P. attributes the Spartan defeat at the Arginousai to Kallikratidas' death in battle—a view for which there is, of course, no evidence whatsoever. P. eliminates in the *Life* the wrong prediction of victory for the Spartan fleet, which is reported in the second version in the *Apoph. Lac.* 222F, because he would be hard pressed to explain why certain omens are not reliable, and therefore they should not be trusted.

Cicero uses Kallikratidas as an example of a commander who, though he sacrificed his life for his country, would not make even the slightest sacrifice for personal glory (*De off.* 1.24.84). He also cites Fabius Maximus as a counter-example to the Spartan's attitude in war. Kallikratidas' response in the *Pelopidas* is offered as an example of misguided modesty.

μὴ παρ' ἕνα τὰ Σπάρτας: an expression which became proverbial; see Breitenbach, 1873: 61n. 32. Kallikratidas' response is a case of misguided modesty.

τὰ Σπάρτας: a Doric form for τὰ Σπάρτης; "the affairs of Sparta", "the fate of Sparta"; cf. the similar use of the genitive at 2.5 (τὸ τῆς ἀρχῆς), which may have been the reason for the emendation (Wyttenbach, following Xylander); the mss. read τὰ σπάρτα (L^1K *mor.*), τὰ σπαρτά (P), τὰν σπάρταν (QL^2).

2.3 οὐκ ἦν εἷς... συναπώλλυτο: P. reformulates what he has already said at 2.1 about the responsibilities of a general to his soldiers. Note also P.'s use of the same source-material in order to make different points in his different works: in *Lys.* 7.1, he praises Kallikratidas for being able to compete with the very best of the Greeks because of his greatness of mind and courage. Although he mentions his death at Arginousai, he avoids giving any of the negative details offered in the *Pelopidas*.

2.4 βέλτιον δ' 'Αντίγονος: the comparative adverb provides a straightforward contrast with the preceding anecdote; see note below on 2.5.

'Αντίγονος ὁ γέρων: whom P. meant here is not immediately clear. Momigliano, arguing against Antigonos Doson on the ground that he died at the age of forty-two, narrows down the possibilities to either Monophthalmos or Gonatas (1950: 107). In P. (*De fort. Alex.* 330E) 'Αντίγονος ὁ γέρων is almost certainly Gonatas (Tarn, 1969: 251). The nickname γέρων was given most likely to Antigonos Gonatas since at the end of his life he had outlived every one of his contemporaries, namely the two Antiochi of Syria, Pyrrhos of Epiros, Areus of Sparta, Magas of Cyrene, Demetrios the Fair, Philetairos of Pergamum, Alexander of Corinth and finally Ptolemy Philadelphos of Egypt. He alone remained the 'old man' of the political world. He was king of Macedonia from 276 to 239 BC and was about seventy-three years old, when he appeared in the Aegean in the spring of 246 at the head of his fleet. See Fellmann, 1930, Tarn, 1969: 378, 461-466, E.L. Brown, "Antigonos Surnamed Gonatas", in G.W. Bowersock, W. Burkert, M.C.J. Putnam, eds., *Arktouros, Hellenic Studies presented to Bernard M.W. Knox* (Berlin New York 1979) 299-307, S. Cioccolo, "Enigmi dell'ἦθος: Antigono II Gonata in Plutarco e altrove", *Studi Ellenistici III*, Biblioteca di Studi Antichi 64, Pisa (1989) 135-190.

ὅτε ναυμαχεῖν... ἀντιστήσεις;: the same saying is repeated twice in P. at *De laud. ips.* 545B and *Reg. et imp. apoph.* 183C, where instead of the battle of Andros, he has Kos, and instead of ὁ γέρων he writes 'Αντίγονος ὁ δεύτερος. Momigliano claims that P.'s confusion between Kos and Andros is not likely to have arisen if the two battles had been fought by two different kings (1950: 108; cf. Tarn, 1909: 267). The battle of Kos was clearly Antigonos Gonatas' great victory over Egypt (Athen. 209e; cf. Momigliano, 1950: 113); as for that of Andros, Tarn argues that it was a victory won by Gonatas over Sophron (Opron) (Trog. *Prol.* 27, Seel), the admiral of Ptolemy Evergetes (1909: 264, 273-274). Momigliano, relying on the more recent evidence of *Pap. Haun.* 6 (Larsen, 1942), arrives at the same conclusion; he assumes that the reference in line 9 of the papyrus is to the sea-battle of Andros, in which the son and co-regent of Ptolemy Philadelphos probably fought against his father at the side of Antigonos Gonatas. Although on Momigliano's own admission the lacunose condition of the papyrus does not lend any substantial evidence for Gonatas' actual participation in the battle of Andros (ib. 111), basing his arguments on the chronological sequence of events included in the papyrus—presumably of the Ptolemaic dynasty—he suggests cautiously that a likely date for the battle at Andros is c. 258 BC, about four years after the sea-battle at Kos (ib. 111-115). If Momigliano's hypothesis about the dates of the two battles is right, P.'s comment on Antigonos' ἐμπειρία and ἀρετή at 2.5 becomes more meaningful: it could mean that Antigonos had acquired experience and had been victorious in a previous battle, namely that of Kos. His earlier victory and expertise could, therefore, excuse his arrogance at the later battle at Andros.

For further discussions of the dates and the protagonists of these two battles, see G. De Sanctis, "La ribellione d' Alessandro figlio di Cratero", *Klio* 9 (1909) 1-9, Tarn, 1909: 264, id., 1910: 209-222, id. 1969, W.S. Ferguson, "Egypt's loss of Sea Power", *JHS* 30 (1910) 189-208, Fellmann, 1930: 66, D. Peremans, "De Zeeslag Van Kos en de Opstand Van Ptolemaios Van Ephese", *Rev. Belge de Philol. et d' Hist.* 12 (1933) 49-58, id. "La Date de la Bataille Navale de Cos", *Antiquité Classique* 8 (1939) 401-408, E. Bikerman, "Sur les batailles navales de Cos et d' Andros", *REA* 40 (1938) 369-383, A.D. Momigliano and P.M. Fraser, "A New Date for the Battle of Andros? A Discussion", *CQ* 44 (1950) 107-118, Gallo, 1975: 77-78, 86-94, F.W. Walbank in *CAH* 7.1, 1984: 221-256.

ἐμὲ δ' αὐτὸν... ἀντιστήσεις;: the point of Antigonos' rhetorical question seems to be that his presence and previous successful record were adequate to guarantee the victorious outcome of the sea-battle at Andros. In *De laud. ips.* 545B, P. recounts the same anecdote but in a different light; he is puzzled with Antigonos' boastful remark, because he thinks that it is not in keeping with his moderate character. He carefully chooses, however, to omit in the *Life* any disquieting comment on Antigonos' arrogant words, in order to provide a clear-cut contrast with Kallikratidas' thoughtless response (2.2); note that the concessive clause (καίπερ ὢν τἆλλα μέγας, 2.2), which is suggestive of a forthcoming negative evaluation of Kallikratidas, does not appear in *Apoph. Lac.* 222F—if we assume, of course, that this compilation represents P.'s raw material for the *Lives*—, on which see Ziegler, 1964, cols. 228-230.

W. Tarn suggested that if ἀντιστήσεις means "to range in order of battle against", then ἐμέ may refer to Antigonos' famous Delian ship ("The dedicated Ship of Antigonos Gonatas", *JHS* 30 [1910] 218); this conjecture would allow us to interpret Antigonos' saying as "against how many ships do you range this of mine?" It is doubtful, however, that P. would expect his readers to make instantly a figurative association between the personal pronoun ἐμέ with Gonatas' ship. In addition, since P.'s emphasis is on Antigonos' personal experience and valor as a commander, it would make more sense to take ἐμέ literally referring to Antigonos himself.

2.5 μέγα... σῴζοντα: Antigonos' words set the tone for P.'s moralization upon the model of ideal leadership, that which is followed by ἐμπειρία καὶ ἀρετή exemplified by Antigonos (cf. Arist. *EE* 1229a14-15, 1230a4-10, *EN* 1116b3-16, Plut. *Apoph. Lac.* 214C, *De laud. ips.* 545C-D). P. appears to have based his criteria for evaluation on the final outcome of each commander's enterprise; while Kallikratidas' military operation resulted in total disaster, Antigonos'—so we are led to understand—was victorious; cf. Pelopidas' jocular response to a frightened soldier at 17.2 before the battle at Tegyra; had Pelopidas been defeated by the Spartans at Tegyra, the self-assured tone of his answer might have been the object of P.'s reproach.

2.6 Τιμόθεος: son of Konon of Anaphlystos and pupil of Isokrates; he was elected strategos in 378, when the Second Athenian League was founded; see Isokr. *Antid.* passim, Diod. 15.29.7, 88.2, 16.85.7, Plut. *Reg. et imp. apoph.* 187C, Polyain. *Strateg.* 3.10, Nepos, *Timotheos*. See also Kirchner, *PA,* 13700, Rehdantz, 1844, Klee, *RE* s.v. "Timotheos, 3", 6.2, 1937: cols. 1324-1329, Cloché, 1934: 66-69, 76-78, 126-128, id. 1951: 318, G. Barbieri, *Conone*, *Studi pubblicati dall'istituto italiano per la storia antica*, 13 (Rome 1955), Davies, *APF*, 13700, pp. 506-512, Kallet, 1983: 239-252, Develin, 3112.

ἐπιδεικνυμένου... τοσαύτης: the anecdote is given in less detail and dramatic effect at *Reg. et imp. apoph.* 187C, where Chares is not named, but is referred to as a daring general. For the exhibiting of war wounds, cf. *Brut.* 20.4. On Chares' typical boastfulness, see *schol. Demosth. Orat.* 4.19 (84b, Dilts), Plut. *Arat.* 16.3. Timotheos' admission of carelessness is meant to create a sharp contrast with Chares' bragging over his war-wounds; on Timotheos' contemptuous opinion of Chares' political abilities, see *An sen. resp. ger.* 788D-E (Stob. 4.13.43).

τοῦ Χάρητος: Chares, son of Theochares of Angele, was an Athenian general (c.400-c.325). The *De Pace* of Isokrates is replete with covert attacks upon him; see also Arist. *Rhet.* 1418a32, Nep. *Chabrias* 3.4, Plut. *Reg. et imp. apoph.* 187C, *An sen. resp. ger.* 788D-E, *Phok.* 14.3-4, *Arat.* 16.3, Arsen. 398 (Walz); Kirchner, *PA*, 15292, id., *RE* s.v.

"Chares, 3", 3.2, 1899: cols. 2125-2128, Parke, 1933: 122-123, 144-145, Davies, *APF*, 15292, pp. 568-569, Pritchett, 1974: 77-85, Develin, 610.

πολιορκοῦντος Σάμον: in 367/366 BC Timotheos was given command of a force of thirty triremes and 8,000 mercenary peltasts, and was instructed to support the satrap of Hellespontine Phrygia, Ariobarzanes, without openly violating the King's Peace. Timotheos started his operations by attacking Samos, which was then held by a Persian garrison under the authority of Tigranes. The island fell after a ten-month siege; cf. Demosth. 15.9, Isokr. *Antid.* 111, Deinarch. 1.14, Ps.-Arist. *Econ.* 2.1350b4-15, Diod. 16.21.2, Nepos, *Timoth.* 1.1, 3.1, 3.3, Polyain. *Strateg.* 3.10.5,9; also Beloch, 1923, 3.2: 255, Parke, 1933: 108-109, Buckler, 1980a: 166-167, S. Hornblower, *Mausolus* (Oxford 1982) 135-136, 170-172, 197, G. Shipley, *A History of Samos 800-188 B.C.* (Oxford, New York 1987) 138-140.

μειρακιωδέστερον... τοσαύτης: this part of the saying is omitted from *Reg. et imp. apoph.* 187C. It is likely that P. himself elaborates on the saying, as a part of his technique of reshaping anecdotes according to the effect he wants to produce. The idea that Timotheos admits to behaving more like an impetuous youth than like a general and commander of a large force recurs again when P. comments on Marcellus' uncontrollable anger against Hannibal (*Marc.* 28.5) and his death, *synkr.* 33(3).6.

μειρακιωδέστερον: in the comparative degree the word occurs only three times in Greek literature (Deinarch. fr. 73.2.4, Phil. *Legat.* 42.6, Galen, *Hippocr. Progn.* 18 b.237.3), while the positive μειρακιωδῶς is used only once by P. (*Mar.* 34.5); cf. the cognate ἀντιμειρακιεύεσθαι (*Su.* 6.4), which is a hapax and is used by P. also with reference to Timotheos.

2.7-8 ὅπου μὲν... στρατηγοῦ: P. is probably thinking here of Aristotle's definition of the μεγαλόψυχος, who "does not run into danger for trifling reasons, and is not a lover of danger because there are few things he values; but he will face danger in a great cause, and when so doing will be ready to sacrifice his life, since he holds that life is not worth having at every price" (*EN* 1124b23; cf. also Cic. *De off.* 1.23.81). P. highlights the significance of a commander's decisions in battle, which should be determined not by his own impulse, as happened in the case of

Kallikratidas (2.3), or in the case of the protagonists under study, Pelopidas and Marcellus (2.11), but by reason. P. occasionally allows room for risk-taking activities only after the general has weighed the pros and cons of his deliberation. He is prepared, however, to make more concessions and excuses for the rashness of Pelopidas, than for that of Marcellus, even if the former's death was clearly detrimental for Thebes. In the concluding *synkrisis* (33(3).4), while he concedes that the best thing is that a general gain a safe victory, in an attempt to exalt the noble motivation of the Theban's rashness and death, he adds that the next best thing is what is proposed by Euripides, namely θανεῖν εἰς ἀρετὴν καταλύσαντα βίον (fr. 994, Nauck; cf. Plut. *De aud. poet.* 24D).

χειρὶ... ἀφειδῶς: a man giving his life for his country is a notion familiar in patriotic speeches, and particularly in funeral speeches; see Loraux, 1986: 98-118, Wilkins, 1990: 179-180.

χρὴ... θνῄσκειν: "it is best for a good general to die *of* old age and if not to die *in* old age". There is a contrast involved between the two parts of the saying: in the first one the death of an old general is caused by old age, and not by an enemy, while in the second, his death is caused in old age by a reason other than old age; in the context of P.'s argumentation about the possible advantages and disadvantages of the loss of a general in battle, the only other cause of the latter form of death seems to be occasioned by the battle itself. The key idea is that a good general should avoid death in battle and die naturally, of old age, not on the battlefield; less good, but still acceptable, is the death of an old general, as opposed, apparently, to that of a young general. P. seems to be disagreeing with the second part of the saying, for he disapproves of rashness in battle regardless of age, as in the case of Marcellus, *Marc.* 28.6.

οὐδεὶς ἀπαιτεῖ... στρατηγοῦ: for the same concept, see *synkr.* 33(3).6, *Sert.* 13.6 (Theophr. fr. 140, Wimmer).

2.9-12 P. offers here a preliminary justification for the choice of his protagonists and the decision to compare the two lives, before embarking on the main narrative. In this section the essential similarities in the character and achievements of Pelopidas and Marcellus are briefly enumerated, to be reviewed at greater length in the concluding *synkrisis* (31(1)-33(3). The only trait, however, that ensures a recognizable likeness between the two heroes is their recklessness and disregard for personal safety, an imperfection which P. explored in its various ramifications in the preceding narrative. Pelopidas' hot-temper, channelled into an uncontrollable anger against the tyrant Alexander, and Marcellus' natural bellicosity, which manifests itself in an obsessive passion to vanquish Hannibal, are seen as weaknesses, partially because of their inherent natures and partially because of the external circumstances by which they are triggered (*Pel.* 25.4, 32.9, *Marc.* 1.2, 28.4-6, 29.3, *synkr.* 33 (3).3-6). Considering P.'s rigid, moralistic tone and his readiness to apportion equal praise and blame to both heroes, one would expect these weaknesses to emerge in the main narrative, as well as in the concluding *synkrisis*. No such consistency is, however, felt. For instance, we observe that the two protagonists' equally blameworthy recklessness gradually loses its initial cohesion and deteriorates finally into dissimilarity (cf. *synkr.* 33(3).3-7). For further discussion, see [**2d**], and Georgiadou, 1992a: 4222-4257. Also Pelopidas' boldness, when tempered with reason, led him to victory at Tegyra and to a successful rush forward at Leuktra.

προαναφωνῆσαι: an uncommon word in P.; cf. *De ex.* 607C, *De es. carn.* 996B.

παραλόγως: cf. σὺν οὐδενὶ λογισμῷ at 2.11, *synkr.* 33(3).1; also Polybios' characterization of Pelopidas' recklessness at 8.35.8, 36.1.

προέμενοι... ἀρχόντων: P. does not dwell on the damaging effects the two protagonists' death had on their respective countries. In the case of Pelopidas, we learn only of how the news of his death at Kynoskephalai was received by the Thebans' allies, the Thessalians (33.8-10, 35.1). Polybios, on the other hand, probably drawing on Kallisthenes (see [**2c**] s.v. "Polybios"), is harshly critical of Pelopidas' imprudence toward the tyrant Alexander (8.35.6-8) and is outspoken about the harmful consequences this aspect of his character had on Thebes; cf. Diodoros, who, reflecting most likely Ephoros' views, connects the loss of Theban hegemony in Greece with Epameinondas' death, 15.88.4.

Chs. 3-4: Introduction
Pelopidas and Epameinondas

P. has little to say about Pelopidas' lineage, family, character and education. The lack of personal details may be attributed to the fact that he had no biography of his Theban hero to rely on, and was, therefore, obliged to confine himself to information supplied to him chiefly by historical sources [**2a**]. To compensate for this shortage he provides a portrait of Pelopidas' moral virtues by way of comparisons with his friend and collaborator Epameinondas [**2e**]. P. switches swiftly from one hero to the other and outlines their outstanding moral qualities, which on several occasions anticipitate future attitudes and events. On P.'s political and compositional reasons for letting Pelopidas' personality emerge through Epameinondas, see [**2e**], and Georgiadou, 1992a: 4222-4257.

3.1 Πελοπίδᾳ... Θήβαις: our knowledge of Pelopidas' ancestry is limited almost entirely to P.'s brief reference to it; cf. the inscription published by Bousquet (1939: 126, and 1963: 207, line 5: Πελοπίδαν Ἱπ[πόκλου]) which was attached to the statue dedicated at Delphi by the κοινόν of the Thessalians; note also P.'s comment on the lack of genealogical information on Epameinondas, *Ages*. 19.6.

ὥσπερ Ἐπαμεινώνδᾳ: on Epameinondas' family, see *De gen.* 583C-D, 585D-E, *Ages*. 19.9, Paus. 9.13.1, 4.31.10, Ael. *V.H.* 2.43, 3.17. See further, Pomtow, 1870, Vischer, 1877, Swoboda, 1900: 460-475, id., *RE* s.v. "Epameinondas", 5.2, 1905: cols. 2674-2707, Peper, 1912, K. von Fritz, *Pythagorean Politics in Southern Italy* (New York 1940) 78, Carrata Thomes, 1952, Fortina, 1958, Shrimpton 1970, Κουμανούδης, 1979, No. 634, Bengtson, 1983. On Epameinondas' patronymic, see the inscription published recently by W. Blümel, "Two New Inscriptions from the Cnidean Peninsula: Proxeny Decree for Epameinondas and a Funeral Epigram", *Epigraphica Anatolica* 23 (1994) 157-159.

οἶκον: Marasco notes that this term, as opposed to the κλῆρος, is more vague and denotes in P. "property" ("La Retra di Epitadeo e la Situazione Sociale di Sparta nel IV secolo", *Antiquité Classique* 49 [1980] 131-145, esp. 139); cf. *Arist*. 1.2, *Lys*. 30.2, *Philop*. 4.5, *Su*. 25.4, 33.3, *Pomp*. 36.6. For a discussion of the different meanings of οἶκος—"house", "property", "lineage", "nuclear family", "progeny" or "persons within a household, including slaves"—, see B.S. Strauss, *Fathers and Sons in Athens: Ideology and Society in the Era of the Peloponnesian War* (Princeton 1993) 33-36; id. 1990: 101-127, esp. 101-104, MacDowell, 1989: 10-21.

ὥρμησε... βοηθεῖν: Pelopidas is portrayed here as following the precept (perhaps Pythagorean; Swoboda, *RE* s.v. "Epameinondas", 5.2, 1905: col. 2678) according to which "friends' goods are common property" (Plut. *Quom. adul.* 65A, Diog. 8.10, Porph. *De vit. Pyth.* 33), on which see J.A. Philip, *Pythagoras and Early Pythagoreanism* (University of Toronto Press 1968) 142-143. On the unselfish sharing of possessions with one's own friends, see Arist. *EE* 1238a16-17, *EN* 1159b31-32, Men. *Adelph.* fr. 9 (Kock), Dio Chrys. *Orat.* 3.110, Plut. *De frat. am.* 490E, *Praec. ger. reip.* 808B, Ter. *Adelph.* 803; cf. J.A.F. Delgado, "Nueva contribución al estudio de los proverbios en *Moralia*", in J.G. López and E.C. Dorda, eds., *Estudios sobre Plutarco: paisaje y naturaleza*, Actas del II Simposio Espanol sobre Plutarco, Murcia 1990 (Madrid 1991) 264.

ἵνα κύριος... δοῦλος: P. characterizes Agesilaos' attitude to money in similar terms: *Praec. ger. reip.* 809B. He has much to say on the kind of corruption which stems from great wealth, though he does not consider that moral degeneracy is a necessary consequence of money (Swain, 1990: 127); cf. *Artax.* 24.9, *Cam.* 2.6, *synkr. Nik./Crass.* 34(1).4, *De cupid. div.*, frr. 'On Wealth': 149-151, Sandbach, 1969: 276-278. P. himself was quite wealthy.

χρημάτων: Ziegler's suggested addition of a definite article (⟨τῶν⟩ χρημάτων) is unnecessary; cf. *Praec. ger. reip.* 809B, Eur. *Suppl.* 877.

δοῦλος: (sc. χρημάτων); the same metaphor is found in Eur. *Suppl.* 877, *Hek.* 865, Arist. *Rhet.* 1394b6. Figurative expressions describing subjugation to money or other passions are not uncommon: Eur. fr. 282 line 5, Nauck, Plat. *Resp.* 329c, Plut. *An sen.* 788E, *Non posse* 1094E, Ael. *V.H.* 2.41. The related notion of φιλοπλουτία is described in analogous terms as a βαρεῖα καὶ πικρὰ δέσποινα in *De cupid. div.* 525B. See further, O'Neil, 1978: 289-362.

3.2 τῶν γὰρ πολλῶν... διατελοῦσιν: (fr. 56, Rose); while Ross (1955: 56), followed by Laurenti (1987, 2: 653-663), assigns it together with *De cup. div.* 527A to Aristotle's lost work περὶ πλούτου, Rose suggests (1967: 66) that this quotation probably comes from the philosopher's πολιτικὸν βιβλίον. It is impossible to define the exact length of Aristotle's moral comment (R. Klaerr and Y. Vernière, *Plutarque, Oeuvres Morales, Traités 37-41*, vol. 7.2 [Paris 1974] 176n. 15). The same idea about the use and abuse of wealth is discussed at length and described with the same verbal pun (χρῶνται... παραχρῶνται) in the *De cupid. div.* 527A-B, where again P. mentions Aristotle as his source. Either version could very well be a paraphrase of the philosopher's own words; see Schuhl, 1968: 5, 12-13. On views regarding P.'s direct knowledge of Aristotle, see note in Proem, chs. 1-2.

τῶν γὰρ πολλῶν: if we assume that Aristotle, from whom P. probably draws here directly, refers specifically to wealthy people, and not generally to the majority of people, the πολλῶν here should be taken to mean the majority of wealthy people; cf. Arist. *EN* 1121b16, where οἱ πολλοί clearly refers to the affluent people (οἱ γὰρ πολλοὶ φιλοχρήματοι μᾶλλον ἢ δοτικοί). In any case, even if we disregard P.'s indebtedness to Aristotle in this passage, the πολλοί makes sense if P. has the wealthy people in mind, since the abuse of wealth can only result from affluence and not from poverty.

οἱ μὲν οὐ χρῶνται [αὐτῷ]: P. is most likely quoting Aristotle verbatim, and in that case αὐτῷ (codd.) refers to πλοῦτος, in the context of Aristotle (cf. *De cupid. div.* 527A) and must be deleted. Ziegler disregards the possibility of a direct quotation, and omits αὐτῷ to avoid the hiatus, although this is not uncommon in the case of quotations in P. (*Cat.Ma.* 9.10). On P.'s views about amassing possessions which are incommensurate with one's own needs, see *synkr. Arist./Cat.Ma.* 31(4).

μικρολογίαν: "parsimony". A common term in the Aristotelian corpus, which is often associated in P. with ἀνελευθερία and φιλαργυρία; cf. *De cupid. div.* 525F and 526C. On P.'s criticism of μικρολογία, see *Cat.Ma.* 5.

οἱ δὲ παραχρῶνται δι' ἀσωτίαν: P.'s model of prodigality is the notorious demagogue Demades (*De cupid. div.* 525C). The abused wealth is compared to a 'captive' and a 'barbarian' in *Luc.* 41.7. For a discussion of ἀσωτία, see Arist. *EN* 1107b10-11, 1119b23-1122a17. For the use of χρῶμαι and παραχρῶμαι in Aristotle, see Laurenti, 1987: 656-657.

3.3 ἐλευθεριότητι: the observance of the mean in relation to wealth (Arist. *EE* 1231b27-30, *EN* 1107b10-11, 1119b22-26). Although P. adopts Aristotle's definition of ἐλευθεριότης (cf. *De virt. mor.* 445A) and portrays Pelopidas here as being endowed with this virtue, he appears to be contradicting himself at 3.7-8, where he talks about the Theban's mismanagement of his ancestral wealth. Kimon, on the other hand, seems to have represented for P. the ideal example of generosity, since he is described as having employed his private resources for the benefit of those of his fellowmen who were in need, and for other charitable purposes (*Kim.* 10); see A. Blamire, *Plutarch, Life of Kimon*, *BICS Supplement* 56, University of London (London 1989) 128-130. For the association of ἐλευθεριότης with money, see Plat. *Resp.* 402c, *Theaet.* 144d, Xen. *Symp.* 4.15.

φιλανθρωπίᾳ: P. uses this term exclusively in the sense of charity, humanity (Aalders, 1982: 46, 65). Martin (1961: 173), commenting on the relationship between generosity and philanthropy observes that the ἐλευθεριότης of Pelopidas refers to the actual sharing of his wealth with his friends, while φιλανθρωπία has more the extensive function of designating the gracious and kindly spirit with which the gifts were made. He suggests, therefore, that ἐλευθεριότης is a particular kind of φιλανθρωπία. The latter quality also characterizes Marcellus (1.3, 10.6, 20.1), who was the first one among the Romans to show such political virtue (*Marc.* 20.1) On the concept of φιλανθρωπία, see Tromp de Ruiter, 1931: 271-306, esp. 295-300, H. Bolkestein, *Wohltätigkeit und Armenpflege im vorchristlichen Altertum* (Utrecht 1939 [1979]), De Vogel, 1966: 82-83, bibliography quoted in Aalders, 1975, 23n. 69, also Brenk, 1992: 4400n. 117.

μόνον δὲ τῶν φίλων... μεταλαμβάνειν: Epameinondas is shown here to be the only one among Pelopidas' friends who did not share in his liberality. In *Arist.* 1.4, P. specifies, however, that "Pelopidas provided the funds that enabled Epameinondas to serve as choregos for a chorus of men with flute accompaniment"; μεταλαμβάνειν seems then to imply in the context of the *Pelopidas* the 'taking advantage of one's wealth' for personal use, and not for the public benefit. Nepos discusses extensively Epameinondas' poverty and lays much stress on his attitude to money and his disinterestedness in acquiring it (3.4-4.6); he specifies that he used his rich friends' wealth merely for the benefit of those who needed it (*Epam.* 3.4-6). Also in the *De genio* P. devotes a lengthy section of his narrative to Epameinondas' refusal to accept Theanor's financial aid (583D-585F), which was offered to him as a compensation for the services that he and his family had bestowed to the

Pythagorean Lysis during the latter's sojourn in Thebes (583C); on Epameinondas' indifference to money, cf. *Philop.* 3.1, *synkr. Arist./Cat.Ma.* 31(4).4, *Arat.* 19.3, *Reg. et imp. apoph.* 193B-C, *De gen.* 579E, 583F-584B, Stob. 3.1.162.

3.4 αὐτὸς μέντοι... καλλωπιζόμενος: taking the interpersonal relationship of the two Thebans as a paradigm of friendship P. grasps the opportunity to trace some of its basic traits. He idealizes Pelopidas' genuine affection for Epameinondas by portraying him as spurning his material goods for the sake of friendship and appropriating voluntarily his friend's mode of humble living. The idea of treating one's friend as "another self", which is reminiscent of Aristotle's definition of friendship (*EN* 1166a31-32, *EE* 1245a29), is widely attributed to the Pythagoreans (cf. the scholiasts in *Il.* 18.82). P. may be drawing on Aristotelian ethics in his delineation of the varying ways in which Pelopidas' friendship towards Epameinondas manifests itself. On the mutual duties between friends, see Arist. *EE* 1240a23-1240b13; on Aristotle's discussion of friendship, see Tracy, 1979: 65-75, Price, 1989: 103-130, Stern-Gillet, 1995.

τραπέζης λιτότητι: Epameinondas' frugal diet (cf. *Reg. et imp. apoph.* 192D, *Lyk.* 13.6, Athen. 419a) was probably the result of his Pythagorean education, on which see De Vogel, 1966: 232-233. Phokion, like Epameinondas, is another hero who typifies, according to P., self-sufficiency and is also praised for his ascetic diet (*De cupid. div.* 525B); cf. also the case of Cato the elder (*Cat.Ma.* 4). The idea that friendship is tested through the enjoyment of the same type of food over a lengthy period of time is discussed by Aristotle; cf. the proverbial consumption of salt among friends (*EN* 1156b27-28, *EE* 1238a2, Cic. *De amic.* 19.67).

τῷ πρὸς τοὺς πόνους... καλλωπιζόμενος: the ἀδόλῳ is unjustifiably deleted by Ziegler, but is kept by Flacelière. The sentence should read: τῷ πρὸς τοὺς πόνους ἀόκνῳ καὶ κατὰ στρατείας ἀδόλῳ καλλωπιζόμενος ("and took pleasure in the combination of unflinchingness towards his toils and guilelessness on campaigns"; τῷ applies both to ἀόκνῳ and to ἀδόλῳ.). The ἀδόλῳ anticipates the absence of envy and rivalry in the two Thebans' political and military cooperation (4.3-4, 24.3, 26.1-2) and reinforces the genuineness of their friendship, which is a recurrent theme throughout the *Life*. The ἀδόλῳ may have also been prompted by the ἀψευδὲς ἦθος of Kapaneus (Eur. *Suppl.* 869), whose personality is compared to that of Pelopidas at 3.5 (cf. *Suppl.* 861-862); see further Flacelière and Chambry (1966: 248) and note on 3.5.

3.5 ὥσπερ ὁ Εὐριπίδου... ἦν: the portrayal of Pelopidas' ethos was possibly triggered by P.'s recollection of Kapaneus' character, as it is described by Adrastos (Eur. *Suppl.* 861-871). The similarities between the two figures, namely their lack of arrogance (despite their wealth) and their frugality, are not limited to the two lines quoted here. The tripartite description of Kapaneus' humility and moderation in life (φρόνημα-ὄλβος-βορά) corresponds evenly to the threefold delineation of Pelopidas' personality (ἐλευθεριότης and φιλανθρωπία, πλοῦτος (3.3), ἐσθῆτος ἀφέλεια and τραπέζης λιτότης, (3.4)). On P.'s use of Euripides, see Tagliasacchi, 1960: 124-142, C.S.J. Mitchell, 1968, Digregorio, 1976, id. 1979, id. 1980, Carrara, 1988: 447-455; generally, on his use of the poets, see B.X. De Wet, " Plutarch's Use of the Poets", *Acta Classica* 31 (1988) 13-25.

βίος: "property".

3.6 'Επαμεινώνδας μὲν... πενίαν: on Epameinondas' poverty, see *Pel.* 5.4, *Fab.* 27.3, *De gen.* 579E, 583E, Paus. 9.13.1, Just. 6.8.6, Nep. *Epam.* 2.1, 3.4, Ael. *V.H.* 2.43, 11.9. P. seems, however, to exaggerate somewhat the extent of it in *Arist.* 1.4 (Sansone, 1989: 178), for we learn in the *De genio* of Epameinondas' fine education (585D) and the rich provisions given to the Pythagorean Lysis by his family during the philosopher's sojourn at Thebes (583C). We can assume therefore that the Theban's poverty is magnified because it was voluntary (ἑλόμενος); see Fortina, 1958: 7.

εὔζωνον καὶ κοῦφον: "unencumbered and less oppressive".

φιλοσοφῶν: P. introduces a theme which he develops at length in the *De genio* (583D-585D). In the treatise Epameinondas develops his theory against the acquisition of wealth (compare P.'s views in the *De cup. divit.*; O'Neil, 1978: 289-362) and justifies his indifference to material goods on philosophical grounds. The Theban's philosophical education and training, on which P. lays especial emphasis in the treatise, are effectively announced here in order to prepare the ground for his reluctance to participate in the conspiracy against the Theban oligarchs (*Pel.* 5.4, *De* *gen.* 576D-E, 594B-C). On P.'s predilection for the philosophically oriented leaders, see [**2e**] and bibliography therein, also De Blois, 1992: 4601-4611.

μονότροπον... ἑλόμενος: Epameinondas' choice of a solitary life is linked, like his indifference to money, with his philosophical inclinations and training; cf. also Diod. 15.87.6, Nep. *Epam.* 5.5, 10.1, Stob. 4.22.62. Polyainos' anecdote about Epameinondas' wife seems to be his own invention (2.3.1); cf. Swoboda, *RE* s.v. "Epameinondas", 5.2, 1905: col. 2677; on his lovers Asopichos and Kaphisodoros, see Plut. *Amat.* 761D. On the anecdotal tradition of his erotic activities, see note on 7.5 and Shrimpton, 1970: 24-26.

μονότροπον βίον: "solitary life".

3.7 Πελοπίδᾳ... παῖδες: P. provides no specific information about Pelopidas' wife and children [**2a**], although he seems to have gone to some lengths to learn the names of his subjects' wives and daughters, on which see Bremmer (1981: 425-426). P. is equally uninformative and vague at 20.2 and 34.4. Nepos informs us of a son of Pelopidas, who had a bad reputation (*Epam.* 10.1).

ἀλλ' οὐδὲν... οὐσίαν: P. made much of Pelopidas' generosity and liberality at 3.3-4 probably in order to make up for his neglect and ineffective management of his wealth mentioned at 3.7-8. Despite the noble intentions which underlie Pelopidas' conduct, negligent attitude to one's own private interests constitutes a reprehensible defect, which detracts from a character's virtue as a 'political' man; on P.'s views about the proper use to be made of wealth, cf. *synkr. Arist./Cat.Ma.* 31(4); see further, Wardman, 1974: 79-86.

3.8 τῶν δὲ φίλων... τυφλόν: Pelopidas probably meant by this that Nikodemos, being blind and lame, was not capable of fine actions, but for those who are, like him, there are better things to do than pursuing the acquisition of money; cf. οὐ γὰρ ἔστι πράττειν μεγάλα φροντίζοντα μικρῶν, *synkr. Arist./Cat.Ma.* 31(4).2. For a discussion of P.'s views about the pursuit of riches and its impact on one's political ἀρετή, see Wardman, 1974: 79-86. The same anecdote with only minor discrepancies appears in *Reg. et imp. apoph.* 194C and Aelian, *V.H.* 11.9, on which see C. Schmidt, *De Apophthegmatum quae sub Plutarchi nomine feruntur Collectionibus*, Diss. Greifswald 1879 69-74.

τοῦ χρήματα ⟨συν⟩άγειν: cf. *Reg. et imp. apoph.* 194C, χρημάτων συναγωγῆς. On the accumulation of wealth, see *De cup. divit.* 524E, *synkr. Arist./Cat.Ma.* 31(4).3, *Cor.* 10.8; see further, Wardman, 1974: 80-86, Barigazzi, "Una declamazione contro la brama della ricchezza: *De cupiditate divitiarum*", and "Un altra declamazione contro la brama della ricchezza: *De vitando aere alieno*", in *Studi su Plutarco*, Università degli Studi di Firenze, Dipartimento di Scienze dell'Antichità (Florence 1994) 75-98 and 99-114.

4.1 ἦσαν... ὁμοίως: "both seemed also equally fitted by nature for all sorts of excellence"; cf. Diodoros (15.88.2-3), who, comparing Epameinondas to all other heroes, including Pelopidas, finds, however, that the former "far surpassed them all". P. continues the delineation of Pelopidas' character by way of comparisons with Epameinondas. He carefully develops elements of similarity in the two characters which, despite certain differences, will provide the basis of their successful political cooperation and reinforce the theme of friendship initiated at 3.3. On P.'s especial emphasis on the ἀρετή of his protagonists, see Valgiglio, 1992: 3999-4010.

πλὴν ὅτι... ἐποιεῖτο: the two Thebans exemplify in their diverse practices the two ways of life known as active and contemplative. Pelopidas' lack of zeal in intellectual matters and his preference for bodily exercises and hunting are contrasted with Epameinondas' dedication to the study of philosophy (cf. Nep. *Epam.* 3.1). Although this difference in their pursuits is too fundamental to be left uncommented upon, P. makes surprisingly little effort to elucidate his statement. He deliberately does not explore the unpleasant implications of Pelopidas' disinclination for learning, nor does he suggest at any time that his inability to control his anger, which was responsible for his recklessness and early death, should be linked chiefly with his lack of commitment to serious intellectual pursuits; on this matter, see [**2e**]. P.'s models in his description of Pelopidas' wordly activities and Epameinondas' philosophical pursuits may have been Amphion and Zethos, who played an important role in the early history of Thebes, and represented two different ways of life; see further Georgiadou, 1995: 187-200.

ὁ δ' ἀκούων τι: Polymnis, Epameinondas' father, talks about his son's keenness on listening and learning rather than discussing his views on philosophical issues, τὸ ἦθος… ἄπληστον δὲ τοῦ μανθάνειν καὶ ἀκροᾶσθαι (*De gen.* 592F); cf. *studiosus audiendi*, Nep. *Epam.* 3.2). Iamblichos mentions that there were at least two levels of exposition and understanding within the Pythagorean regimen, the 'Hearers' and the 'Learners' (Burkert, 1972: 192-198, G. Clark, *Iamblichus: On the Pythagorean Life*, Translated Texts for Historians, Greek Series, vol. 8 [Liverpool 1989] 2-34-35), the lowest of which was the ἀκροατής, which he equates with the ἀκουσματικός. Both he (*De vit. Pyth.* 81-82) and Porphyry (*De vit. Pyth. 37*) claim that the ἀκουσματικός received only the essentials of the Pythagorean doctrines. Iamblichos' reference to Epameinondas' philosophical training (ἐγένετο Ἐπαμεινώνδας ἀκροατής, 35.250) could refer to the Theban's limited exposure to Pythagoreanism. Whether he uses the term ἀκροατής as a Pythagorean term, like ἀκουσματικός or in the more common sense of "student" is unclear. P. seems, however, to be referring here not to the depth of Epameinondas' acquaintance with the philosophy, but to the Theban's pursuit of the Pythagorean regimen of education; see Buckler, 1993: 104-108 and bibliography therein.

φιλοσοφῶν ἐποιεῖτο: on Epameinondas' philosophical training and his adherence to the tenets of Pythagoreanism, see Plut. *De lib. educ.* 8A, *De gen.* 576D-E, 585D, *Ages.* 27.6 (ἀνὴρ ἔνδοξος ἐπὶ παιδείᾳ καὶ φιλοσοφίᾳ); also Diod. 10.11.2, 15.39.2-3, 52.7, 16.2.3, Dio Chrys. *Orat.* 22.2, Ael. *V.H.* 3.17, 7.14, Justin. 6.8.9. He is mentioned as having been taught by the Pythagorean Lysis of Tarentum, on whom see Diod. 10.11.2, Plut. *De gen.* 583C, 585E, Paus. 9.13.1, Nep. *Epam.* 2.2, Cic. *De orat.* 3.139, *De off.* 1.155, Ael. *V.H.* 3.17, Iambl. *De vit. Pyth.* 35.250, Diog. Laert. 8.7, Porph. *De vit. Pyth.* 55, Dio Chrys. *Orat.* 49.5. On the influence of Pythagoreanism on P. see Swoboda, *RE* s.v. "Epameinondas", 5.2, 1905: col. 2676, Cloché, 1952: 131, Fortina, 1958: 5-6, Shrimpton, 1970: 22n. 8, Brenk, 1977: 65-84, Hardie, 1992: 4781-4783, Buckler, 1993: 104-108. P., being himself a serious student of philosophy, is especially interested in public figures like Epameinondas, Cicero, Dion, who are versed in it; cf. *Nu.* 20.8-12, *Max. cum princ. philos.* 776C-777A, 777C-E, 778B, *Ad princ. inerud.* 779F; see further [**2e**], Herbert, 1957: 86, J.R. Hamilton, 1969: xxxvii-xxxix, Wardman, 1974: 211-220, De Blois, 1992: 4601-4611.

4.2-4 P.'s glorification of the two Thebans' unconditional friendship and harmonious cooperation in their public functions is a theme underlying the entire narrative and provides a further link between the two *Lives* [see **2d**, **e**]. Especially bothered by the jealousy and animosity of Greek politicians (Aalders, 1982: 26-27, 48-53) P. offers here a little lecture, replete with crude moralism and patriotic fervor, on the blessings of amicable collaboration in leadership. The entire section, which reads like a miniature of the *Praecepta gerendae reipublicae*, demonstrates not only the moral superiority of the Theban model of leadership—embodied in Pelopidas and Epameinondas—over the Athenian, despite his laudatory comments on Athens' military achievements and his admiration for the great deeds of Athenian generals and statesmen in *De glor. Athen.*, but also anticipates a sharp contrast with the Roman *Life*, on which see [**2d**].

P. strongly and repeatedly disapproves of political discord and antagonism in the *Praecepta* (809B-809F, 813A, 825E) and sees them as evils proceeding from personal ambition and jealousy (805C, 806C, *An seni* 787 C, 788E). His practical advice to future politicians is to avoid unnecessary competition (*De tuenda sanit.* 135D-E) and urges them to strive for κοινωνία (*Praec. ger. reip.* 812E), ὁμόνοια and φιλία (816A-B, 819D. 824C-D, 827D). He commends Phokion and Cato, neither of whom would ever allow their personal enmity to have any bearing upon their political differences (809D-E). P. does not disapprove of all types of rivalry and ambition; he praises the Sacred Band for their ἅμιλλα and φιλόνικον, because they inflamed their courage and inspired them for the accomplishment of noble actions (19.5).

4.2 οἱ νοῦν ἔχοντες: "the thoughtful ones"; P. refers to his ideal readers (cf. *De gen.* 575B-D), apparently those who will not share the false views of the many. This is a sophistic mode to convey what he regards as important truth and expects his select readers to agree; on P.'s audience, see Wardman, 1974: 37-48; Georgiadou, 1995: 187-200, esp. sect. 3: "Plutarch's Ideal Audience".

διὰ τοσούτων... πολιτειῶν: "through all their struggles, military commands and civil services"; it anticipates the negative triptych ὅσων... διαφορῶν καὶ φθόνων καὶ ζηλοτυπιῶν, at 4.3. Examples of their harmonious military and political cooperation are given at 12.6, 20.3, 23.2-5, 24.

εὔνοιαν καὶ φιλίαν: P. includes Pelopidas and Epameinondas among the legendary pairs of Achilles/Patroklos, Theseus/Pirithoüs, Orestes/Pylades, Damon/Phintias, which are generally accepted as model-friendships of the old times (*De am. mult.* 93E). The theme of exemplary friendship is also discussed by Cicero (*De amic.* 4.15), but we have no evidence that the two Thebans were included in his list (cf. also a list of seven pairs of Φιλέταιροι by an anonymous author in A. Westermann, *Παραδοξογράφοι, Scriptores Rerum Mirabilium* [Brunswig 1839] 219). Young statesmen are recommended to show "good will and friendship" towards the established leader to whom they choose to attach themselves (*Praec. ger. reip.* 806F), and colleagues are urged to show ὁμοφροσύνη καὶ φιλία among themselves (816A). Fortina suggests (1958: 9) that the account of the signs of friendship of the two Thebans may derive from P.'s desire to present their friendship like the one of Alkibiades and Sokrates (*Alk.* 7.3-4, 6; cf. Swoboda, *RE* s.v. "Epameinondas", 5.2, 1905: col. 2678). For the expression φιλίαν... συμμείνασαν, cf. Plat. *Phaedr.* 232B: χαλεπὸν φιλίαν συμμένειν. On the notion of friendship in Greek literature, see Fraisse, 1974, esp. ch. 3, sect. 2: "Plutarque: Le charme de l'affection", 434-441. On the history of friendship in classical antiquity, see Konstan, 1997, and bibliography therein.

συμμείνασαν: ἀναμείνασαν (L^1KPQ), ἐμμείνασαν (L^2 editt.). Reiske's emendation (συμμείνασαν) harmonizes with P.'s emphasis on the two men's collaboration and friendship through a series of compounds with συν; cf. συνάρχοντας καὶ συστρατήγους (4.3), συνεστρατεύσαντο, συμμάχοις (4.5), συνασπίσαντες (4.6); for a similar use, cf. *Praec. ger. reip.* 816B. For P.'s particular interest in stressing the immutability and duration of Pelopidas' and Epameinondas' amicable relationship, see Georgiadou, 1992a: 4224-4229.

4.3 εἰ γάρ τις... πολιτείαν: having introduced the unparallelled friendship of the two Thebans P. goes on to draw a sharp contrast between three pairs of outstanding Athenian politicians of the fifth century, which he uses as cautionary examples of dissension and political rivalry, and the Theban pair, which provides a worthy example for emulation; cf. Diodoros' comparison of Epameinondas with Greek heroes of the past, 15.88. The reader can easily detect in this contrast P.'s method at work, namely to elicit from comparisons rather than from straightforward descriptions of character those qualities which will be shown in the *Life* to guide and determine his heroes' actions.

The contrasts in enmity and in friendship between two protagonists are seen as part of the comparative principle and the antitheses deriving from the Peripatos (T.W. Hillard, "Plutarch's Late-Republican *Lives*: Between the Lines", *Antichthon* 21 [1987] 19-48, esp. 35, Scardigli, 1995: 10-11), although, I would suggest that P. was far too eclectic to adopt this or that school's precepts wholesale. The use of historical examples as a means of persuasion goes back to Homer, when characters such as Nestor and Phoenix use examples drawn from what is *for them* a historical past in an attempt to persuade. Thukydides, Plato and Aristotle (τὸ λέγειν πράγματα προγεγενημένα, *Rhet.* 1393a28-29) were also commonly using historical examples; on P.'s use of historical *exempla*, see Pelling, 1995: 205-220, esp. 215-216.

P. wrote individual *Lives* for each of these Athenian politicians; see Levi, 1955, Podlecki and Duane, 1992: 4053-4127. On the rivalry of Aristeides and Themistokles, see Hdt. 8.79, Arist. *Ath. pol.* 23.3, Plut. *Arist.* 2.2-6, *Them.* 5.7, 11.1. He omits any mention of the occasional cooperation between Aristeides and Themistokles; cf. Arist. 8.3, *Praec. ger. reip.* 809B; see further, F.J. Frost, *The Scholarship of Plutarch,* Ph.D. diss. University of California, 1961, 38-57, id. 1980: 72. For a discussion of P.'s different treatment of the rivalry between Aristeides and Themistokles in the two *Lives*, see Larmour, 1992: 4179-4200. On the political opposition between Kimon and Perikles, see *Per.* 7.4, 9.5; on Nikias' and Alkibiades' rivalry, see *Nik.* 11. 1-3, *Alk.* 13.1, 14.1-3.

διαφορῶν... ζηλοτυπιῶν: the cumulative effect of the nouns enhances the contrast. The theme of envy among politicians is brought up again at 25.

συνάρχοντας καὶ συστρατήγους: the two Thebans were genuine fellow-officers and fellow-generals, united in their services to Thebes by their virtue and untroubled by envy of each other's achievements (Wardman, 1974: 27-9). The συστρατηγεῖν καὶ συνάρχειν are often causes of enmity among political and military leaders (*Praec. ger. reip.* 816B). On P.'s binary structures, see Ambrosini, 1991: 21.

4.4 αἰτία... γενομένην: P. may be thinking of Aristotle in his definition of virtue, who says that "those acts that one does not perform with the self in mind are fine or at least preferable; those that one performs for one's country, in contempt of one's own interest, are absolutely good" (*Rhet.* 1366b35-38).

Despite the two Thebans' different personalities, P. attributes the success of their cooperation exclusively to their virtue and their common aspirations regarding their country. They contributed jointly towards the same objective and fought for the same noble purpose (cf. the counter-examples of men like Agesilaos, Lysander, Nikias, and Alkibiades, who had been great warriors, but had not known how to use their victories to noble and glorious ends, as P. notes in *Flam.* 11.3-6). P.'s moral message has clearly political overtones; he reformulates the idea suggested earlier (4.3), that "political activity... should be a school of cooperation rather than a struggle of rivals each aiming at his own dominance" (Wardman, 1974: 27). P.'s precepts are in line with his advice to future political leaders: τοῖς δὲ πολιτικοῖς ἔξεστι μὴ σώματα μηδὲ χρήματα μόνον, ἀλλὰ καὶ τύχας καὶ δυνάμεις καὶ ἀρετάς, ἂν ὁμονοῶσιν, εἰς μίαν χρείαν συντιθέντας εὐδοκιμεῖν μᾶλλον ἄλλου περὶ τὴν αὐτὴν πρᾶξιν (*Praec. ger. reip.*, 819D).

Wardman claims (1974: 29) that P.'s glorification of Pelopidas' and Epameinondas' unrivalled collaboration is made without reference to the political situation in Thebes at the time. This is partly true. P. is clearly more concerned to invest their amicable synergy with idealistic motives rather than present a more realistic political picture of Thebes, which was emerging then as a leading city-state in Greece, and needed more than ever political leaders who would fight, untroubled by envy, for the same purpose. He is more interested in extracting "points of general, timeless significance" (Pelling, 1995: 210-211) from particular past events and applying them to new contexts than looking into the specific political and social circumstances apposite to one period. Occasionally, however, his comments strongly suggest that the general political atmosphere in Thebes at the time made the two Thebans' amicable cooperation imperative (24-25). On P.'s idealization of their motives, see also Aalders (1982: 16); for a study of friendship and politics, see H. Hutter (*Politics as Friendship: The Origins of Classical Notions of Politics in the Theory and Practice of Friendship* [Waterloo, Ontario 1978]) Connor, 1971: 33-84.

ἡ ἀρετή: P. exalts here a particular type of virtue, the political, which he defines as the accomplishment of fine actions in the interest of one's own country (cf. 19.5, 25.13), and not for the sake of personal gratification; cf. P.'s emphasis on the individualistic nature of the Roman triumph, *Marc.* 8, 21-22. P. may be thinking of Aristotle, for whom the good of the state is greater and far more complete than the good of the individual (*EN* 1094b8-9, *Pol.* 1252a4-8; cf. Plut. *synkr. Arist./Cat.Ma.* 30(3).1), although this notion is explicitly present in Perikles' last speech in Thukydides, in Tyrtaios, and in the entire Spartan system; on the political *arete* in P., see Wardman, 1974: 49-78.

οὐ δόξαν, οὐ πλοῦτον... μετιόντες: P.'s point is that ἀρετή should be the right aim of political life, not the achievement of a glorious name, or the acquisition of wealth. Although it is understood that the value to be ascribed to fame or reputation is not to be disregarded altogether (cf. 4.2; *Agis* 2.1-4) and that it even contributes, to some extent, to the achievement of noble purposes (27.4, 29.2, 11, 30.1, 4, 13), there is a danger that δόξα may get out of control and be pursued for its own sake, as he suggests at 8.2, when he refers to Pelopidas' associates; note also Diodoros' comment on Epameinondas' desire for everlasting fame without attaching to it any noble motivation (15.66.1). Likewise, the pursuit of wealth is not considered a 'fine' action, since it will prevent one from performing his political tasks properly; see Wardman, 1974: 44-45, 79-86.

ἀλλ' ἔρωτα... γενομένην: P. applies different standards in evaluating the driving force behind his heroes' actions; while the two Thebans' desire to see their country most glorious by their efforts—cf. also the similar case of the Theban Theagenes, *Mul. virt.* 259D—is ascribed to their 'divine eros', Marcellus' passion to fight a decisive battle with Hannibal (*Marc.* 28.4) is described in derogatory terms and is said not to fit his age (28.6). The notion of 'divine passion' associated with ἀρετή is reminiscent of the Platonic ἔρως for virtue which makes someone ἔνθεον (*Phaedr.* 255B); see also note below on *Pel.* 18.7. In the *Praecepta*, P. maintains that the strongest and most divine kind of love is that which is engendered in states and individuals for an individual because of his virtue (821F). Similarly, Thukydides' Perikles uses the word ἐραστής to incite the Athenians to become lovers of their city (2.43.1).

ἔρωτα... ἐρασθέντες: cognate accusatives are common in P.; cf. *Pel.* 15.2, 33.4.

ὥσπερ ἰδίοις... κατορθώμασιν: "they used to this end one another's achievements as their own". P. has aptly fashioned his argument about the two men's good-willed collaboration into an appropriate vehicle for introducing us subtly to an issue which is vital to the *Life*, namely their joint undertakings and the sharing of their accomplishments. It is questionable how equal their roles were, as well as their shares in the various exploits, during the Theban hegemony; in fact, in the *synkrisis* he raises doubts about Pelopidas' equal share in the first invasion of the Peloponnese and the Theban victory at Leuktra (32(2).2). Elsewhere in P.'s works, these two events are presented as personal triumphs of Epameinondas (*De glor. Athen.* 346b, *De garr.* 514C, *De laud. ips.* 540D-E, 542C, *An sen. resp.* 786D, *Non posse* 1098A); cf. Nepos' assignment of the leading role to Epameinondas, and the secondary to Pelopidas (*Pel.* 4.3); for a discussion of the possible existence of diverging traditions regarding the events of the Theban hegemony see [**2e**], and Georgiadou, in *Boeotia Antiqua.*

4.5-6 τῆς ἐν Μαντινείᾳ... ἐπιφερομένους: P. refers to the military expedition led by the Lacedaemonians against Mantinea in 385 under the leadership of Agesipolis (cf. Xen. *Hell.* 5.2.1-7, Diod. 15.12.1-2; also Isokr. *Paneg.* 126, Polyb. 4.27.6, Polyain. 2.25; on the campaign, see Cartledge, 1987: 370-371); the reason was to punish the Mantineans for failing to fulfill their commitments during the Corinthian War (395-386 BC). P. and Pausanias (9.13.1)—both probably drawing from the same tradition at this point—claim that a force was dispatched to Mantinea in which Pelopidas and Epameinondas served as hoplites. Pausanias' account presupposes the existence of a Theban-Spartan alliance. P. and Isokrates, who was well-known for his antipathy towards Thebes, are the only authors who mention explicitly the existence of such an alliance after the Peace of Antalkidas (*Pel.* 4.5, *Plataikos* 27-28); more specifically Isokrates accuses the Thebans of joining the Spartans. Xenophon, on the other hand, who would have no motive for suppressing knowledge of such an alliance—he could, in fact, use it later in his account, in order to portray the Spartan seizure of the Kadmeia as a retribution for violating the terms of alliance with Sparta—has nothing to say about it. For recent arguments against the existence of such an alliance, see Buckler, 1980c: 179-185, Urban, 1991: 113 and n. 439, 136n. 524, Buck, 1994: 63; for the contrary view, see Cawkwell, 1972: 254-278, esp. 257, Hack, 1975: 10-12, id. 1978: 216-219; C.D. Hamilton, 1991: 142 and n. 82.

ἔτι φίλοις καὶ συμμάχοις: it anticipates P.'s comment at 5.1 (τῶν Σπαρτιατῶν λόγῳ μὲν ὡς φίλοις καὶ συμμάχοις προσφερομένων τοῖς Θηβαίοις) and refers to the alleged brief Theban-Spartan alliance (see previous note) established by the Peace of the King. Xenophon, however, in his account of Leontidas' address to the Lacedaemonians, after the seizure of the Kadmeia, does not leave any doubts about the continued hostility of the Thebans towards them (*Hell.* 5.2.33); likewise, Nepos mentions (*Pel.* 1.3) that after the Peloponnesian War and the defeat of Athens the Lacedaemonians thought that the Thebans were the only ones who dared to resist them. That the Thebans rejoined the Peloponnesian League in 386—perhaps when Leontidas and his faction came into power—can only be deduced from the accounts of P., Pausanias and Isokrates.

τεταγμένοι... ἐγεγόνει: P. talks about a pitched battle, in which the Spartan wing, where Pelopidas and Epameinondas served as hoplites, was routed by the Arkadians. Xenophon (*Hell.* 5..2.4-5) and Diodoros (15.5.3-5) mention only siege operations and no Spartan defeat.

τὸ κατ' αὐτοὺς κέρας: αὐτούς, sc. τοὺς 'Αρκάδας.

4.7 Πελοπίδας... κείμενον: this anecdote is also found in a concise form in Paus. 9.13.1. Stern (1884: 36-37n. 8) has pointed out that the scene involving Epameinondas and Pelopidas has noticeable similarities with an incident at Potidaia where Sokrates saved his beloved Alkibiades (Plut. *Alk.* 7.3-6; Plat. *Symp.* 220e-221b). Pausanias' reliance on P.'s lost *Epameinondas* for the sections 9.13.1-15.6 has been a matter of dispute among scholars; see recent discussion by Tuplin, 1984: 346-358. There remains the possibility that both writers are deriving this episode from the same source, one which would have invested Epameinondas with Socratic traits; as such it may have inspired P. for his portrayal of the Theban in the *De genio*, on which, see Georgiadou, 1996: 113-122 and bibliography therein.

Epameinondas is shown to save Pelopidas' life once more at 29.4, and his effort to rescue the latter is presented again as a matter of the highest priority for him. P. concludes then his speech on comradeship and synergy among colleagues with an extreme example of Epameinondas' demonstration of genuine friendship towards Pelopidas. The former's daring act in Mantinea may be linked with his philosophical training; cf. the hospitality and friendship of Epameinondas' family to the Pythagorean Lysis which is linked with the family's philosophical training. On philosophy as conducive to friendship, cf. *Max. cum princ. philos.* 777C-E); on Pythagoreans' high esteem of friendship, cf. *De gen.* 582D-584D, *De frat. amor.* 488B-C; see further, De Vogel, 1966: 150-159.

τραύματα... ἐναντία: the word ἐναντία is not used in a hostile sense, but refers to wounds suffered while facing the enemy; for a similar use, see *Pel.* 18.7, *Aem.* 20.4, *Cat.Ma.* 1.7; see further, Georgiadou, 1990: 81-82.

πολλοῖς... πολεμίοις: a similar description is found at *Cat.Ma.* 20.11 (ἐν πολλοῖς σάγμασιν ὅπλων καὶ σώμασι νεκρῶν ὁμοῦ φίλων τε καὶ πολεμίων κατασεσωρευμένων). P. exploits this powerful and dramatic scene with a particular psychological empathy; having death unite both friends and enemies he builds up an emotional and personal scene with Epameinondas trying to defend his friend's supposedly dead body at the risk of his own life; cf. an analogous scene in *Marc.* 2.2, in which the description is made in a remarkably dry and matter-of-fact manner.

4.8 'Αγησίπολις: the son of the exiled Pausanias (*Ages.* 20.5) and king with Agesilaos. He descended from the Agiades family and reigned from 393 till 381 BC. He died at the campaign of Olynthos in 380. Agesilaos declined the command against Mantinea on the ground that Mantinea had helped his father Archidamos II to put down a Messenian Helot revolt in the 460s (Xen. *Hell.* 5.2.3), and so Agesipolis had to undertake the operation. On his personality, see Diod. 15.19.4, Plut. *Ages.* 20.7; see further Niese, *RE* s.v. "Agesipolis, 1", 1.1, 1894: cols. 805-806, Poralla, 1985, No. 14.

ἀνελπίστως: P. probably points out the unexpected character of the Spartan king's aid to the two Thebans revealing thereby his general mistrust of the Spartans (cf. λόγῳ μὲν... ἔργῳ δέ, 5.1).

Chs: 5-13. The seizure of the Kadmeia by the Lacedaemonians (382) and the liberation of Thebes (379)

The events which led to the seizure of the Kadmeia: The Corinthian War (394-387) was followed by the signing of the King's Peace, also known as the Peace of Antalkidas (387/6). This peace guaranteed nominally the autonomy of every Greek city (Xen. *Hell.* 5.1.31); the autonomy clause, meant, however, in the context of Spartan interests, the breaking up of any political combinations in Greece that had formed a coalition against Sparta, such as the Boiotian Confederacy (Plut. *Ages.* 23.5), and secured Spartan supremacy in Greece (Bengtson, 1975, No. 242). The Thebans resented King Agesilaos' insistence that they should abide by the terms of the Peace, which would not allow them to take the oath to the Peace on behalf of the Boiotian cities (Xen. *Hell.* 5.1.32). Although they were at first opposed to this provision, threatened by imminent Spartan invasion, they acceded to the treaty and relinquished control of Boiotian foreign policy (*Hell.* 5.1.33, Diod. 14.110). With the Peace of Antalkidas the Boiotian Confederacy, which was under Theban control, was dismantled, and each Boiotian city was now autonomous with regard to both local and foreign affairs.

The peace terms, particularly beneficial to the Spartan interests, put an end, at least temporarily, to the revival of the Athenian imperialism, Theban control of the Boiotian cities and the dominion of Argos over Corinth, and confirmed Spartan supremacy in Greece. After the signing of the Peace, Agesilaos put into effect a policy of chastising some of Sparta's allies who had not been adequately cooperative to her demands during the Corinthian War, and imposing docile, pro-Spartan oligarchies. The splitting of Mantineia (*Hell.* 5.2.7, Diod. 15.12.2) in 385/4 (Harding, 1985, No. 30) and the establishment of aristocratic government by Agesipolis was followed by the restoration of a group of pro-Spartan exiles in Phlius (*Hell.* 5.2.8-10; Legon, 1967: 324-337). An opportunity for Spartan expansion to the north was given by a delegation from Akanthos and Apollonia, cities in the Chalkidike, and from Amyntas III of Macedon. Their envoys asked for Spartan intervention against Olynthos, which was leading the Chalkidian League (*Hell.* 5.2.11-23). A potential coalition of Thebes, Athens and Olynthos could pose a severe threat to the Spartan interests. As a result, together with her

Peloponnesian allies, and under the pretext of enforcing the violated principle of autonomy, Sparta voted to send out an expeditionary force to aid the Akanthians, Apollonians and Macedonians. Thebes refused, apparently, to participate in this campaign, and issued a proclamation forbidding any Theban from serving with the Spartans against Olynthos (*Hell.* 5.2.27). A hostile Thebes and its strategic position in central Greece impeded, of course, Sparta from intervening effectively in the north.

In 383 the Spartan Eudamidas was sent on a preliminary expedition to the coast of Thrace with the agreement that he be followed later by his brother Phoibidas with the rest of the army (*Hell.* 5.2.24). In 382 Phoibidas, in the course of his march to Olynthos and, probably acting on secret instructions given to him by Agesilaos (*Ages.* 24.1, Diod. 15.20.2), seized the Theban acropolis with the support of Leontidas, a pro-Spartan polemarch in Thebes (*Hell.* 5.2.25-29, *Pel.* 5.2). The seizure took place while the council was in session away from the Kadmeia because of the celebration of Thesmophoria by the Theban women (*Hell.* 5.2.29, *Pel.* 5.3). This act was considered as a blatant violation of Thebes' own autonomy, and even Xenophon, with his pro-Spartan bias, criticized it severely (*Hell.* 6.3.9 and 11).

For a treatment of the events prior to the seizure of the Kadmeia, see: Accame, 1936: 337-355, Wilcken, 1941: 3-20, Smith, 1953-4: 277-280, Ryder, 1965: 25-45, Rice, 1971: 1-71, id. 1974: 164-179, Seager, 1974: 36-63, id. *CAH* 1994: 156-186, Hack, 1975: 1-27, id. 1978: 210-227, Sealey, 1976: 386-406, Cawkwell, 1976: 62-84, Cartledge, 1979: 278-288, id. 1987: 194-199, 293-296, C.D. Hamilton, 1979: 301-329, id. 1991: 120-141, Seibert, 1979, vol. 1: 110-112, vol. 2: 478-480, Lanzilotta, 1980: 129-178, Cargill, 1981: 7-10, De Voto, 1982: 167-187, Hornblower, 1983: 202-206, Urban, 1991: 109-160, Buck, 1994: 58-64, Tuplin, 1993: 65-96, Jehne, 1994: 31-51.

On the seizure and liberation of the Kadmeia, see Ryder, 1965: 46-53, Rice, 1971: 71-85, id. 1974: 179-182, Seager, 1974: 36-45, Hack, 1975: 16-61, id. 1978: 221-227, Borthwick, 1976: 148-151, Sealey, 1976: 406-410, Seibert, 1979, vol. 1: 112-116, vol. 2: 481-482, De Voto, 1982: 188-222, id. 1989: 101-116, Hornblower, 1983: 206-209, Mastrocinque, 1983: 237-247, Gehrke, 1985: 175-180, González, 1986:

69-83, Hammond, 1986: 469-470, 482-483, Cartledge, 1987: 296-299, C.D. Hamilton, 1991: 141-159, Buck, 1992: 103-109, id. 1994: 64-80, Tuplin, 1993: 96-100, Jehne, 1994: 51-54, Dillery, 1995: 216-231.

Sources: The Spartan seizure of the Kadmeia and its liberation by the Theban democrats is also recounted by P. in the *De Genio*. In the *Life* P. is primarily interested in Pelopidas' role in the liberation of the Theban acropolis, while in the treatise this event, described in much greater detail, constitutes the framework of the narrative around which a number of philosophical issues are discussed. Certain discrepancies between the two accounts (on which see Lattanzi, 1933: 115-117, Corlu, 1970: 29-31, Hani, 1980: 39-40) should probably be attributed to the different nature and scope of each work and not necessarily to the use of different sources; see [**2g**], Georgiadou, 1995: 187-200, ead. 1996: 113-122. On P.'s sources for chs. 5-13, see [**2c**], esp. s.v. "Theopompos and Xenophon", Bock, 1910: 52-68, Accame, 1936: 341-342n. 4, Westlake, 1939: 21-22, Corlu, 1970: 31-39. On the chronological sequence in the composition of the *Pelopidas* and the *De genio*, see [**2g**]; for bibliography on the different interpretations of the treatise, see same section. Briefer accounts of the events recounted in chs. 5-13 are preserved in Xenophon's *Hellenika* (5.4.1-13), Nepos' *Pelopidas* (2.1-4.1), Diodoros (15.25-27) and in Polyainos' *Strategemata* (2.3.1, 2.4.3). For a discussion of the discrepancies among the accounts of P., Xenophon and Diodoros, see Stern, 1887: 15-69, Schaefer, 1930: 12-21, Accame, 1936: 337-345, Corlu, 1970: 31-39, Hani, 1980: 40-43, Buck, 1994: 64-69.

5.1 λόγῳ μὲν... ὑφορωμένων: P. shifts from Pelopidas' and Epameinondas' indisputable friendship, which was put to the test on various occasions, to the superficial friendship and unsteady alliance between Sparta and Thebes at that time. The possibility of a Theban-Spartan alliance referred to at 4.5 is a matter of dispute; see also note ad loc.

φίλοις καὶ συμμάχοις: cf. ἔτι φίλοις καὶ συμμάχοις, 4.5; the repetition is probably intended to show more forcefully the forthcoming reversal of their supposed alliance.

τὸ φρόνημα... ὑφορωμένων: P. echoes verbally the argument used by the ambassadors of Akanthos and Apollonia to Lacedaemon regarding the threat that the Olynthians posed to the Spartan hegemony in Greece

(Xen. *Hell.* 5.2.16-18). The ambassadors lay stress on the Olynthians' increasing power and ambitious spirit (δύναμις... φρόνημα; cf. τὸ φρόνημα... τὴν δύναμιν, *Pel.* 5.1), while at 5.2.27 Xenophon suggests that the power of Thebes is more threatening than that of the Olynthians, since by refusing to join the Spartans against Olynthos (*Hell.* 5.2.27) and through her discussions and negotiations with Olynthos over the terms of an alliance (*Hell.* 5.2.15), Thebes is beginning to assert her strength again. Diodoros also mentions the Spartans' fear lest Thebes claim the leadership of Greece (15.20.1).

τὴν Ἰσμηνίου... ὁ Πελοπίδας: referring to the period after the King's Peace P. makes a clear distinction between the democratic (δημοτικὴν εἶναι δοκοῦσαν) group of Ismenias and Androkleidas and the oligarchic group of Archias, Leontidas and Philip (ἄνδρες ὀλιγαρχικοί). Xenophon attaches nothing ideological either to Ismenias' faction or to Leontidas', and makes no mention of democrats against oligarchs in Theban politics, but considers them simply as political factions that were at variance with each other (*Hell.* 5.2.25). From the *Hell. Oxyrh.* (Bruce, 12.1 and Chambers, 406-415) we learn that Thebes was split into two factions, at least as early as 395 (Bruce, 1967: 110-111, note on 12.1; on Theban politics in the years between 404 and 395, see Cloché, 1918: 315-343). Of these factions, the one was philo-Lakonian and was led by Leontidas, Asias and Koiratadas and was opposed by the faction led by Ismenias, Androkleidas, and Antitheos, who were responsible for helping the exiled Athenian democrats in 404 (*Hell. Oxyrh.* 12.1; see Bruce, *ib.*: 112, Chambers, 412-414, McKechnie and Kern, 162; Diod. 14.32.1). This picture of the opposed factions and their leaders is also confirmed by Xenophon (*Hell.* 5.2.25). The author of the *Hell. Oxyrh.* makes a vigorous refutation of the accusation that Ismenias' circle was pro-Athenian, and despite the corruption of this sentence, he seems to be criticizing the selfish motives of Ismenias' faction; see Bruce, 1967: 112-113, Cook, 1981: 10-11, Mckechnie and Kern, 1988: 162. Their acceptance of the Athenian refugees, apparently against the opposition of their pro-Spartan opponents, may have been responsible for their label as pro-Athenians. For a discussion of the factional division in Thebes in this period, see Cook, 1981: 11-17, González, 1991a: 121-135, esp. 121-125.

On the eve of the capture of the Kadmeia by Phoibidas, Ismenias and Leontidas are still leading members of their respective factions; the rivalry between the two political groups is mentioned in somewhat similar terms by Xenophon (*Hell.* 5.2.25). For P.'s dependence on Xenophon at this point, see Accame, 1936: 340.

P.'s comment on the Spartan dislike of Ismenias' and Androkleidas' faction suggests the rivalry between the two political groups and also points to the special circumstances of the two political leaders' death (*Pel.* 5.3). It is maintained that the major difference between the two factions lay in their foreign policy (Cook, 1981: 162). For a discussion of the democratic sympathies of Ismenias' group, see ead. 159-161. Buckler raises doubts about Ismenias' democratic leanings (1980a: 36-37; cf. also González, 1991a: 126-129). On the antagonism of the factions within Thebes, see Buck, 1985: 292-293; throughout the Boiotian Federation, see Bruce, 1967: 113; on Ismenias' group, in general, see Buckler, 1980a: 36-45. P. adds Pelopidas to the latter group not because of his leading position in it but presumably in order to create a symmetry between the two rival factions, by listing three leaders from each side.

'Ισμηνίου: his name appears aspirated (Hismenias) in *IG* VII 2892; cf. A. Führer, *De Dialecto Boeotica* (Göttingen 1876) 5-6, Head, 1881: 65. His faction seems to have acquired much of its power after the Peloponnesian War in opposition to the pro-Spartan group of Leontidas and his followers (on which see Bruce, 1960: 76-80). He is mentioned as having given assistance to the Athenian exiles at the time of the Thirty Tyrants (Just. 5.9.8). Ismenias was polemarch of Thebes along with his opponent Leontidas at the time of the seizure of the Kadmeia. He received Persian money from the Rhodian Timokrates (Xen. *Hell.* 3.5.1, Paus. 3.9.8), was tried by Spartan judges on charges of befriending the Persian satrap, accepting bribery from the Persians (on Ismenias' medism, see Dusanic, 1985: 227-235) and precipitating the Corinthian war together with Androkleidas (*Hell. Oxyrh.*, Bruce, 13.1 and Chambers, 461-471); for these reasons he was condemned to death (Xen. *Hell.* 5.2.35-36, *Pel.* 5.3, *De gen.* 576A); see further, Swoboda, *RE* s.v. "Ismenias, 2", 9, 1914: cols. 2139-2140, Κουμανούδης, 1979, No. 1007, Bruce, 1967: 110, Buckler, 1980a: 36-38. He was known for his wealth and generosity (Plut. *De tranq.* 472D, *De cup. div.* 527B, *Praec. ger. reip.* 823E, Plat. *Men.* 90a, *Resp.* 336a) and should be distinguished from Ismenias, Pelopidas' friend (*Pel.* 27.1, 29.12, *Artax.* 22.8), who was probably his son.

Ἀνδροκλείδου: Xenophon mentions that Androkleidas persuaded the Thebans to assist the Lokrians against the Phokians in 395 (*Hell.* 3.5.3-5, Paus. 3.9.9, *Hell. Oxyrh.*, Bruce, 13.1 and Chambers, 476-479) and at 5.2.35 makes him jointly responsible with Ismenias for the events that led up to the Corinthian war (cf. P.'s note on the Spartan dislike of Ismenias' faction, τὴν Ἰσμηνίου καὶ Ἀνδροκλείδου μισούντων ἑταιρείαν, *Pel.* 5.1). Both Xenophon and Pausanias associate him with Ismenias in the receipt of Persian money from Timokrates (*Hell.* 3.5.1, Paus. 3.9.8), while P. does not mention Ismenias in this connection (*Lys.* 27.3). He fled together with other Theban democrats to Athens (*IG* II^2 37, *SEG* 32, 1982: 47) and was murdered by members of Leontidas' faction while being in exile; see *Pel.* 6.3, *De gen.* 596B; see further Toepffer, *RE* s.v. "Androkleidas", 1.2, 1894: col. 2147, Κουμανούδης, 1979, No. 123, Bruce, 1967: 110-111, Buckler, 1980a: 39-40, 286-287n. 58.

μισούντων: P.'s mention of the Spartans' hatred for Ismenias' and Androkleidas group echoes Xenophon's report of Ismenias' loathing (διὰ τὸ μῖσος τῶν Λακεδαιμονίων, *Hell.* 5.2.25) for the Lacedaemonians.

Πελοπίδας: the best known member of Ismenias' faction, thanks to P.'s and Nepos' biographies of him. He was apparently too young (ἐν τοῖς νεωτάτοις ὤν, *Pel.* 7.1) to have played any influential role before the events of 382.

φιλελεύθερον... δοκοῦσαν: "it was thought to be friendly to freedom and to a democratic form of government"; P. is here referring to the period after the King's Peace in 387/386; on the change of the political philosophy and outlook of Ismenias' group after the return of the Theban exiles, see Buckler, 1980a: 36-37, 43-45.

δημοτικήν: "of (relating to, for) the demos" or in the sense of political labeling, "taking the side of the demos", "favoring the rule of the demos"; see discussion of the term in Stadter, 1989: 94-95. P. uses the term here not only to denote the democratic principles of Ismenias' faction (on which, see Buckler, 1980a: 36-45), but also with an eye to the tyrannical regime of the faction of Archias and Leontidas (6.2); see also Gehrke, 1985: 175n. 69.

5.2 Ἀρχίας: he may be identical with Ἀσίας (*Hell. Oxyrh.*, Bruce, 12.1 and Chambers, 410), also called Ἀστίας (*Hell. Oxyrh.*, Bruce, 12.2 and Chambers, 426), both of which are attested by epigraphical evidence (*IG* VII 2428, 1.8, VII 3068, ll.7-8; on the spelling of the name, see Bruce, 1967: 111, Κουμανούδης, 1979 s.v. Ἀστίας. Archias was one of the leaders of the philo-Lakonian group (*Hell. Oxyrh.*, Bruce, 12.1 and Chambers, 410), one of the polemarchs in 382 (Xen. *Hell.* 5.4.2, *Pel.* 7.4, *Ages.* 24.2, *De gen.* passim). He was murdered by the group of Charon and Melon (*Pel.* 11.2-4); see Κουμανούδης, No. 297.

Λεοντίδας: Λεοντίδης in the *De gen.*; Λεοντιάδης in Xenophon's *Hellenika* (Λεοντιάδας by Flacelière in Plut. *Ages.* 23.11). He may have been the son of Eurymachos (Εὐρυμάχου τοῦ Λεοντιάδου, ἀνδρὸς Θηβαίων δυνατωτάτου, Thuk. 2.2.3; Gomme, 1956: 3). He is mentioned along with Asias (Archias) and Koiratadas (Korrantadas) in the *Hell. Oxyrh.* (Bruce, 12.1 and Chambers, 410) as one of the leaders of the philo-Lakonian group in the period around 395. In 382 he is still mentioned as a leading member of his group. He was murdered in 379 by the group of Pelopidas and Damokleidas (*Pel.* 11.1-8, *De gen.* 597D-F). See further, Lenschau, *RE* s.v. "Leontiades, 2", 12, 1925: col. 2040, Κουμανούδης, 1979, No. 1236.

Φίλιππος: one of the polemarchs in Thebes in 379 (*Pel.* 7.4, *De gen.* passim). Xenophon mentions his tyrannical rule during the occupation of the Kadmeia (*Hell.* 5.4.2). He died at the hands of Charon and Melon (*Pel.* 11.2-4, *De gen.* 597A-B). See further, Treves, *RE* s.v. "Philippos, 5", 19.2, 1938: cols. 2264-2265, Κουμανούδης, 1979, No. 2046.

πλούσιοι: P. draws attention to the wealth owned by these oligarchic leaders implying a contrast with the democratic faction. At 3.1, however, he refers to the affluence of Pelopidas' family, and at 5.4 he specifies that Epameinondas did not go into exile along with the other Thebans on philosophical grounds, but also because of his poverty, on which see note on 3.6. This remark allows us to infer that the Thebans who fled to Athens, and apparently belonged to Ismenias' and Androkleidas' group, were generally wealthy. See also *Hell. Oxyrh.* (Bruce, 12.1 and Chambers, 406-407), where the leaders of the two opposing Theban groups in the period around 395 are said to be οἱ βέλτιστοι καὶ γνωριμώτατοι τῶν πολιτῶν, which indicates that they were all of an upper social class. We also know that Ismenias was famous for his wealth (Plut. *De tranq.* 472D, *De cup. div.* 527B, *Praec. ger. reip.* 823E).

ἀναπείθουσι... Καδμείαν: in late summer 382, when Phoibidas was on his way to join his brother Eudamidas, in the north, he encamped outside the city, near the gymnasium (*Hell.* 5.2.24-25). Xenophon indicates that there was stasis in Thebes at that time between the two factions and that Ismenias and Leontidas, who were polemarchs, were at variance with each other (5.2.25). It is assumed that while the balance of power between the two factions in 382 was pretty even, it was the group of Ismenias which was in control of Theban affairs in Thebes (Rice, 1971: 72-73, Hack, 1975: 20-21, Buck, 1985: 293). According to both P. and Xenophon, Leontidas sought to convince Phoibidas to occupy the Theban acropolis; P. reports briefly Leontidas' effort to persuade Phoibidas (similar account is given in *De gen.* 575F), while Xenophon relates in detail the former's arguments for surrendering the Kadmeia to the Spartans (5.2.26-28). At 6.1, one might argue that P. suggests subtly that the Lacedaemonians may have been privy to the plan of the occupation of the Kadmeia, when he says that "the Greeks were amazed at the Lacedaemonians' inconsistency, since they punished the doer, but approved his deed". In *Ages.* 24.1 he makes it clear that the Spartan general was not acting on his own authority, and states that suspicion of Agesilaos' complicity in Phoibidas' operation arose immediately after the seizure of the Kadmeia (cf. also 23.6, 11). Diodoros also testifies to the secret staging of Phoibidas' operation by the Spartans (15.20.2), while Xenophon, either not wishing to show Agesilaos in a poor light (H. Montgomery, *Gedanke und Tat: zur Erzählungstechnik bei Herodotos, Thucydides, Xenophon und Arrian* [Lund 1965] 114, Lanzilotta, 1980: 160, Cartledge, 1987: 156) or fearing that "it would damage the impression of general Spartan acceptance of what had been done at Thebes" (Tuplin, 1993: 99; cf. Schaefer, 1930: 46, C.D. Hamilton, 1991: 146-148), expressly says that ὅτι οὐ προσταχθέντα ὑπὸ τῆς πόλεως ταῦτα ἐπεπράχει (5.2.32); furthermore, it is not incidental that he lays stress on Leontidas' treason and Phoibidas' rashness. At 5.2.31 he refers to the group of Thebans "who had been acquainted with the affair" (οἱ μὲν δὴ εἰδότες), implying those of Leontidas' group, and the ones that were "unaware" (τῶν δὲ μὴ εἰδότων), those of Ismenias' faction. We can therefore assume, following Xenophon's account, that the supporters of the oligarchic group were at least informed of the secret arrangements between Leontidas and Phoibidas. Nepos' statement at *Pel.* 1.2 (*idque suo privato, non publico fecit* [sc. Phoebidas] *consilio*) implies knowledge of the rumors regarding the complicity of the Spartan authorities in this incident.

ἀναπείθουσι: "they try to convince"; conative present, with the result of the effort mentioned later on, at 5.3: πεισθέντος δ' ἐκείνου (sc. τοῦ Φοιβίδου); cf. *Ages.* 24.1, where P. tells us about a prearranged Spartan occupation of the Kadmeia authorized by the Spartan authorities. The reason for which P. silences the possibility of Spartan secret plotting against the Theban democrats in the *Pelopidas* should be sought in his biographical methods: he first, drawing attention to the factional division within Thebes, speaks positively of the group of Ismenias and makes a negative criticism on that of Leontidas; he then strengthens the notion that the latter faction acted unscrupulously against Theban interests by simply omitting all mention of the role that the Spartan authorities may have played in the capture of the Theban acropolis. By casting blame solely on the group of Archias, the act of surrendering Kadmeia voluntarily to the Spartans emerges as pitifully treacherous.

Φοιβίδαν: brother of Eudamidas and father of Isadas (*Ages.* 34.8), possibly a kinsman of Agesilaos (Cartledge, 1987: 147-148). While being harmost in Thespiai in 378 he was killed in a battle fighting against the Thebans (Xen. *Hell.* 5.4.41-46, Plut. *Pel.* 15.6, Diod. 15.33.6). Polyainos gives a sensational, probably unhistorical, account of his death (2.3.1); see Ehrenberg, *RE* s.v. "Phoibidas", 20.1, 1941: cols. 347-348, Poralla, 1985, No. 734.

μετὰ στρατιᾶς διαπορευόμενον: P. uses the participle to denote vaguely Phoibidas' action of "passing through a place" without mentioning his intended final destination. We only learn that Phoibidas was passing through Thebes with an army, which indicates that he was on a military mission. P. is apparently not interested in questioning and investigating the Spartan's march through the unfriendly city, even at a time when Thebes' attitude towards Olynthos met with the Spartans' disapproval. Why Phoibidas chose to take the longer road from Plataia to

Thebes, in order to go to Olynthos, instead of the shorter and easier route, which led from Mount Kithairon to Plataia, then through the Boiotian plains to Thespiai and Haliartos, and from there continue northward along the shores of Lake Kopais, neither P. nor Xenophon explains. As Buckler notes (1980a: 15), Phoibidas' detour, although it was seemingly inexplicable, it was "hardly an unauthorized one".

Καδμείαν: the city of Thebes centers around a rather flat acropolis, which is a pear-shaped area; it was named after Kadmos, who was the city's mythological founder. Still in classical times, the name Kadmeia was used to designate the old city of Thebes, while Thebes was reserved for the entire city, which had expanded around the acropolis; see Symeonoglou, 1985, esp. 8-13, Demand, 1982: 8-9. We learn from the *Hell. Oxyrh.* (Bruce, 11.4 and Chambers, 404-405) that the Federal Council of Boiotia met on the Kadmeia; see Symeonoglou, *ib.*: 138. Naturally, after the dissolution of the Boiotian Confederacy with the King's Peace only the Theban council met there; see Ziehen, *RE* s.v. "Thebai, 1", 5.2, 1934: cols. 1424-1451.

πρὸς τὸ Λακεδαιμονίων... τὴν πολιτείαν: "to make the government suitable for subjection to the Lacedaemonians by establishing an oligarchy", "to make the government subject to the power of the Lacedaemonians"; see 6.2, when the promises of Leontidas' faction are finally put into effect; cf. Xen. *Hell.* 5.4.1.

5.3 πεισθέντος δ' ἐκείνου: Xenophon elaborates on Phoibidas' reaction to Leontidas' promising invitation to seize the Kadmeia and lays especial emphasis on the latter's responsibility for the conception and operation of his plan (*Hell.* 5.2.28). His negative portrayal of Phoibidas is intended to demonstrate the outright irrationality of his decision to take the Theban acropolis, and, perhaps, to remove any doubts about Agesilaos' share in this plot (see Tuplin's reservations, however, for Xenophon's attempt to exculpate Agesilaos, 1993: 98-99). Diodoros, on the other, clearly attributes the event to a Spartan initiative (15.20.2).

καὶ μὴ προσδοκῶσι... Θεσμοφορίων ὄντων: Phoibidas' coup took the Thebans unawares (ἐξαίφνης) not only because it was an unexpected violation of the autonomy clause in the King's Peace, but also because of the celebration of the festival of Thesmophoria (ἐν σπονδαῖς καὶ εἰρήνῃ τὴν Καδμείαν καταλαβών, *Ages.* 23.6). Although a festival was a time of peace, and not even legally sanctioned violence was admissible, it appears to have constituted the ideal occasion for a surprise attack (Hdt. 6.87, Thuk. 3.65.1, Aen. Tactic. 4.8) or for uprisings within the state (Aen. Tactic. 22.17, 17.3, Diod. 13.104.5, Xen. *Hell.* 4.4.2-4); see further, Parker, 1983: 154-159.

The Thesmophoria, the principal rite in honor of Demeter, were a women's festival and the most widespread cultic observance in Greece (Versnel, 1992: 31-55). Since the women celebrated it on the Kadmeia, where the sanctuary of Demeter the Lawgiver (Θεσμοφόρος) is said to have been, the Theban Thesmophoria seem to have been connected, at least topographically, with the cult of Kadmos; see Paus. 9.16.5, Ael. *V.H.* 12.57; see further Keramopoulos, 1917: 353-356, Ziehen, *RE* s.v. "Thebai, 1", 5.2, 1934: col. 1550, Arbesmann, *RE* s.v. "Thesmophoria", 6.1, 1936: cols. 15-28, Y. Béquignon, "Déméter, Déesse Acropolitaine", *RA* 2 (1958) 149-177, esp. 152-153, Vian, 1963: 135-137, Schachter, 1981: 165-168, Symeonoglou, 1985: 123-125, Valgiglio, 1988: 257-259. Xenophon notes that the Theban women celebrated this rite in the summer (*Hell.* 5.2.29), which indicates that the Theban summer festival did not take place at the same time as the Attic Thesmophoria, which were in Pyanepsion—equivalent to October/November. Munn[1] basing his argument on a reference in Lysias' speech *On the Scrutiny of Euandros* (26.23-24) to Thrasyboulos' involvement with "a Boiotian revolution" believes that Lysias is referring here to the seizure of the Theban acropolis; he also argues that the occupation of the Kadmeia must be placed sometime in June 382, a few weeks before Lysias' speech was delivered, and not in August of 382, as Beloch does (1923^2, vol. 3.2: 232-233); the latter follows Aelius Aristides' testimony (*Eleus.* 258), who associates the seizure of the Kadmeia with the Pythian festival.

While the women were celebrating the festival on the Kadmeia the council could not be convened on the acropolis, as usually, but had to meet in the lower city, "in the stoa in the agora" (*Hell.* 5.2.29). The entire setting, then, in Thebes at the time of Phoibidas' march to the north offered him a particularly favorable opportunity for a surprise attack; cf. *Marc.* 18.4, where P. tells us that Marcellus took advantage of a festival

[1] I wish to thank Prof. M. Munn for showing me a draft of an article dealing with Lysias 26 and the Date of Thesmophoria at Thebes.

celebrated at that time in the city in honor of Artemis to conquer Syracuse. Xenophon strongly disapproved of attacks made during religious events (*Hell.* 4.4.3).

τῆς ἄκρας κυριεύσαντος: Xenophon informs us that after Phoibidas seized the Theban acropolis without any opposition, Leontidas proceeded to the assembly and alerted the Thebans to the new situation (*Hell.* 5.2.29-30; cf. Nep. *Pel.* 1.2). Diodoros, on the other hand, talks about a subsequent fight between the Thebans who resented Phoibidas' act and his army (15.20.2). On the seizure of the Kadmeia, see also *Ages.* 23.6, Androtion, *FGrHist* 324F50, Isokr. *Paneg.* 126, *Plat.* 19, Polyb. 4.27.4, Ael. Arist. *Panath.* 293, *Leuktr.* 1.10-11, 2.2, 3.2 (Behr).

'Ισμηνίας... ἀνῃρέθη: following Xenophon's account, Leontidas ordered the arrest of Ismenias, after appointing another polemarch in his place. Ismenias was extradited to Sparta, tried before a tribunal of Spartan and allied judges (on which see Cawkwell, 1973: 53n. 3) on a charge of medism (Dusanic, 1985: 227-235, Tuplin, 1993: 98-99) and for having been responsible along with Androkleidas for all the disturbance in Greece. Although Ismenias was allowed to defend himself, he failed to convince the court, and was put to death (*Hell.* 5.2.30, 32, 35-36; Plut. *De gen.* 576A); on Ismenias' trial, see Hack, 1975: 24-27. A series of coins stamped HIΣME and ΑΝΔΡ date to the period from 378 to 336 and probably commemorate the two Thebans; see Seltman, 1955: 159, Kraay, 1976: 113-114.

συναρπασθείς: "seized and carried off", "robbed away".

Φερένικος: son of Kephisodotos, who had assisted the exiled Athenian exiles in Thebes during the tyranny of the Thirty at Athens (Lys. fr. CXX.78.2, Thalheim). He became Androkleidas' heir, after the latter's death (Lys. fr. CXX.78.4, Thalheim). He was in charge of a contingent of exiles during the liberation of the Kadmeia, *Pel.* 8.1, *De gen.* 576C; see further Lenschau, *RE* s.v. "Pherenikos, 1", 19.2, 1938: cols. 2034-2035, Κουμανούδης, 1979, No. 2031, Buckler, 1980a: 40.

Πελοπίδας... ἐξεκηρύχθησαν: both Xenophon (*Hell.* 5.2.31) and Diodoros (15.20.2) agree on the number of exiled Thebans, which they estimate as three hundred, while Nepos says that they were no more than a hundred, *Pel.* 2.3. According to Androtion (*FGrH* 324F50) they were four hundred. It is not known whether exiles from other Boiotian cities joined the Thebans of Ismenias' faction. Buck observes that it is reasonable to assume so given the large force of exiles deployed in Attica on the Boiotian frontier (1994: 71), but there is no concrete evidence to substantiate his suggestion. Diodoros notes that many eminent Thebans went into exile (15.25.1). Xenophon's account suggests that apart from those who had been legally sentenced to exile, other Thebans also, who shared the same views as Ismenias and Androkleidas, must have voluntarily fled from Thebes (*Hell.* 5.2.31). P.'s description of the measures taken against the faction of Ismenias (ἀνῃρέθη/φεύγοντες ἐξεκηρύχθησαν) recalls Diodoros' account of the exile and killing of Athenian democrats at the time of the Thirty Tyrants at 14.32.1.

5.4 'Επαμεινώνδας... ἀδύνατος: after having elaborated at length on Pelopidas' and Epameinondas' common aspirations (*Pel.* 4), P. deems it necessary to comment briefly on the latter's decision to remain in Thebes and not join the other exiles in Athens or assist them in their conspiracy. We learn in the *De genio* that he had not totally disengaged himself from the exiles' plot against the tyrants, but that he was privy to the day of their return to Thebes and had arranged with Gorgidas to act when an opportune occasion presented itself (594B). Epameinondas' refusal to participate in the exiles' plot against the Theban tyrants (cf. also Nep. *Epam.* 10.3, *Pel.* 4.1) is a leading issue in the *De genio* (576D-F, 594B-C), on which see [**2e**], Riley, 1977: 257-273, Barigazzi, 1988 [1994] 409-425, and Georgiadou, 1995: 187-200, 1996: 113-122. On Epameinondas' philosophy, see note on 4.1; on his poverty and P.'s exaggeration of it, see note on 3.6.

τῷ καταφρονηθῆναι... ἀδύνατος: "being looked down upon, as one whom philosophy had made inactive, and poverty powerless".

6.1 'Επεὶ δὲ... κατέσχον: P.'s account of the same events in the *Agesilaos* (23.6-24.2) and the *De genio* (575F-576A) sheds additional light on the aftermath of Phoibidas' operation. We learn that the Lacedaemonians, and especially Agesilaos' opponents, were displeased at Phoibidas' act and that Agesilaos came forward to defend Phoibidas' action on grounds of expediency, especially since he almost certainly had a hand in the affair. He succeeded in convincing the ephors and the displeased citizens that Phoibidas' operation was not intrinsically wrong, since it was performed with a view to Spartan interests. Although Phoibidas was relieved of his command and was fined (cf. Nep. *Pel.* 1.3), the Spartans not only continued to hold the Kadmeia but also reinforced the garrison (ἐγκρατέστερον ἐφρούρουν τὴν ἄκραν, *De gen.* 576A). They also appointed three harmosts, Lysanoridas, Herippidas and Arkesos (*Pel.* 13.3, *De gen.* 576A, 586E, 598F) and entrusted the administration of Thebes to the philo-Lakonians Archias and Leontidas; cf. also Xen. *Hell.* 5.2.32-36 (there is no mention of Phoibidas' punishment, while in *Pelopidas* 15.6 P. places Phoibidas' death as a harmost in Thespiai before the battle of Tegyra), Diod. 15.20.2, Nep. *Pel.* 1.3; see further, Rice, 1971: 74-77, Hack, 1975: 22-24, De Voto, 1982: 189-191, Tuplin, 1993: 98-100.

δέκα δραχμῶν μυριάσιν ἐζημίωσαν: cf. Diod. 15.20.2; the extravagant fine of 100,000 drachmas, which corresponds approximately to 17 talents (436.6 kg of silver), suggests that Phoibidas was wealthy; see Ziegler, 1955: 372n. 265. Cartledge thinks that Agesilaos may have paid Phoibidas' fine, as Philistos had done for Dionysios I of Syracuse (1987: 156). Xenophon makes no mention of the fine, on which see C.D. Hamilton, 1991: 146-147.

εἰ τὸν μὲν πράξαντα... δοκιμάζουσι: the conditional sentence here expresses the public opinion and has a causal force as at 9.11; it corresponds roughly to the causal clause in *Ages.* 23.6 ('Επεὶ γὰρ Φοιβίδας ἔργον εἰργάσατο δεινὸν... πάντες μὲν ἠγανάκτουν; cf. Smyth 2246; LSJ s.v. εἰ, B.VI). Although the outcome of Phoibidas' trial was a great surprise to the rest of the Greeks, there is no evidence that any city expressed its sympathy in any official and tangible way. Isokrates' *Plataikos* 19 and *Panegyrikos* 126 (Thompson, 1983: 75-80, Usher, 1990: 183-184) merely attest to the ill repute the Spartans suffered from the blatant violation of the principle of autonomy, but they bear no compassion for the plight of the Thebans. Diodoros mentions that "the Lacedaemonians were discredited in the eyes of the Greeks" (15.20.2) and Xenophon regarded the Spartan seizure of the Theban citadel as the action of a rash man (οὐ μέντοι λογιστικός γε οὐδὲ πάνυ φρόνιμος, *Hell.* 5.2.28); he denounced this act at 5.4.1 and saw its political consequences at 6.3.9, 11 (cf. also Plut. in *Praec. ger. reip.* 808B, Polyb. 4.27.4). For a discussion of Xenophon's disapproval of the seizure of the Theban citadel and the defence of the act by Agesilaos, see Cartledge, 1987: 296-297, Dillery, 1995: 216-218, 221-225.

6.2 τοῖς δὲ Θηβαίοις... πολιτείαν: we have no evidence of a change in the Theban constitution in 382 introduced by the philo-Lakonian ruling government other than P.'s assertion that the Thebans were deprived (ἀποβεβληκόσι) of their ancestral constitution. His comment probably suggests that the ascendancy of Leontidas' faction to power produced a 'new' type of oligarchy in Thebes which was much more repressive than the moderate oligarchies the city had experienced in the past. P. mentions in the *De genio* that Archias and Leontidas ruled unconstitutionally and violently (576A), while Diodoros tells us of the loss of Theban independence (15.20.3); see Buckler, 1979: 51, Buck, 1994: 70-71. The constitution of the Boiotian League was, most likely since 447 and until the dissolution of the League in 387/6, oligarchic; cf. Cloché, 1952: 71-74, Bruce, 1967: 102-103.

ὑπὸ τῶν περὶ 'Αρχίαν καὶ Λεοντίδαν: "by Archias, Leontidas and their associates". On this construction, see Dover, 1960 [1988] 69, Schwyzer, 1975: 417, vol. 2:1, Dubuisson, 1976-77, S.L. Radt, "Noch Einmal Aischylos, *Niobe* fr. 162 N.², 278 M", *ZPE* 38 (1980) 47-58, esp. 51. The plural formulas οἱ περί τινα and οἱ ἀμφί τινα are common in P. (*Pel.* 7.4, 9.10, 10.5, 12.2, 14.4, 27.6) and can refer generally to a person and his associates or simply to the persons explicitly mentioned: *Pel.* 14.4, *Su.* 12.13, *Sol.* 29.6, *Them.* 7.6, *Cat.Ma.* 22.1, *Alk.* 36.6, *Aet. Rom. et Gr.* 268F.

τῇ Σπαρτιατῶν δορυφορουμένην ἡγεμονίᾳ: "guarded by the Spartan hegemony"; the new type of oligarchy established in Thebes is in line with the initial proposal of Archias and his supporters (πρὸς τὸ Λακεδαιμονίων ὑπήκοον ἁρμόσασθαι... τὴν πολιτείαν, *Pel.* 5.2). The word δορυφορούμενην suggests a tyrant's bodyguard. P. is probably making an ironic statement about Spartan power protecting tyrants. It may also suggest a parallel with the 30 tyrants installed by the Spartans at Athens in 404.

εἰ μή τις ἄρα παύσειε... ἄρχοντας: the prerequisite for the re-establishment of the ancestral constitution in Thebes—contained in the protasis of the conditional clause and expressed impersonally and in general terms—is fulfilled partly at 13.6-7; cf. 8.5, where the scheme of shaking the Spartan supremacy becomes more concrete in the conspirators' minds. P. tells us that the twelve Theban men—although Pelopidas takes most of the credit for the exploit—"loosed and broke the fetters of the Lacedaemonian supremacy", in other words they challenged the Spartan hegemony.

After the Peace of Antalkidas and for the following few years until the liberation of the Kadmeia Sparta was actively involved in the increase and consolidation of its power in Greece. The διοικισμός of Mantinea into its constituent villages (Xen. *Hell.* 5.2.1-7), the imposition of an extreme oligarchy on democratic Phleius (*ib.* 5.3.10-17, 21-25), the break-up of the Chalcidian sympolity led by Olynthos (5.2.11-24, 5.2.37-3.8) and the occupation of Thebes were some of her important achievements during this period (cf. Nep. *Pel.* 2.4, Isokr. *Paneg.* 125-127); see also introduction to chs. 5-13.

κἀκείνους: the emphatic καί alludes to the Athenians, when they were defeated by the Lacedaemonians at the end of the Peloponnesian war and lost their hegemony in Greece.

6.3 τοὺς φυγάδας... κρύφα: the Theban exiles found a favorable reception in Athens from those Athenians who were opposed to the Spartan policy and the succesive violations of the King's Peace. The exiles went as ἰδιόξενοι (*De gen.* 576A), that is "private friends in a foreign state", and were probably exempted from the tax on aliens (*IG* II^2 37, *SEG* 32 [1982] 47; Seibert, 1979, vol.1: 113 and vol. 2: 481n. 901, Gehrke, 1985: 176). Their presence in Athens gave rise in Thebes to numerous rumours of conspiracies. The Theban oligarchs had, of course, good reasons to suspect that the exiles would attempt to return and overturn their government—note Charon's effort to appease Archias' fears of the rumors regarding the exiles' plots against the oligarchic regime (*De gen.* 596B, *Pel.* 9.5), while the analogous attempt of Thrasyboulos against the Thirty Tyrants in Athens is still fresh in their minds—and they sent assassins to Athens to forestall the exiles' action. They only succeeded in doing away with Androkleidas, but not with the rest. On the hospitality of Athenian circles to the Theban exiles, see Plut. *De gen.* 576A, Aeschin. 3.138-139, Deinarch. *contra Dem.* 38.

πέμψαντες ἀνθρώπους ἀγνῶτας: "sending some unknown fellows"; P. lays emphasis on the secretive aspect of the operations (κρύφα) deployed by the faction of Archias against their political opponents.

'Ανδροκλείδαν... διαμαρτάνουσιν: fearing the return of the exiles and the overthrow of the oligarchic regime Leontidas plotted the murder of the political associates of Ismenias. He was successful with Androkleidas, but failed with the rest. P. refers incidentally to Androkleidas' death in *De gen.* 596B and for this reason it has been used by Lattanzi, against Christ (1901: 81), as an indication that the treatise must have been composed after the *Life* (1933: 117n. 3). For coins probably issued in commemoration of Androkleidas' death, see Seltman, 1955: 159, Kraay, 1976: 113-114.

6.4-5 ἧκε δὲ... Θηβαίους: although the Spartans demanded the expulsion of the Theban exiles from Athens on the grounds that they had been declared common enemies by the allied cities, the Athenians did not yield to Spartan pressure in return for Theban hospitality to the Athenian democrats in 404/403 (Xen. *Hell.* 2.4.1-2, Diod. 14.32.1). What happened at that time with the Athenian exiles was, in fact, very similar to the situation in Thebes in 382: despite the fact that the Spartans had forbidden the allies of Athens to harbor Athenian exiles, the latter were given refuge in Argos and Megara, while the Theban government refused to comply with the orders. According to Diodoros' and P.' accounts the Thebans issued counter-decrees according to which no Theban was to see or hear any Athenian passing through Boiotia under arms against the tyrants and that any Theban not assisting an exile would be liable to financial punishment (Diod. 14.6.1, Plut. *Lys.* 27.2-3). P. also mentions that the Thebans provided money, arms, secrecy and a base of operations to the Athenian democrats (*Lys.* 27.4), while Justin refers explicitly to private, not public, assistance (5.9.8; cf. Lys. fr. 78.2 [120] Thalheim, Diod. 14.32.1). See further, Cook, 1981: 7-26, C.D. Hamilton, 1979: 149-150, Cartledge, 1987: 282-283. Fuscagni (1975: 32) ascribes P.'s tone regarding the Spartan measures against the Theban exiles to the anti-Spartan mood of his source; for objections to this view, see [**2c**].

ὑπὸ τῶν συμμάχων: the cities of the Peloponnesian League.

οἱ μὲν οὖν 'Αθηναῖοι... φιλάνθρωπον: on the ancient philanthropy of the Athenians, cf. Plut. *Arist.* 27.7, *Kim.* 10.6, Diod.

15.29.7, 15.63.2, Podlecki, 1988: 231-234; for the notion of philanthropy in P., see earlier note on 3.3.

ἐάν τις 'Αθηναίων... μηδ' ὁρᾶν: there are very few inscriptions from the early fourth century and there is no evidence of the existence of such a decree in 382. The decree P. is referring to here may be a fabrication inspired by the purported Theban decree in 404/3 on behalf of Thrasyboulos and his followers, which is reported in *Lys*. 27.3: ἂν δέ τις 'Αθήναζε διὰ τῆς Βοιωτίας ἐπὶ τοὺς τυράννους ὅπλα κομίζῃ, μήτε ὁρᾶν τινα Θηβαῖον μήτε ἀκούειν.

7.1-2: The Theban exiles in Athens and Pelopidas' leading role in the conspiratorial scheme against the Theban oligarchs: P. gradually moves on in his narrative to Pelopidas and focuses exclusively on his activity during the period of his exile in Athens from 382 to 379. He is far from being explicit about the factional politics in Athens during the sojourn of the Theban democrats. In the light of the Plutarchean account we can assume that the exiles must not have had an easy time in Athens, since they were trying to ingratiate themselves with the demagogues, who, being able speakers could manipulate the crowds. The Thebans' attitude suggests the existence of at least two factions in Athens: one, which did not view favorably their presence in Athens (it was led by Kallistratos of Aphidna, who was hostile to the Thebans, *Praec. ger. reip.* 810F; see further, Sealey, 1956: 178-203, Rice, 1975: 97-100), as well as another one, which was friendly to the Theban cause (cf. καλοὶ καὶ ἀγαθοί, *Pel.* 6.3) and opposed to the Spartan claim for their expulsion from Athens. The presence of these opposed attitudes seems to be confirmed by what happened before and after the *coup de main* of the Theban exiles. At the beginning, after the Spartan occupation of the Kadmeia, Athens had a benevolent, but a rather passive role with respect to the exiles who took refuge to Athens—perhaps so that it would not disturb Sparta by encouraging other cities, as well, to defend their freedom. Later, when Pelopidas and his fellow-men entered Thebes secretly, Athens, supporting a ψήφισμα proposed by Kephalos (Deinarch. *contra Dem.* 38-39, Worthington, 1992: 192-197), offered to assist the rebels to expel the Spartan garrison from the Kadmeia. Finally, when the acropolis was liberated by the Theban exiles and the Lacedaemonians invaded Boiotia a year later, the threat of Spartan invasion appears to have assured the prevalence of the anti-Theban faction in Athens (*Pel.* 14.1); see further, Ferone, 1992: 276-278. The period of exile in Athens must have had a strong impact on the redefinition and reshaping of the policies of Pelopidas' group, since they had the opportunity to observe directly and in depth the principles and workings of the Athenian constitution (Buckler, 1980a: 44-45).

P. informs us that Pelopidas kept urging the other exiles not to neglect their garrisoned city and to stop fawning on the smooth-tongued orators and relying on Athenian decrees. Accame argues (1936: 340) that Nepos' account of the stay of the Theban exiles in Athens and their coup against the Theban oligarchs (*Pel.* 2-3),—which, he claims, derives from Ephoros—provides us with the nucleus on which P. moulded his narrative (cf. *Pel.* 7.2 and Diod. 15.25.4). He observes, however, that the two writers differ substantially in their presentation of the Theban exiles: in P.'s account the exiles appear to be busy flattering the Athenian demagogues and seem to be undecided about the line of policy they should follow, while in Nepos' narrative they appear well determined upon the liberation of the Kadmeia from the first moment of their exile in Athens. In the former's account Pelopidas assumes the role of the Athenian Thrasyboulos (cf. also the association of the two heroes in *Non posse* 1098A), and takes both the military and the political initiative for the liberation of Thebes, while in Nepos' narrative Pelopidas' role is drastically downplayed. If we assume then, following Accame's suggestion, that P. did, in fact, depend for the entire episode, at least in part, on Ephoros (on which see also De Voto, 1989: 104, Buck, 1994: 148n. 48) or on some other source deriving ultimately from him, we can attribute certain discrepancies between the two accounts to P.'s re-elaboration of his source-material to suit his biographical purposes (Ferone, 1992: 278; cf. Sordi's suggestion about P.'s use of Theopompos in conjunction with Xenophon for the whole episode [chs. 5-13], 1995: 415-423, and discussion in [**2c**] s.vv. "Theopompos and Xenophon"). At any rate, the emphasis on Pelopidas' protagonistic role in instigating his fellow-exiles is in line with P.'s general tendency in the *Life* to bestow more praise on his hero's patriotic initiative than on any of the other Thebans.

ἐν τοῖς νεωτάτοις ὤν: after the death of Ismenias (5.3) and Androkleidas (6.3) Pelopidas emerged, despite his young age, as the most prominent among the exiled Thebans. Although we have no precise information about his age at the time of the liberation of the Kadmeia we can assume that given his participation in the battle of Mantinea six years earlier (on which see note on 4.5) he must have been in his early twenties.

ἰδίᾳ τε... πλῆθος: Pelopidas is here presented as the inspiring force in the organisation of the liberating movement of the Kadmeia. Xenophon, on the other hand, makes no mention of Pelopidas in connection with the entire episode, but emphasizes, instead, the coordinating role of Phyllidas (*Hell.* 5.4.2); see Accame, 1936: 340.

θεραπεύειν... τὸν ὄχλον: "subserviently flattering those able to speak and to persuade the people"; ἐκκρέμασθαι + genitive is balanced by ὑποπεπτωκότας + dative (τοῖς... δυναμένοις is perserved by the codd., on which see following note). The verb ἐκκρέμασθαι is being used metaphorically here.

τοὺς λέγειν δυναμένους: Ziegler's emendation of the reading of the codd. (τοῖς... δυναμένοις) is unnecessary, since the dative of the participle, which depends upon ὑποπεπτωκότας, is also supported by *PHeid.* 209 (Siegmann, 1956: 88-90, H. Gärtner, *Plutarchi Vitae Parallelae*, vol. 2, fasc. 2 [Stuttgart 1994] 341).

πείθειν τὸν ὄχλον: cf. μ[ετ]απ⟨ε⟩ίθειν τὸν δῆμον in *PHeid.* 209, col. II. 9-10 (Siegmann, 1956: 88-90). The use of the compound verb in the *PHeid.* 209 highlights both the struggle between the two opposing factions and Pelopidas' efforts to convince his compatriots to undertake the noble enterprise of liberating the Kadmeia. On P.'s use of ὄχλος in the sense of δῆμος, see Wyttenbach, s.v. ὄχλος; see Ferone, 1992: 278-9.

παράδειγμα... Θήβας: in his endeavour to stir up his fellow-men's fading patriotism Pelopidas, P. tells us, reminds them of Thrasyboulos, the hero of the democratic restoration in Athens, whom he portrays as a model of boldness and virtue worthy of imitation. P. is anxious to demonstrate that the Theban exploit, which he is about to recount, is comparable to the Athenian achievement, and makes, therefore, Pelopidas appear as equally capable and significant as the Athenian hero. The same comparison recurs at 13.4, where P. likens the grandeur of the Theban exploit to Thrasyboulos' successful *coup de main* against the Thirty Tyrants at Athens. Echoes of the Athenian exiles' struggle to overthrow the oligarchy are felt throughout P.'s account of the entire episode: note, for instance, P.'s emphasis on the small number of the Theban exiles that succeeded in demolishing the oligarchy (13.5-7; cf. the few Athenian exiles led by Thrasyboulos, Xen. *Hell.* 2.4.2, Arist. *Ath. pol.* 37.1). His comment on the Thebans' loss of their ancestral constitution (6.2) recalls Thrasyboulos' determination to see the ancestral constitution restored in Athens (*Hell.* 2.4.42, Diod. 14.32.6). Even the weather conditions seem to be similar in both exploits and equally favorable to the returning exiles (9.2; cf. *Hell.* 2.4.2, 2.4.14, Arist. *Ath. pol.* 37.1, Diod. 14.32.2).

ἵν' ὡς ἐκεῖνος... τυράννους: Thrasyboulos, son of Lykos of Steiria was banished by the Thirty Tyrants and fled to Thebes, whence he set off in 404/403 with seventy exiles and occupied the fortress of Phyle. He seized the Peiraeus and defeated the troops of the Thirty. He finally led the armed democrats to the Acropolis and the democracy was restored in Athens. On the liberation of Phyle by Thrasyboulos, see Xen. *Hell.* 2.4, Andok. 1.81, Arist. *Ath. pol.* 37-40, Diod. 14.32-33, Plut. *Lys.* 27.7; see further, Cloché, 1915, Berve, 1967, vol. 1: 208-212, vol. 2: 631-632, Krentz, 1982: 69-101. On Thrasyboulos, see Nep. *Thrasybulus*, Kirchner, *PA*, 7310, Schwahn, *RE* s.v. "Thrasybulos, 3" 6A1, 1936: cols. 568-574, A. Andrewes, "The Generals in the Hellespont 410-407 B.C.", *JHS* 73 (1953) 2-9, esp. 3-4, Seager, 1967: 95-115, S. Perlman, "Athenian Democracy and the Revival of Imperialistic Expansion at the Beginning of the Fourth Century B.C.", *CP* 63 (1968) 257-267, Davies, *APF*, 7310, pp. 240-241, Cawkwell, 1976b: 270-277, Develin, 3033.

προελθόντες: cf. ἐλθ[ό]ντες, *PHeid.* 209, col. II, 20.

7.3 ὑπολελειμμένους τῶν φίλων: cf. ἀπ[ολελειμμέ]ν[ους, *PHeid.* 209, col. II. 24-25. P. suggests here that friends of the exiles, obviously people who shared the political ideology of Ismenias' and Androkleidas' group, had not left for Athens. Their presence in Thebes was, in fact, indispensable for the success of the coup for various reasons. They could inform the exiles of the political situation in Thebes, assist them in organizing their return to Thebes and making it more secure and, also, participate in the plot of the overthrow of the oligarchy. In the *De gen.* 576A we learn that the exiles learned about the conditions in Thebes through letters sent to them by Gorgidas. Among the friends of Ismenias' group who remained in Thebes and are mentioned in the treatise we find Charon (595F; cf. *Pel.* 7.4), Amphitheos (598B), Galaxidoros (577A) and Kephisodoros (595E). No less than thirty Thebans are mentioned in *De genio* 586C to have been privy to the exiles' plans. Xenophon, who seems to suggest that all those who held the same views as Androkleidas and Ismenias fled to Athens (*Hell.* 5.2.31), does not account for the presence of people like Phyllidas or Charon in Thebes (5.4.2-3), who had a principal role in the coup.

7.4 Χάρων: Charon's leading role in the uprising and the murder of the oligarchs is especially emphasized by P. in both accounts: *Pel.* 8.3-4, 9.6-10.5,11.2, 13.1, *De gen.* 576D, 586B-587B, 594E-597A. Xenophon mentions him once (*Hell.* 5.4.3). After the liberation of the Kadmeia he was elected a boeotarch together with Melon and Pelopidas (*Pel.* 13.1). He led a successful cavalry attack on Plataia probably in 375 (see note on 15.6) and was still a major figure in Theban politics in the mid-360s, when Menekleidas tried to bring him into conflict with Pelopidas (*Pel.* 25.8). He was killed in 353/2 at the battle of Koronea; see Kirchner, *RE* s.v. "Charon, 4" 3.2, 1899: col. 2179, Κουμανούδης, 1979, No. 2153.

Φυλλίδας... πολεμαρχούντων: the wording is similar in Xenophon (*Hell.* 5.4.2; cf. also *De gen.* 577B), but the effort made by Phyllidas to have himself appointed secretary to Archias, despite his hostility to the regime of Archias and Philippos, appears only in the *Life*. P. mentions that Phyllidas "contrived to be appointed secretary to the polemarchs" after the arrangements regarding the exiles' return to Thebes were made. Hack notes that it is more likely that he already held this position of trust (1975: 52n. 75; Xen. *Hell.* 5.4.7).

γραμματεὺς... πολεμαρχούντων: the polemarchs and a secretary are referred to by Xenophon as executive officials at Thebes in the years 386-379 (*Hell.* 5.2.25, 32, 5.4.2, 7-8; Nep. *Pel.* 3.2). With the abolition of the office of Boeotarchy after the dissolution of the Boiotian Confederacy, the polemarchs seem to have been the chief local executive magistrates in Thebes and in other Boiotian cities; see Schaefer, *RE* s.v. "Polemarchos", 8, 1956: cols. 1097-1134, Salmon, 1978: 144-145, Buck, 1979: 160, Buckler, 1979: 50-64.

Φυλλίδας: Φιλλίδας in *PHeid.* 209, col. III. 7-8; Φυλλίδας in *De gen.* 577B and passim, Xen. *Hell.* 5.4.2; see further, Lenschau, *RE* s.v. "Phyllidas", 20.1, 1941: cols. 1020-1021, Κουμανούδης, 1979, No. 2055. The fact that P. mentions Phyllidas' role in the conspiracy abruptly, without any introductory information about his status, as he does with Charon, or with other characters, seems to reflect his selective use of sources. It is conceivable that P.'s source laid much stress on Phyllidas' principal part in the conspiratorial movement, which P. tried to play down by way of omission, in order to magnify Pelopidas' own contribution; see Accame, 1936: 345n. 3.

τῶν περὶ 'Αρχίαν καὶ Φίλιππον: for this construction, see note on 6.2.

7.5 'Επαμεινώνδας... διαφέρουσιν: Epameinondas was among the Thebans who remained in Thebes after the seizure of the Kadmeia (see note on 5.4). P. refutes implicitly the accusation of ἀπραγμοσύνη brought against him by his fellow-men for refusing to support the pro-democratic Thebans in their conspiracy (cf. *De gen.* 576E-F, 594B-C; cf. *Pel.* 5.4, Nep. *Pel.* 4.1, *Epam.* 10.3) by presenting him here, but not in the *De genio*, as actively involved in inflaming the Theban youth in the palaestra to challenge the Lacedaemonians and keeping alive their hatred for them (for his sentiments on physical training, see Nep. *Epam.* 2.4-5, 5.3-4). The entire scene with Epameinondas being depicted as the moral instigator of the young Thebans has striking parallels in phraseology with Polyainos' description of the same incident (2.3.6) and functions as a doublet of the preceding episode (7.1-2), where Pelopidas has been trying to strengthen his men's fading patriotic fervor. This incident is omitted in the *De genio* (see Accame, 1936: 346n. 1); we learn, however, that the palaestra functioned as the meeting place of the Thebans who were privy to the plot (577B, 594C); see Pritchett, 1974: 218.

Shrimpton (1970: 24-25) mentions the possibility of erotic overtones in this scene. Although there is nothing suggestive of Epameinondas' sexual behaviour here, there was clearly a tradition which emphasized Epameinondas' erotic exploits, on which see Shrimpton, *ib.*, 23-26.

ἐν τοῖς γυμνασίοις... παλαίειν: the Thebans' training in the palaestra is seen as the principal reason for their victory at Leuktra in

Quaest. conv. 639F; cf. Diod. 15.50.5, Ephoros in *FGrH* 70F97; see further Pritchett, 1985: 64-65.

ἐπιλαμβάνεσθαι: means grappling while wrestling, but suggests also attacking in battle; see LSJ III.2 s.v.

8-12 The return of the Theban exiles and the overthrow of the tyrants in 379 is recounted at length in the *De genio* (576C-577D, 586B-588B, 594B-598F). Despite, however, the prominence of the philosophical debates interspersed in the account of this Theban exploit in the treatise, the novelistic touches of its narrative and certain factual discrepancies between the two works, the outline of the events recounted in the *De genio* is historically correct. Shorter versions of the same events are found in Xenophon's *Hellenika* 5.4.3-13, in Diodoros (15.25.1-27), in Nepos' *Pelopidas* (3) and in Polyainos' *Strategemata* (2.3.1, 2.4.3); for a detailed bibliography, see above in the Introduction to chs. 5-13 and [**2g**].

8.1 'Ημέρας... ὁρισθείσης: the conspiratorial plan of overthrowing the tyrants started to materialize when a day agreed upon by the exiles in conjunction with their associates in Thebes was set for their return. Xenophon sheds some light on the arrangements made between Melon and Phyllidas during the latter's visit in Athens (*Hell.* 5.4.2) and also on the exact time chosen for the surprise attack of the exiles against the tyrants (5.4.4). It was the Aphrodisia, a festival celebrated at the end of the official Boiotian year by the annual magistrates upon the expiration of their term of office before transferring their powers to the new officials (*IG* VII 3172, ll.63-65, Roesch, 1965: 97, 100). P. fixes the event in the winter-time, but, with the exception of the drinking party of Archias and his associates (*Pel.* 9.4, *De gen.* 577C), he mentions neither Aphrodisia nor any other formal celebration taking place at that time. For a discussion of the celebration of Aphrodisia in connection with the conspirators' coup, see Vian, 1963: 143, F. Sokolowski, "Aphrodite as Guardian of Greek Magistrates", *HThR* 57 (1964) 1-8, Croissant and Salviat, 1966: 464-471, Schachter, 1981: 38-40, Pirenne-Delforge, 1994: 281-286.

ἔδοξε τοῖς φυγάσι... πόλιν: since rumors about the exiles' secret plotting had spread in Thebes (*De gen.* 586E-F, 587C, 596A-C) the coup was planned to proceed in two distinct phases so that they would not attract attention and also minimize the amount of possible losses. On the route taken by the two groups, see De Voto, 1989: 108-109. P. does not specify the size of the main body of Theban exiles, which was assigned to wait in the Thriasian plain under Pherenikos until news about the success of the coup allowed them to join the remainder of the exiles (12.5). In his much shorter version of the same events, Xenophon mentions nothing about the presence of any exiles awaiting on the Boiotian frontier, but informs us of the presence of Athenian troops on the Boiotian border under the command of two unnamed generals—a detail which is omitted by P. in both the *Life* and the dialogue; see Buck, 1992: 105-106, id. 1994: 149n. 56—; he also implies that this force was not officially decreed: cf. *Hell.* 5.4.9, 19, Isokr. *Plat.* 29, Deinarch. *contra Dem.* 38-39, Diod. 15.25.1-27.4, Aristeid. *Panath.* 283D. On the problems connected with the divergent sources regarding the role of Athens in the liberation of Thebes, see Stern, 1884: 44, E. Fabricius, "Die Befreiung Thebens", *RhM* 48 (1893) 450-451, Cloché, 1934: 55-57, Accame, 1941: 23, Burnett, 1962: 15-16, Rice, 1971: 88-92, id. 1975: 100-103, Cawkwell, 1973: 56-60, Hack, 1975: 58n. 88, Ober, 1980: 227-230, Cargill, 1981: 56, Kallet-Marx, 1985: 140-147, De Voto, 1989: 108, 114, Worthington, 1992: 193-197.

Φερένικον: see note on 5.3 and *De gen.* 576C. Lattanzi, prompted by the abrupt and incidental mention of Pherenikos (and also of Androkleidas, 596B) in the *De genio*, has suggested that the *Life* was written before the dialogue (1933: 117-8n. 3). Christ (1901: 81), on the contrary, believes that P.'s casual reference to men like Pherenikos or Androkleidas in the dialogue demonstrates his desire to focus mainly on the main conspirators and not on the auxiliary forces.

ἐν τῷ Θριασίῳ: the Thriasian plain is close to the small region of Thria, northeast of Eleusis; see Wrede, *RE* s.v. "Θριάσιον πεδίον", 6A1, 1936: cols. 599-601. It was the starting point of Pelopidas' operation. The proximity of this gathering place to Eleusis, Clinton suggests (1974: 16-17), may have facilitated the hierophant Archias' discovery of the plot (*Pel.* 10.7).

8.2 ὑφίσταται... πρῶτος: P., Diodoros (15.81.1) and Nepos (*Pel.* 2.5, 4.1) highlight the protagonistic role of Pelopidas in the whole enterprise, while Xenophon makes no mention of him and gives to Melon most of the credit for the restoration of the Theban freedom (*Hell.* 5.4.3-7, 19); on Xenophon's anti-Theban bias, see bibliography in [**2a**].

Μέλων: in the *Life*, Melon appears as one of the main figures of the rebellion (11.2, 12.1), while in the *De genio* he is briefly mentioned as belonging to the group of Pelopidas in the assassination of the polemarchs (596D). He was elected boeotarch in 378, immediately after the liberators had assassinated Leontidas and his colleagues (*Pel.* 13.1), and he may have held the office again sometime in the late 360s or early 350s, on which see Buckler, 1980a: 41. Xenophon presents Melon as the leader of the rebels (*Hell.* 5.4.3-7, 19).

Δαμοκλείδας: he fled to Athens after the occupation of the Kadmeia and played a leading role in the coup against the tyrants (*Pel.* 11.1, *De gen.* 594D, 596D). He was one of the seven boeotarchs at the battle of Leuktra (Paus. 9.13.6; cf. 4.32.6); see further, Kirchner, *RE* s.v. "Damokleidas, 2", 4.2, 1901: col. 2068, Κουμανούδης, 1979, No. 455, Buckler, 1980a: 130-131.

Θεόπομπος: a member of a prominent Theban family. He was among the exiles who found refuge in Athens after the seizure of the Theban acropolis. He is probably the same Theopompos whose name appears on the inscription set up shortly after the battle of Leuktra: *IG* VII 2462, Larfeld, 1883, No. 308, Tod, 1950, No. 130, Wickersham and Verbrugghe, 1973, No. 33, Tuplin, 1987: 96 and n. 72; he is mentioned twice in the *De genio* (594D, 597C) and was an ambassador in Athens in 377 (*IG* II^2 40, l.7); see Κουμανούδης, 1979, No. 917, Buckler, 1980a: 131.

ἄνδρες... φιλονίκως ἔχοντες: P.'s keen interest in the topic of friendship, especially among men in leading positions, is apparent. He does not romanticize the three Thebans' friendly relationship, as he had done earlier, when he was dealing with Pelopidas and Epameinondas (4.2-4), and explicitly says that their friendship was sitting uncomfortably with their desire for glory and heroic achievements, suggesting, thereby, that their rivalry could, in fact, hinder them, in certain circumstances, from achieving noble results. P. probably had good reasons for making this disquieting remark, but refrains from providing any examples to substantiate his claim. We are never told throughout the account of the liberation of the Kadmeia, or even after it, not even in the vaguest manner, that their failings cast any shadow over the greatness of the exploit.

P. has delineated three distinctive groups of political partners: the first comprises the three Athenian pairs which were characterized by mutual dissensions and envy (4.3); the second includes the Theban triad of Melon, Damokleidas and Theopompos, who were friends, but rivals for glory (8.2), and the third one, which embodies the ideal and unconditional friendship and partnership, is represented by the pair of Pelopidas and Epameinondas (4).

8.3 γενόμενοι... δώδεκα: a small band of twelve Thebans (cf. 13.7, *De gen.* 576C, Nep. *Pel.* 2.5) chosen from the youngest exiles (τῶν νεωτάτων, 8.1) set out for Thebes, while the rest under Pherenikos would remain in the Thriasian plain. Xenophon differs from P. in setting the number of this group of exiles at seven, *Hell.* 5.4.1, 5.4.3 (on Xenophon's narrative, see Gray, 1989: 65-70, De Voto, 1989: 109). Within the group of the twelve conspirators only six are mentioned by name in the *Life*: Melon, Pelopidas, Damokleidas, Theopompos (8.2), Kephisodoros (11.7) and Menekleidas (25.5). A few others, such as Eumolpidas, Samidas, Hismenodoros and Bacchylidas, are mentioned only in the *De genio* (577A, 582D) and may or may not have been in exile (see Buck, 1994: 73).

τοὺς ἀπολειπομένους: those remaining under the leadership of Pherenikos (8.1).

προῆγον ἐν χλαμυδίοις... δοκοῖεν: P.'s description of the exiles' disguise bears striking similarities with that of Nepos' (*Pel.* 2.5): *cum canibus venaticis exierunt, retia ferentes, vestitu agresti, quo minore suspicione facerent iter.* Accame attributes the almost word by word correspondence between P.'s version and that of Nepos to their common source, Ephoros; he notes, however, that the *vestitu agresti* does not correspond to P.'s ἐν χλαμυδίοις, but to the subsequent description of the exiles' disguise as peasants (ἐσθῆτας γεωργῶν μεταλαβόντες, 9.1; cf. *De gen.* 576C) and proposes that Nepos "fused two distinct incidents in Ephoros' account, the peasants' dress and the hunting uniform" (1936: 351; cf. also Bradley, 1968: 156-157). Xenophon reports that the group of seven "came to the city as if on their way back from the country", thus implying that some type of disguise in field-workers' or hunters' clothes had taken place (*Hell.* 5.4.3).

8.4 αὐτὸς μὲν... οἰκίαν: P. is particularly interested in creating sharp contrasts between his heroes' characters. The unfailing courage and patriotism of Charon upon which P. will dwell later in his account (9.9) are set in opposition to Hipposthenedas' cowardice (*Pel.* 8.5) and Chlidon's hesitation to perform his duties (8.8).

8.5 Ἱπποσθενείδας: nothing is known about Hipposthenedas' activity in Thebes during this period other than P.'s references to him in both the *Life* and the *De genio* (586B-588B).

οὐ πονηρὸς... ἐλπίσιν: the continuity of the theme of bravery and its variations through the *Life* is clear (1-2, 7.2, 7.5). P. criticizes Hipposthenedas for not possessing daring to the degree required by the present circumstances, but does not condemn his sceptical attitude entirely. Hipposthenedas' uncertainty about the successful outcome of the conspiratorial operation—depicted as neither unpatriotic nor unrealistic (cf. *De gen.* 586E-587B)—, gives P. an opportunity to redefine the limits of bravery and cowardice. He is perfectly aware of the great risks involved in this enterprise, since he himself acknowledges the part "fortune" played in it (13.4; cf. *De gen.* 596D-F; see further, Hack, 1975: 54). For an elaborate account of Hipposthenedas' arguments in support of the postponement of the enterprise, see *De gen.* 586B-588B; his self-defense against Phyllidas' and Charon's accusations is reminiscent of the key-theme in P.'s *chreiai* in the Proem (*Pel.* 1, 586B, D). P. mentions in the dialogue only that the conspirators in Charon's house suspected Hipposthenedas of having betrayed their plans (595A).

μόλις ποτὲ τῷ λογισμῷ συμφρονήσας: "with difficulty having come to the realization"; for this meaning of συμφρονέω, which is common in P., see LSJ, s.v.

ὅτι τρόπον τινα... ἐλπίσιν: explanatory apposition to λογισμῷ.

τὴν τῶν Λακεδαιμονίων σαλεύουσιν ἀρχήν: the power of the Lacedaemonians is metaphorically portrayed as a ship (cf. Soph. *O.T.* 23, Plut. *Dion* 8.5, *Quom. adul.* 68F, *De Alex. magn. fort.* 337C); an analogous vivid illustration of the liability of the Lacedaemonian empire is found at 13.6. The verb σαλεύω corresponds to Nepos' *perculsa* in his description of the shaking of the Spartan hegemony (*Pel.* 2.4), which may suggest the use of a common source (Ephoros?) by both writers at this point; it should also be noted that the entire section in P.'s *Pelopidas*

(13.4-7) on the magnitude of the Theban accomplishment by so small a number of exiles bears striking similarities with Nepos' comments on the same exploit (*Pel.* 2.4), on which see Bradley, 1968: 155-156.

8.6 Χλίδων: in the *De genio* he is introduced as the overseer of Melon's charioteers and privy to the plot at Thebes (τῶν Μέλωνος ἁρματηλατῶν ἐπιστάτης, 587D); see Accame, 1936: 347n. 2, 349, Κουμανούδης, 1979, No. 2158.

8.7-8 καὶ κατὰ σπουδὴν... τραπέσθαι: a much more detailed account of the same incident is given in the dialogue (587E-588A), which Accame views as mere invention by P. The tragicomic scene between Chlidon and his wife is merely meant, according to Accame, to create a symmetrical antithesis with the following episode with Charon and his son (1936: 349); for a different view, see De Voto, 1989: 110.

ἀπορουμένης... τραπέσθαι: compare this scene with a similar one between Pelopidas and his wife, in which the latter does not succeed in convincing her husband to stay away from the battle (20.2).

<καὶ κεχρηκέναι>: Ziegler's emendation is supported by the corresponding passage in *De gen.* 587F.

9.1 Οἱ δὲ περὶ τὸν Πελοπίδαν... μεταλαβόντες: P. describes a double disguise, first as hunters (8.3) and now as peasants. De Voto (1989: 109) concedes that this is inherently plausible in view of the suspicions of possible reprisals by the exiles (cf. *De gen.* 586E, 587C, 596A-C) and the precautions taken by the Theban laconizers (586E-F).

διελόντες... αὐτούς: the twelve exiles (8.3) were divided into two bands (in the *De genio* the splitting took place just after the exiles had passed Kithairon, 594E) in order to enter the city from different parts being as little conspicuous as possible.

παρεισῆλθον... οὔσης: P. does not allow any time for delay between the twelve men's departure from the border and their arrival at Thebes; he reports that the twelve conspirators entered Thebes furtively when it was still daylight, while Xenophon assigns a day's interval between the massing of the exiles at the frontier and their return to Thebes (*Hell.* 5.4.3); see De Voto, 1989: 108-111, Buck, 1994: 74.

9.2 ἦν δέ τι πνεῦμα... ἀέρος: cf. *De gen.* 596C. The weather conditions in Thebes are reminiscent of those existing at the time of the liberation of Athens by Thrasyboulos (Xen. *Hell.* 2.4.2, 2.4.14, Arist. *Ath. pol.* 37.1, Diod. 14.32.2). This and a few other similarities between the two exploits (see note on 7.2) are clearly meant to magnify the Thebans' achievement (see also Buck, 1994: 149n. 56).

⟨ᾧ⟩: Ziegler's addition of ᾧ is unsatisfactory; ⟨καὶ διὰ τοῦτο⟩ or simply ⟨διὸ⟩, which is very common in P., is preferable and is supported by the καὶ διὰ τοῦθ' of 18.1 and the reading in *De gen.* 594D. The first καί is connecting, the second, in καὶ μᾶλλον, is intensive; ⟨καὶ διὰ τοῦτο⟩ καὶ μᾶλλον ἔλαθον should be translated then as "and for this reason they escaped notice even more".

9.3 ἐγένοντο... δέοντες: a total of forty-eight conspirators had assembled at Charon's house: the twelve exiles who came secretly over Kithairon from Athens were joined by thirty-six other dissidents from within the city; cf. also *De gen.* 594E. Xenophon does not specify the number of the conspirators that gathered in Charon's house; he only mentions Phyllidas and Charon (*Hell.* 5.4.2-3).

9.4 Φυλλίδας... ἐπιτιθεμένοις: Phyllidas, who was earlier reported to have been privy to the exiles' plans (7.4), had arranged that the conspirators' assault on the Theban oligarchs should coincide with the officials' entertainment. While P. or his source fixes the drinking-party of Archias and his associates in the winter-time (9.2) and lays particular stress on its pre-arranged date (εἰς δὲ τὴν ἡμέραν ἐκείνην ἐκ παλαιοῦ κατηγγελκώς), he makes no mention of a formal celebration taking place at that time; see also note on 8.1. Xenophon, on the other hand, drawing most likely on a different source (on the two different traditions represented by Xenophon, Nepos and Polyainos, on the one hand, and P. on the other, see Schachter, 1981: 39-40), explains that the occasion for such a party was the Aphrodisia, a festival celebrated at the end of the polemarchs' term of office (*Hell.* 5.4.4; cf. Nep. *Pel.* 2.2, Polyain. 2.4.3). Schachter notes that according to P.'s version the only reason for the proposed debauchery was Phyllidas' suggestion to the polemarchs that they should have a drinking-party with women; but, as he suggests, an official celebration of Aphrodisia seems to be a more plausible occasion, which could be used as a pretext for the conspirators to gain entry to the polemarcheion (1981: 39); for further discussion, see bibliography cited at 8.1.

γύναια τῶν ὑπάνδρων: "a few married women"; testimony varies as to the morals of these women. In referring to them P. uses the terms γύναιον and γυνή interchangeably and in no derogatory manner (cf. 10.4, 11.3, *De gen.* 577C: τινὰ τῶν ἐν ἀξιώματι γυναικῶν); elsewhere in his works, he uses the word γύναιον apparently without negative connotations (*Alk.* 39.9, *Ant.* 2.4, 10.5). Xenophon, on the other hand, emphasizes the upper-class status of these women (τὰς σεμνοτάτας καὶ καλλίστας τῶν ἐν Θήβαις, *Hell.* 5.4.4), but also has the drunken men call them ἑταῖραι (5.4.5; cf. also Polyain. 2.4.3). Schachter (1981: 40) suggests that these women may have been the polemarchs' wives, thus representing in human form Ares and Aphrodite and symbolizing the institutions of the state; see further Pirenne-Delforge, 1994: 283-284.

ἔπραττεν: "he schemed".

9.5 οὔπω δὲ... κρυπτομένων: suspicions of the exiles' return to Thebes are repeatedly mentioned in the *De genio* (586E-F, 587C, 596A-C) in order to heighten the dramatic tension of the narrative. Xenophon mentions nothing of the rumour that the conspirators were concealed in the city, nor of the letter sent by the hierophant Archias to the polemarch Archias (*Pel.* 10.6-10, Nep. *Pel.* 3.1-2).

ἀβέβαιος... μήνυσις: cf. περιφώνησιν ἀσαφῆ καὶ τυφλὴν ὑπόνοιαν ἡ πρᾶξις λαβοῦσα, *De gen.* 587C.

9.6 ὅμως 'Αρχίας... πρὸς τὸν Χάρωνα: a slightly different version appears in the *De genio* (594E-F), where both Archias and Philippos are mentioned to have sent two attendants to Charon's house.

⟨πρὸς⟩ αὐτόν: without the addition of πρός the pronoun αὐτόν refers to Charon, already mentioned. With the preposition it refers to Archias, and makes the sentence clearer. Cobet's emendation is supported by the text in *De gen.* 594F: σὲ ἥκειν πρὸς αὐτούς.

9.7 ἦν δ' ἑσπέρα... ἀνειληφότες: for a list of night attacks attested in Greek historiography, see Pritchett, 1974: 162-171; cf. also Sphodrias' nocturnal invasion of Attica, *Pel.* 14.6.

9.8-13 P.'s dramatization of the entire scene contributes to the suspense which started building up from the day of the exiles' secret return to

Thebes (*Pel.* 8.1). While Charon is earlier described as οὐδ' ὑπὸ δεινοῦ πλησιάζοντος ἔτρεψέ τι τῆς γνώμης, ἀλλ' ἀνὴρ ἀγαθὸς ἦν (8.4; cf. *De gen.* 576D-E) he is portrayed here as showing great fear not for his personal safety, but because his reputation appeared to be in jeopardy. As a token of his good faith he offered his young son as a hostage in case he should subsequently appear to betray them. P.'s renewed exploration into the theme of bravery is partly to be explained by the daring aspect of the operation against the tyrants; the great risks involved in the coup in conjunction with the small number of men participating in it made its success an even more memorable event in the Theban history. Bravery, and the varied degrees in which it is manifested, is a recurrent theme in the *Pelopidas* (1-2, 4.7, 7.2, 5, 8.4-8, 10.5, 17.11-13, 18.7, 19.3, 27.7, 29.4-5). For a fuller account of the Charon episode, see *De genio* 595A-F.

9.9 παρασχεῖν ἑαυτὸν ἀνυπόπτως: Sintenis' emendation to παρασχεῖν ἑαυτὸν ⟨δεῖν⟩ ἀνυπόπτως (after ἑαυτὸν all codd. except K add δή) is preferable since it intensifies Charon's sense of duty recounted elaborately in the treatise: φύσει πρὸς τὸ καλὸν ὑπὸ τῶν νόμων ἀγόμενος τὸν μέγιστον ὑποδύεται κίνδυνον ἑκουσίως ὑπὲρ τῆς πατρίδος, 576D-E.

9.10 παραλαβὼν... καθ' ἡλικίαν: the bravery of Charon's unnamed son is described elaborately in the treatise and parallelled with that of Achilles' son, Neoptolemos (595D-E).

9.11 πάντες δ' ἠγανάκτουν... αἰτιᾶσθαι: C. Kahle (*De Plutarchi ratione dialogorum componendorum,* Diss. Göttingen 1912, 73) draws attention to the different sequence of emotions described by P. in the *Life* and the treatise; while in the former the exiles' tears and indignation seem to be expressed simultaneously, in the latter, he points out that the two scenes are "artificially separated" (595C); claiming that the description in the dialogue is meant to create a sharp contrast between Charon and his companions, in the manner of Sokrates and his friends (cf. Pl. *Phaed.* 117d), he attributes, rather unconvincingly, the insignificant variation between the two versions to the later composition of the *De genio*; for a different view, see Corlu, 1970: 25.

εἰ δειλὸν... παρόντος: conditional clause with a causal force; see also note on 6.1.

9.13 ὁ δὲ Χάρων... τελευτῆς;: Charon's refusal to accept his son back before his integrity is proven gives P. another opportunity to refer to the idea of an honorable death, which he discussed at length in the prooemium. The emotional tone which pervades the whole scene is probably P.'s own creation in order to highlight Charon's character. Charon's rhetorical question "if any kind of life or safety could be more honorable than a decorous death" is partly answered by P. in *Pel.* 1-2, esp. 1.7-8; cf. also 34.5.

ἀνυβρίστου τελευτῆς: "a death not insulted", "a decorous death"; cf. *Luc.* 18.8.

10 Archias summons Charon for questioning about the rumor regarding the exiles' activity in Thebes. The latter allays the oligarch's fears and then informs Pelopidas of the incident. The plot is again threatened with exposure when a letter containing details of the exiles' plans is brought from Archias, the Athenian hierophant, to his Theban namesake.

10.1 τινὰς... τῶν πολιτῶν: the lack of precision in Archias' questions confirms the vagueness of the reports concerning the exiles' return to Thebes (cf. 9.5, *De gen.* 595B).

10.2 καὶ ὁ Χάρων διαταραχθεὶς τὸ πρῶτον: since P. mentioned earlier that Charon composed his countenance and modulated his voice so as not to betray what he was really doing (9.13), we could assume that his disturbed reaction to Archias' questions was part of this dissimulation; cf. also 9.9, where P. portrays Charon as determined not to raise any suspicions in his encounter with Archias (καὶ παρασχεῖν ἑαυτὸν ⟨δεῖν⟩ ἀνυπόπτως τοῖς ἄρχουσιν). This assumption, however, is not supported by either text; cf. *De gen.* 596A.

10.3 οὐ μὴν ἀλλὰ: "nevertheless", adversative; for this conjunction of particles, common in P., see Denniston, 28-30.

10.4 καὶ τὸν 'Αρχίαν... πότον: P. suggests that Charon's visit interrupted the drinking-bout of the polemarchs, whereas in the treatise he makes it clear that Charon found Archias and Philippos already incapacitated by drink (595F-596A).

διεπαιδαγώγει τὸν πότον: "he was protracting the drinking"; unusual meaning of διαπαιδαγωγῶ; cf. *De gen.* 596F, *Sert.* 16.11,

Agis/Kleom. 34.7, and Stadter's note on διαπαιδαγωγῶν, *Per.* 11.4 (1989: 137); on P.'s neologisms, see Corlu, 1970: 104-105.

10.5 μετὰ φόνου πολλοῦ τῶν πολεμίων: the conspirators' preparation for a great slaughter of their enemies, namely Archias and his associates, recalls Epameinondas' concern about certain conspirators' vindictive tendencies and his fear that the coup end up in civil bloodshed (*De gen.* 576E, 594B-C).

τὸ μὲν ἀληθὲς... πλασάμενος: P. preserves contradictory testimony in his account of this incident; he tells us in the *Life* that Charon concealed the real reason for Archias' summons from his fellow-conspirators and shared it only with Pelopidas, whereas in the treatise he has Charon break the news to all the men (595F-596C). Stern (1887: 53; cf. Bock, 1910: 57-58), noting this contradiction, takes the expression τοῖς περὶ τὸν Πελοπίδαν as refering to Pelopidas and his very close friends, but I see no reason why the group of Pelopidas only, and not the rest of the conspirators, should be entrusted with Charon's news. The divergence between the two versions lies most likely in the different scope of the works: P. wanting to highlight the importance of Pelopidas in the *Life* as the leader of the enterprise, has Charon share Archias' words only with him ; in the treatise however, as he is more restrained in his praises of Pelopidas, all the conspirators become privy to Charon's message; see Lattanzi, 1933:70; Flacelière and Chambry (1966: 145; cf. also note on 6.2) translate "au seul Pélopidas".

10.6 ἔτι δὲ τοῦ πρώτου παραφερομένου: sc. χειμῶνος.

χειμῶνα: is here used metaphorically. P. uses it freely in the *Life* to denote either the harsh weather conditions which existed on the day of the exiles' return to Thebes (9.2) or the disaster which might ensue from the damaging effect of the rumours regarding the exiles' secret activity in Thebes.

10.7-10 ἧκε γάρ τις... τοῖς Ἕλλησι: cf. also *De gen.* 596E-F, Nep. *Pel.* 3.2. A parallel incident is recounted in *Caes.* 65.1-4, where Artemidoros the sophist, who was privy to Brutus' designs, brought to Caesar in a small roll disclosure of future events and met with Caesar's indifference. In both cases the characters' disregard for, and wrong evaluation of, serious matters proved to be fatal for their lives. Bradley (1968: 157-8) draws attention to the similarity between P.'s and Nepos' account and ascribes it to the use of the same source by both authors; he notes that the only difference between the two is "the strong epideictic tone adopted by Nepos" in the sentence introducing the episode at 3.1.

'Αρχίου τοῦ ἱεροφάντου: Nepos names the Athenian Archinus (*Pel.* 3.2). Accame (1936: 350) suggests that P. himself altered the name of the hierophant from Archinos to Archias to enhance the irony of the polemarch's response. On the Athenian Archias, see Judeich, *RE* s.v. "Archias, 9", 2.1, 1895: col. 462, Clinton, 1974: No. 3, pp. 16-17, Osborne and Byrne, 2, 1994, s.v. "Archias, 14", Traill, 1995, 3, No. 212235. On Archias the Athenian and Archias the Theban, see Herman, 1987: 21. Demosthenes informs us that the Athenian was later convicted of sacrilege (*contra Neaera*, 116-117). Hierophant is the title of the first and most distinguished priest of the Eleusinian cult, who revealed the *sacra* to the initiates; see Stengel, *RE*, s.v. "'Ιεροφάντης", 8, 1913: cols. 1581-1583, G. Mylonas, *Eleusis and the Eleusinian Mysteries* (Princeton 1961) 85-86, 229-232 and passim, Kerenyi, 1967: 22-25, 90-94, 108-131 and passim, Clinton, 1974: 8-47, esp. 46-47.

ἐπιστολὴν... οὐ κενήν: an ironic allusion to Charon's deceptive words to Archias: κενός τις... λόγος, 10.3.

τὴν ἐπιστολὴν... ὑπέθηκεν: Nepos simply adds that the polemarch placed the letter unopened (*sicut erat signata*, *Pel.* 3.2) under his cushion.

οὐκοῦν εἰς αὔριον τὰ σπουδαῖα: cf. εἰς ἕω τὰ σπουδαῖα, *Quaest. conv.* 619D-E, *Paroem. Gr.*, 1, 404, s.v. 'Εν ἀοῖ τὰ σπουδαῖα.

ὁ μὲν οὖν λόγος... Ἕλλησι: the origin of this popular proverb owes its celebrity to Archias' fatal negligence, an otherwise minor incident in the history of the Greek world (Borthwick, 1976: 150n. 11). The diachronic value of this proverbial expression is also attested in K.P. Kavafy's poem, Μαρτιαὶ Εἰδοί (*Ποιήματα 1896-1918*, vol. 1^4 [Athens 1968] 18), which is based on Caesar's reluctance to heed Artemidoros' advice.

11: The murder of the polemarchs: The time for action arrived and the two groups of plotters left Charon's house. Disguised as revellers one group gained entrance to the polemarcheion and killed Archias and

Philippos, while the other slew Leontidas and Hypates in their respective homes. There are certain discrepancies between the accounts of Xenophon (*Hell.* 5.4.5-7) and P. regarding the number of conspirators participating in the murder of the polemarchs and the mode of assassination, and a few minor differences between P.'s own two accounts (*Pel.* 11, *De gen.* 596C-597F); the episode is only briefly reported by Nepos (*Pel.* 3.3) and Diodoros (15.25.2) and there are also two anecdotal notices in Polyainos' *Strategemata* (2.3.1 and 2.4.3); see further, Borthwick, 1976: 148-151, De Voto, 1989: 110-113, Buck, 1994: 72-78. On the different traditions regarding this episode, see Mastrocinque, 1983: 237-247.

11.1-2 οἱ μὲν περὶ Πελοπίδαν: on this construction, see note on 6.2.

δίχα διελόντες... Φίλιππον: P. has at least three Thebans undertake the murder of Leontidas (Pelopidas, Damokleidas and Kephisodoros, two of whom later murder Hypates) and at least two more, Charon and Melon, who are responsible for the murder of Archias and Philippos. Xenophon ignoring the participation of Pelopidas has the murder committed by a group of seven exiles (5.4.3) with the addition of Phyllidas for the murder of Leontidas (5.4.7). Likewise, Nepos makes no mention of two groups of plotters, but represents the entire operation as unitary (*Pel.* 3.3).

τὸν Ὑπάτην: he was a leading figure among the laconizing Theban oligarchs (cf. Xen. *Hell.* 7.3.7), and probably polemarch in 379 BC; see Sundwall, *RE* s.v. "Hypates", 9, 1916: col. 241, Κουμανούδης, 1979, No. 2001.

Χάρων καὶ Μέλων: "Charon and Melon (and their group)"; cf. Χάρων δὲ καὶ Μέλων καὶ οἱ μετ' αὐτῶν, *De gen.* 596D. P. shifts construction between the μέν and δέ clauses. On Charon, see note to 7.4, and on Melon, note to 8.4.

ἐσθῆτας... πρόσωπα: P. does not distinguish in the *Life* between those dressed in women's clothes and those covered with garlands of pine and fir, as he does in *De gen.* 595D. Commenting on the discrepancy between the two accounts Accame suggests that there is an awkward fusion of two different traditions in the latter work and a harmonious blend of Xenophon's account with an Ephorean tradition in the former (1936: 347-9; cf. Buck, 1994: 73). Pirenne-Delforge regards the two traditions, P.'s and Xenophon's, as irreconcilable (1994: 284; cf. also Mastrocinque, 1983: 241). Besides P.'s version of the conspirators disguised as women (5.4.5), Xenophon seems to know of another tradition in which the conspirators arrived in the guise of komasts (5.4.7); see Accame, ib. 337. The motif of men disguised as women to deceive the enemy is a literary topos; cf. Hdt. 5.20, Paus. 4.4.3, Herakl. Pont. in Müller, *FHG* 2, 222, fr. 32, Plut. *Sol.* 8.5-6 (cf. Polyain. 1.20.2); see further, F. Le Corsu, *Plutarque et les Femmes* (Paris 1981) 143, Mastrocinque, ib., 242-243.

τοῖς θώραξι: P. mentions simple breastplates in the *Life*, but describes them as semi-breastplates (ἡμιθωράκια) in the *De gen.* 596D, on which, see Polyain. 4.3.13, *SIG*[3], 421, Pollux, *Onom.* 1.134, 10.142, M. Feyel, "Le Règlement militaire trouvé à Amphipolis", *Revue Archéologique* 6 (1935) 29-67, esp. 38-40.

11.3 διὸ καὶ ταῖς θύραις... ἥκειν: there is a striking discrepancy between P.'s two accounts of this incident, which is to be attributed probably to the different scope of the two works: the scene of the arrival of the conspirators dressed up as women provoked excitement in the people at the banquet-room and they started shouting and clapping their hands; the same episode is described in different terms in *De gen.* 597A: Ἡ μὲν οὖν τῶν στεφάνων καὶ τῆς ἐσθῆτος ὄψις παραλογιζομένη τὴν ἐπιδημίαν ἡμῶν σιγὴν ἐποίησεν. Embellished with novelistic touches and rhetorical features (καθάπερ δρᾶμα τὴν πρᾶξιν... διαποικίλλουσα κινδυνώδεσιν ἐπεισοδίοις, *De gen.* 596D-E; see further Mastrocinque, 1983: 244-247; P. Desideri, "Il De genio Socratis: un esempio di storiografia tragica?", *Athenaeum* 62 [1984] 569-585), the account in the treatise naturally lends itself to the invention of incidents, such as this one, which enhance the dramatic tension of the plot. Bock suggests that the silence described in the *De genio* is intended to foreshadow the polemarchs' death (1910: 58).

ἃς πάλαι προσεδόκων γυναῖκας: see note on 9.4.

11.4 ἐπεὶ δὲ περιβλέψαντες... ἀπέκτειναν: in the *Life* P.'s account of the assassination of Archias and Philippos by the group led by Charon and Melon lacks the details given in the treatise (596F-597C); this difference of emphasis is to be expected since this phase of the operation is only indirectly relevant to Pelopidas' activity. Both P. and Xenophon mention the scanty attire of the attackers: ἔχοντες οὐδὲν ἕτερον τῶν ὅπλων ἢ μάχαιραν ἕκαστος, *De gen.* 596C; ξιφίδια ἔχοντες καὶ ἄλλο ὅπλον οὐδέν, *Hell.* 5.4.3.

συνεξανισταμένους διὰ τὴν μέθην: "excited on account of their drunkenness", "reduced by drunkenness"; συνεξανίσταμαι is uncommon in classical Greek; see Wyttenbach, s.v. συνεξανίστασθαι; for an analogous description of Archias' condition upon Charon's arrival at the polemarcheion, cf. *De gen.* 596F: τῇ μέθῃ κατακεκλασμένος.

11.5-10 The elaborate description of the conspirators' murder of Leontidas and Hypates by Pelopidas' group is to be explained by the leading role P. assigns to the Theban in the *Life*, but not in the treatise (cf. 597D-F). Borthwick (1976: 148-151, esp. 149-150) suggests that the subject-matter of the scene on the late fourth century Panagjurischte amphora is a historical incident and not a 'genre-scene' from comedy, as Griffith proposed (1974: 38-49, esp. 43-49 and plate V). The scene depicts with some remarkable fidelity, Borthwick claims, the attack made by Pelopidas and his group to assassinate Leontidas.

τοῖς δὲ περὶ τὸν Πελοπίδαν... ἐχώρουν: Pelopidas' activity in P.'s account is substituted for Phyllidas' in Xenophon's narration of the same events (*Hell.* 5.4.7).

ἐργωδέστερον: "with greater difficulty". Accame rightly observes that P. does not fail to remind the reader of the particular difficulty of Pelopidas' task (1936: 347). The mention, for instance, of Leontidas' soberness (νήφοντα), which is omitted in the *De genio*, enhances the greatness of Pelopidas' exploit, as opposed to the easy killing of the drunken Archias and his associates by the other group of conspirators (καὶ συνεξανισταμένους διὰ τὴν μέθην οὐ πάνυ χαλεπῶς ἀπέκτειναν, 11.4). P. stresses anew the additional obstacles encountered by Pelopidas during his fight with Leontidas at 11.8. No systematic attempt to individualize Pelopidas' role in the murder of Leontidas and Hypates is made in the treatise.

δεινὸν ἄνδρα: cf. εὔρωστος δὲ τῇ ψυχῇ καὶ κατὰ χεῖρα ῥωμαλέος, *De gen.* 597E.

11.5-6 καὶ κεκλεισμένην... ὥρμησαν: following clearly a different tradition Xenophon reports that when the plotters arrived at Leontidas' house, he was still reclining on his couch after dinner and was with his wife who was carding wool (*Hell.* 5.4.7). The idyllic setting in his account reinforces, as Corlu notes, the gruesome nature of Leontidas' assassination (1970: 34). P. introduces more details regarding the first phase of the assault in the *De genio* (597D); he recounts that the conspirators, in order to gain entrance to Leontidas' house, persuaded a servant that a message had come for Leontidas from Kallistratos of Aphidna (Cloché, 1934: 67-68). In Xenophon's account, Phyllidas, accompanied by three of the conspirators, proceeded to Leontidas' house, and knocking at the door said that he wished to give a message from the polemarchs. The discrepancies between the accounts of Xenophon and P. are hard to reconcile and it is difficult, if not pointless, to speculate about which of the two versions is closer to the facts (see also, De Voto, 1989: 112). Stern argues that Xenophon's version is more convincing (1887: 29-30); he suggests that the historian may have used as his source conversations with surviving members of the oligarchs' families, such as Leontidas' widow (1884: 57), or with conspirators, such as Phyllidas (1887: 17-18). Buck proposes Menekleidas, who was later strongly opposed to Pelopidas and Epameinondas (1994: 76).

11.7 ἔλαθε δ' αὐτὸν καταβαλεῖν τὰ λύχνα: "he forgot to throw down (i.e. "overturn" or "upset") the lamps"; the accusative and infinitive after λανθάνω is rare: LSJ A3, s.v.; cf. Plut. *Arist.* 17.6-7.

αὐτοὺς ἑαυτοῖς περιπετεῖς ποιῆσαι: "cause them to fall foul of each other"; cf. *Marc.* 26.5, *Fab.* 12.5.

καὶ τὸν πρῶτον... κατέβαλε: a discrepancy between P.'s two accounts regarding the sequence of Kephisodoros and Leontidas' deaths should be attributed to the differences of emphasis in the works. In the *Life*, Kephisodoros dies before Leontidas leaving therefore Pelopidas to fight the tyrant unaided, whereas in the dialogue his death follows Leontidas', after giving Pelopidas his hand and saluting the others (597F); an attempt to enhance the dramatic effect of the entire scene is immediately apparent here. In his much shorter version, Xenophon makes no mention of Kephisodoros' death.

11.8 συνεπέπλεκτο: imperfect (following KCB and Sintenis) instead of pluperfect; the latter is usually accompanied by ἤδη; cf. *Pel.* 12.3, *Tim.* 2.2, 3, 23.6.

Κηφισόδωρος: a common Theban name. He was the son of Dio-<gei>ton (*De gen.* 595E); see also 596D, 597E-F, Κουμανούδης, 1979, No. 1135.

11.9-10 ἐκράτησε δ' οὖν... διέφθειραν: P. avoids mentioning in the *Life* that Pelopidas was wounded during his encounter with Leontidas; cf. *De gen.* 597F. Hypates' assassination, which constituted the second part of Pelopidas' plan, does not appear in the *Hellenika*; there is an incidental reference to his death in a speech delivered in 366 (7.3.7).

παρεισέπεσον: "they stole in".

αἰσθόμενον... διέφθειραν: in the *De genio* P. has Hypates being killed as he was attempting to flee over the rooftop to a neighboring house (597F).

12 Pelopidas' party is joined by the other group led by Melon and also by Epameinondas and Gorgidas. They proclaimed the liberation of the city and called the citizens to assembly. Having sent for the rest of the exiles who were waiting under Pherenikos in the Thriasian Plain (*Pel.* 8.1) they launched an attack on the Kadmeia with the aid of armed citizens. P. especially favors in this episode the powerful effect of visual images.

12.1 ἔπεμψαν... φυγάδων: Xenophon mentions nothing about any Boiotian exiles on the frontier, but reports the presence of deployed Athenian forces who were summoned by the returned exiles as soon as the coup succeeded (*Hell.* 5.4.9); on the presence of Athenian troops on the Boiotian frontier, see note on 12.5.

καὶ τοὺς προσιόντας... ἀναρρηγνύντες: a slightly different version emerges from the account of Xenophon, who has Phyllidas and two of his men provide the liberated prisoners, not the citizens, with arms (*Hell.* 5.4.8). Since the release of the Theban political prisoners is not directly connected with Pelopidas' activity, but with Phyllidas', P. makes no mention of this incident in the *Life*, while he recounts it in vivid detail in the treatise (598 A-C). According to Xenophon (5.4.8) the rebels and the prisoners gathered at Ampheion; see Hani, 1980: 213-214n. 1, Symeonoglou, 1985: 12.

ἀφαιροῦντες... σκῦλα: for a discussion of captured arms dedicated in temples, in stoas, and in various structures within a sacred precinct, see Pritchett, 1979: 293-295; J.J. Coulton, *The Architectural Development of the Greek Stoa* (Oxford 1976) 12-13. Hani (1980: 128n. 1) suggests that the trophies mentioned here were from the Battle of Delion in 424.

τὰ περὶ τὴν † οἰκίαν ἐργαστήρια: a corrupt passage. Ziegler's proposed emendation to τὰ ⟨τῶν⟩ περιοικούντων ἐργαστήρια, which he does not incorporate in the text, finds some support in the corresponding passage in *De gen.* 598D (τὰ τῶν ἐγγὺς οἰκούντων ἐργαστήρια μαχαιροποιῶν). Wilamowitz's suggestion that the temple of Εὔκλεια may have been part of the corrupt line here is not convincing (*Pindaros* [Berlin, 1922 (1966)] 30n. 1), chiefly because P. is, generally, not in the habit of providing in the *Life* any topographical information in his account of the liberation of the Kadmeia. His narrative technique is different, however, in the *De genio*; when the same historical events are reported at a greater length by Kaphisias to his Athenian audience in the treatise, a number of Theban sites, in which the action took place, are mentioned by name (Ἄμφιον, 577B, πολύστυλον, 598 A, τὸ τῆς Ἀθηνᾶς ἱερόν, 598D). The reason for P.'s varied handling of the topographical material—sparing in the one case and more informative in the other—lies clearly in his different aims in these two works. In the *Pelopidas*, as in his other *Lives*, he is not particularly concerned about detailed accounts of the historical events he recounts unless they provide the reader with character-revealing material or are used as a point of departure for his comments. In the *De genio*, however, P. concentrates on the emotive and pictorial aspects of the same events. As a result, topographical clarity in the lengthy accounts of the conspirators' activities in this historico-philosophical treatise becomes a vital ingredient of its narrative. It not only enhances the vividness (ἐνάργεια) of the descriptive passages, but also imparts to them a sense of historical accuracy, which is otherwise blurred by the fictional nature of the philosophical debates and Kaphisias' flair for sentimental incidents (cf. 586B-588B, 587D-588A, 594F-595E). To assume, therefore, that the name of a Theban site is missing here would not be in accord with P.'s characteristic reticence about Theban topography throughout the entire account of the return of the exiles to Thebes and the liberation of the Kadmeia. On P.'s method of autopsy, see Buckler, 1992: 4788-4830, esp. 4807-4808.

12.2 ἦκον δὲ βοηθοῦντες... βελτίστους: cf. *De gen.* 598C-D. P. specifies in the dialogue that the two groups of plotters met at the Stoa Polystylon (598A) and were joined by Epameinondas and Gorgidas near the temple of Athena (598D; see Aloni: 1982: 118-119). Epameinondas joined the plotters after the murder of the polemarchs and those members of the oligarchy who were not attending the banquet abstaining, therefore, from acts that resulted in civil bloodshed to which he was strongly opposed (cf. *De gen.* 576D-E, 594B-C, Nep. *Epam.* 10.3).

Γοργίδαν: a member of the democratic group; he had not fled to Athens when Kadmeia was occupied and is mentioned to have been a hipparch before 382 (cf. *De gen.* 577C) and also in 378 (Polyain. 2.5.2; for a different view regarding the latter office, see Buckler, 1979: 53), when he commanded the Theban troops during Agesilaos' invasion of Boiotia. He is credited by P. as the founder of the Sacred Band (*Pel.* 18.1, Polyain. 2.5.1) and is ranked by Diodoros as among the most illustrious Theban generals together with Pelopidas and Epameinondas (15.50.6). Buckler (1980a: 41) suggests that it is probable that Gorgidas was killed in action sometime between 378 and 376 (his view is refuted by De Voto, 1992: 9), since he is not heard of again after 378, and since by 375 Pelopidas was commanding the Sacred Band. On Gorgidas' Boeotarchy after the liberation of the Kadmeia, see notes on 13.1 and 14.2; more generally, see Swoboda, *RE*, s.v. "Gorgidas", 7.2, 1912: cols. 1619-1620, Κουμανούδης, 1979, No. 394, De Voto, ib., 4-9.

12.3 ἡ δὲ πόλις... περιέμενον: both P. (in the *Life*) and Xenophon comment on the Theban people's momentary inertia (they waited for the following day to join the conspirators) in the face of the recent developments. P. attributes their passive response to their surprise and the lack of precise information, while Xenophon talks about their incredulity (ἀπιστοῦντες, *Hell.* 5.4.9). Whereas both writers draw attention to the lapse of one day between the murder of the polemarchs and the joining of the revolutionaries by the Theban citizens in the liberation of the Kadmeia, in the *De genio* (598D-E) action appears to have taken place right away. Corlu (1970: 34) attributes this discrepancy to P.'s tendency to highlight the Theban patriotism more in the treatise than in the *Life*, while Accame ascribes the lack of chronological precision in the treatise to the overpowering effect of the dramatic episode (1936: 347); cf. J. Rohrmoser, "Zeugenverhör über die Befreiung Thebens 379 v. Chr.", *Zeitschrift für die österreichischen Gymnasien* 41 (1890) 581-593, esp. 591.

ἀνεπτόητο: metaphorical use of ἀναπτοέομαι ("to be in a flutter of excitement").

12.4 ὅθεν ἁμαρτεῖν... ὄντες: P. observes that the Spartan commanders were viewed to have committed a tactical blunder in not engaging in battle with the Thebans at once, taking, therefore, advantage of their delay in joining the rebels and also of the size of the Spartan garrison, which amounted to approximately 1,500 soldiers (cf. also Diod. 15.25.3). P. reports, moreover, that the Spartan garrison was strengthened by the addition of many pro-Spartan Thebans (οἱ λακωνίζοντες), who fled from the city and took refuge to the acropolis, and the so-called "Betters" (κρείττους); the latter usually encamped nightly at the foot of the Kadmeia (*De gen.* 598 E). For a discussion of the strength of the Spartan garrison at Thebes, see Parke, 1927: 162-163.

τὸν ψόφον: Sintenis' emendation for ὄχλον codd.; it is based on the parallel passage in *De gen.* 598E.

12.5 ἅμα δ' ἡμέρᾳ... ὡπλισμένοι: at dawn the exiles left in the Thriasian plain under the leadership of Pherenikos (cf. *Pel.* 8.1) arrived at Thebes to join the rest in the effort to expel the Spartan garrison. This group constituted the major body of the Theban exiles who had taken refuge to Athens; for a discussion of the number of the exiles, see note on 5.3 and Buck, 1994: 78.

The evidence about Athenian assistance to the Thebans at the time of the liberation of the Kadmeia is conflicting. P. says nothing about any official or unauthorized Athenian action and his silence on this matter is uniformly viewed as a deliberate omission in order to magnify the role of the Theban rebels. Xenophon, on the other hand, mentions the presence of Athenian forces under the command of two unnamed generals on the Boiotian frontier, who, being summoned by the Theban rebels, aided in the siege of the Kadmeia (*Hell.* 5.4.9-11); he seems to suggest the essentially private nature of the assistance afforded Thebes by Athens and also mentions the subsequent punishment of the two generals (5.4.19). A slightly different version emerges from Diodoros' account (supported by those of Aristeides and Deinarchos, on which see below), who talks about official Athenian assistance to the Thebans in reply to their request; he even specifies the size of the Athenian contingent (15.25.1-27.4). Diodoros says nothing about any generals being punished. Finally, there is the testimony of three Athenian orators, Isokrates (*Plat.* 29), Deinarchos (*contra Dem.* 38-39) and Aristeides (*Panath.* 283D), all of whom agree in that the Athenians assisted the Thebans to expel the Spartan garrison and liberate their city. For a discussion of the reliability of the Athenian orators' testimony regarding the Athenian aid to the Thebans, as well as the discrepancies among sources, see Cloché, 1934: 55-57, Accame, 1941: 22-23, Burnett, 1962: 1-17, esp. 16, Ryder, 1965: 53, Rice, 1971: 89-92, id. 1975: 95-130, esp. 101, 103, 110-111, Cawkwell, 1973: 47-60, esp. 56-60, Seager, 1974: 45-46, Hack, 1975: 58, Ober, 1980: 227-231, Cargill, 1981: 56, Kallet-Marx, 1985: 140-147, Urban, 1991: 161-163, Buck, 1992: 103-109, id. 1994: 74-75, 83-87, 149nn. 56-58, C.D. Hamilton, 1991: 156-159, Worthington, 1992: 194-195, Munn, 1993: 137n. 12, 216-224.

συνήθροιστο δ' εἰς τὴν ἐκκλησίαν ὁ δῆμος: since P. mentions earlier that the conspirators were arming the Theban citizens (12.1), we can safely assume that most of them, at least, were armed at the time of the assembly; a similar instance of Thebans being armed at a public assembly is reported by Pausanias (9.1.6) for the year 373; see Sordi, 1973: 84.

12.6 εἰσῆγον... βοηθεῖν: Epameinondas and Gorgidas appear here to act unofficially as boeotarchs since we are told that they brought before the assembly Pelopidas and his associates with the recommendation that they be welcomed as deliverers and benefactors; for a discussion of the probouleutic duties of the boeotarchs, see Buckler, 1980a: 25-26.

12.7 εὐεργέτας καὶ σωτῆρας: used in a formulaic manner; cf. *Thes.* 33.1, *Aem.* 39.8, *Cat.Mi.* 71.1, *Demetr.* 9.1. The doublet is frequently found in honorific and dedicatory decrees in Greek inscriptions (cf. *SIG* s.v. εὐεργέτης, 4.2, Dittenberger). C.P. Jones (1971: 115) commenting on P.'s vocabulary notes that it is "echoed by inscriptions that celebrate local magistrates". On P.'s use of εὐεργέτης and σωτήρ in the *Pelopidas*, see Fuscagni, 1975: 31-55, esp. 44-48.

13: The liberation of the Kadmeia and the re-establishment of the Boeotarchy in 378

13.1 Ἐκ δὲ τούτου... ὁ Πελοπίδας: P. records that Melon, Charon and Pelopidas were elected to the Boeotarchy by the assembly of the Theban people (cf. 12.6-7; see Buck, 1994: 150n. 79); in *Ages*. 24.6, only Pelopidas and Melon are mentioned as boeotarchs at that time, while in *Pel.* 14.2 Gorgidas is reported to hold this office together with Pelopidas. Following P.'s different accounts there were four Theban boeotarchs elected in 378. The accuracy of his testimony about the actual election of boeotarchs—the first election, in fact, after the abolition of the Boeotarchy in 386, when the Boiotian Confederacy was dissolved—on the eve of the liberation of the Kadmeia and the number of the elected boeotarchs have been much discussed by other scholars. Cawkwell argues against the premature restoration of the Boeotarchy at that time agreeing with Beloch's reading of Isokrates' statement in *Plat.* 29 (cf. Beloch, 1922^2, vol. 3:1, 145n. 2, on which see Buckler, 1979: 52-53); he suggests that all four men were merely magistrates—with Pelopidas, Melon and Charon as polemarchs and Gorgidas as hipparch, cf. Polyain. 2.5.2—acting in place of the murdered polemarchs (1972: 275-276). The re-establishment of the Boeotarchy in 379/8, which signalled the Thebans' intention of restoring the Boiotian Confederacy (cf. Diod. 15.28.1) and their leadership in it, has been supported by Accame, 1941: 26—albeit with some reservations about the actual date of the re-establishment of the κοινὸν τῶν Βοιωτῶν, 25—, Bersanetti (1949: 49-50, esp. 89-93), Roesch (1965: 100-103, 124), Larsen (1966: 72), Sordi (1973: 79-91), who agrees with Fuscagni's suggestion (1972: 415-433) that two (Melon and Charon, 422), not four, boeotarchs would have come from Thebes at this time, and Hack (1975: 66); finally, a strong case for the accuracy of P.'s account of the constitutional reforms in 378, as well as of the number of Theban boeotarchs elected at this time has been made by Buckler (1979: 50-64) supported by Kallet-Marx, 1985: 144n. 75, and Buck, 1994: 150n. 78.

βοιωτάρχης: the boeotarchs were the highest officials of the Boiotian Confederacy, who had the supreme command of the Boiotian military forces and operations and were also in charge of the supervision of foreign affairs; they were, therefore, instrumental in shaping and implementing the policy of the Confederacy. Their main duty in the legislative sphere was probouleutic, an aspect of which was diplomacy (cf. the introduction of foreign ambassadors to the assembly, *Pel.* 26.1). There is a board of eleven boeotarchs in 424 (Thuk. 4.91) and in 395 (*Hell. Oxyrh.*, Bruce, 11.3 [Chambers, 383-384]) and seven listed for the period between the liberation of the Kadmeia and the Battle of Chaeronea (Paus. 9.13.6-7, *IG* VII 2407 and 2408; cf. *Teiresias: Epigraphica* 13 [1983] no. E.82.50, *ib.* 15 [1985] no. E.85.17, and Roesch, 1984: 45-60, esp. 45-51); for further on the role of boeotarchs, see Cloché, 1952: 72-74, Salmon, 1978: 129-143, Buck: 1979: 124-125, id. 1985: 293-294, id. 1994: 108-110, Buckler, 1980: 24-30, Roesch, 1982: 287-290, Kallet-Marx, 1985: 135n. 35, Salmon, 1995: 365-383.

εὐθὺς ἀπετείχιζε... ἐπελθεῖν: Pelopidas emerges in P.'s account as the protagonist in the blockade and assault of the Kadmeia (cf. ἀπετείχιζε... ἐποιεῖτο... σπουδάζων), while the role of the other boeotarchs and revolutionaries is entirely effaced and the Athenian aid omitted (on which, see note on 12.5). P. reports that Pelopidas precipitated the whole operation of the liberation of the acropolis before a relief Spartan army arrived, thus suggesting that the Spartan commanders had already requested military reinforcement during the night (cf. also Diod. 15.27.1). Xenophon also tells us that upon the proclamation of the Theban rebels the Spartan harmost in the Acropolis sent immediately to Thespiai and Plataia for help (*Hell.* 5.4.10).

13.2 καὶ παρὰ τοσοῦτον... δυνάμεως: ancient sources disagree regarding the interval between the surrender and retreat of the garrison and the arrival of the relief force in Boiotia. P. reports that Kleombrotos and his rescue force responded so quickly to the garrison's request for help that they met the retreating troops in the Megarid; the swiftness of the Spartan response in the *Pelopidas* seems to suggest the short duration of the siege and the undelayed capitulation of the garrison (cf. Buck, 1994: 82-83). Xenophon implies a rapid surrender of the garrison and evacuation of the Kadmeia, while he indicates that there was a lengthy interval between the retreat of the garrison and the mobilization of a rescue force (*Hell.* 5.4.10-12); both P. (*Ages.* 24.3) and Xenophon (5.4.13) agree in their accounts as to who led the expedition against

Thebes. In his more lengthy account of the siege and the surrender of the Spartan garrison, Diodoros says that the Lacedaemonians spent much time mustering rescue forces and narrowly missed arriving at Thebes before the capitulation of the garrison (15.27.1-3); a more prolonged siege is clearly indicated in Diodoros' version. The omission of certain incidents in P.'s account, such as the Athenian assistance—most likely unofficial—to the Thebans to recapture their city (on which see note on 12.5), or the killing of the Laconizers by the enraged Thebans (Xen. *Hell.* 5.4.11-13; cf. *De gen.* 577A, where Epameinondas anticipates such atrocities), is not accidental, nor can it be due merely to the compression of his narrative. P. was clearly familiar with Xenophon's account regarding the Thebans' breaking of promise towards the Spartan garrison after the liberation of the Kadmeia and the subsequent massacre of their political foes, but carefully chose to pass them over in silence, as they would no doubt undermine the Theban exploit, and indirectly Pelopidas' role in it. He maintains the same attitude in the *De genio* (598F; cf. also Diod. 15.27.2). For a discussion of the discrepancies among the sources, see Rice, 1971: 91-92, Hack, 1975: 62-65, De Voto, 1989: 114-115, Buck, 1994: 81-87. On the dispatch of a Theban embassy to Sparta asking for peace terms (Isokr. *Plat.* 29) in the interval between the retreat of the Spartan garrison and Kleombrotos' expedition to Thebes, see Hack, 1975: 64-77. On Kleombrotos' campaign against Thebes, see note on 14.1.

Κλεόμβροτον: Kleombrotos the first was from the royal family of Agiades, the son of king Pausanias and the brother of Agesipolis I, whom he succeeded as a king of Sparta in 380 BC; he fell at the battle of Leuktra in 371 (Plut. *Ages.* 28.8, Diod. 15.55.5, Xen. *Hell.* 6.4.13); see further, Lenschau, *RE* s.v. "Kleombrotos, I", 11.1, 1956: cols. 677-678, Poralla, 1985, No. 434, s.v. Κλεόμβροτος, I.

13.3 οἱ δὲ Σπαρτιᾶται... μετέστησε: P., agreeing with Diodoros (15.27.3), records that two of the harmosts who had been present at the Kadmeia were executed by the Spartans, while the third one having been severely fined—and being unable to pay, as Diodoros notes—left the Peloponnese. In the *De genio* he adds that the third harmost was in Haliartos at the time of the coup (598F); his absence from Thebes appears then to have been the reason for not suffering the same penalty as the other two. Xenophon mentions only one unnamed harmost's trial and death-penalty for having abandoned the Acropolis instead of waiting for the relief force (*Hell.* 5.4.13); see further, Parke, 1927: 164-165, Hack, 1975: 64, Munn, 1993: 218-221, Buck, 1994: 83.

In an attempt to explain the number of harmosts present in Thebes at this time (three instead of the regular one) and reconcile P.'s and Diodoros' divergent accounts, on the one hand, and Xenophon's, on the other, Parke (1927: 164-165) suggests that only one (Lysanoridas) was the regular harmost of the garrison, while the other two (Herippidas and Arkesos) were each the commander of a distinct force temporarily placed together; one of them may have been the acting harmost of the garrison during Lysanoridas' absence in Haliartos.

The harmost was a Spartan military commander, usually with a garrison, exercising some authority in a city outside Sparta; on harmosts, see Parke, 1927: 161-162, id. 1930: 37-79, esp. 78-79, id. 1931: 31-38, Bockisch, 1965: 129-239, S. Link, *Der Kosmos Sparta: Recht und Sitte in klassischer Zeit* (Darmstadt 1994) 11-12.

Ἡριππίδαν: Bryan's emendation finds support in Xenophon; he led an embassy to Tissaphernes in 396 (*Hell.* 3.4.6), commanded the Cyreans at the battle of Koronea (4.3.15, 17) in 395 and was appointed nauarch, probably in 391 (4.8.11); he may be identified with Θηριπίδης mentioned in Diodoros as commander at Herakleia in Trachis in 399 (14.38.4-5, Polyain. 2.21) and in the Spartan expedition to Histiaea (15.30.3-4). The reading Ἑρμιππίδαν (codd.) is found also in *De gen.* 598F, while 586E reads Κρηππίδας; see Lenschau, *RE* s.v. "Herippidas", 8, 1913: cols. 684-685, Parke, 1927: 159-165, esp. 162-163, Poralla, 1985, No. 349, s.v. Ἡριππίδας, Munn, 1993: 219-221.

Ἄρκεσον: we have no other information about Arkesos other than P.'s testimony (cf. also *De gen.* 598F). He may have been leader of a field force in Thebes, or the acting harmost of the garrison in Lysanoridas' temporary absence; see Parke, 1927: 124, Poralla, 1985, No. 143, s.v. Ἄρκισσος.

Λυσανορίδας: a Spartan commander, probably to be identified with the Lysandridas of Theopompos (*FGrH* 115F240) quoted by Athenaios from the fifty-sixth Book of the *Philippika* (Athen. 609b, on which see Parke, 1927: 159-160n. 4); he is mentioned in the *De genio* to have replaced Phoibidas (576A; also 577A-B, 578A-B), see further, Poralla, 1985, No. 503, s.v. Λυσανδρίδας.

13.4 Ταύτην τὴν πρᾶξιν... οἱ Ἕλληνες: P. stresses the analogy between the Athenian's successful overthrow of the Thirty Tyrants and the Theban's liberation of the Kadmeia, in order to magnify Pelopidas' role in it; cf. also 7.2 and note ad loc.; the likening of the two exploits, which P. may have derived from his source, is not limited to this *Life* (cf. *Arat.* 16.3-4, *Non posse* 1098A).

καὶ βραβευθεῖσαν ὁμοίως ὑπὸ τῆς τύχης: Sordi (1995: 415-423) draws attention to the role of τύχη in the Theban exploit, which P. explicitly views as a decisive element in the overthrow of the pro-Spartan tyrants, while Nepos merely suggests it with his emphasis on the unprecedented success of this dangerous enterprise (*Pel.* 2.3, 3.1; cf. Plut. *Pel.* 13.5). Based on this and a few other similarities between the two accounts she proposes that both authors probably derive their account of the occupation and liberation of the Kadmeia independently from Theopompos' *Philippika*, and particularly from Bk. VIII, which was dedicated to the Θαυμάσια (*FGrH* 115F64-76). She also suggests that the author upon which P. and Nepos derived their accounts, Theopompos, drew among other sources on Xenophon's *Hellenika* (417-420); see **[2c]** s.vv. "Theopompos and Xenophon". For a discussion of the meaning of τύχη in P., see Brenk, 1977: 145-183, id. 1987: 305-316 and 345n. 133; in general, see Ziegler, *RE* s.v. "Tyche", 7B, 1948: cols. 1689-96.

13.5-7 Nepos expresses the same sentiments about the venture and the magnitude of the Theban accomplishment, but in a reverse order, that is by way of introduction and not by way of conclusion, as does P. Furthermore, he views the consequences of the success of this undertaking as extending beyond the political rivalry of the two parties to the ascendancy of Thebes over Sparta and the rest of Greece (*Pel.* 2.3-4; cf. Bradley, 1968: 155); see also P.'s praise of the exploit in the *synkrisis* 31(1).6. Diodoros, at 15.81.1, makes Pelopidas the protagonist of the liberation of the Kadmeia, as P. does here.

13.7 ὁ γὰρ καταλύσας... καὶ παύσας... πόλεμος: see note on 6.2 (εἰ μή τις... ἄρχοντας).

ἐξ ἐκείνης ἐγένετο τῆς νυκτός: "began from that night"; the liberation of the Kadmeia marked only the beginning of the collapse of the Spartan hegemony. The subsequent battles at Tegyra and at Leuktra would be the final blow to it.

οὐ φρούριον, οὐ τεῖχος, οὐκ ἀκρόπολιν: anaphora; the use of the tricolon with οὐ instead of οὔτε, is not uncommon in P.; cf. *Pel.* 34.4, *Per.* 12.3, 39.2, *Sol.* 7.1, *Fab.* 17.3. On the use of cumulative negatives by P., see Ambrosini, 1991: 19-34, esp. 19-20. The words φρούριον, τεῖχος, ἀκρόπολις seem to refer in no specific order to the different stages of Thrasyboulos' achievement, namely the occupation of Phyle, Mounichia and Peiraeus.

δωδέκατος: should be translated as "one of the twelve" and not as "twelfth"; the ordinal may be used here by P. like the phrase δωδέκατος αὐτός (Dover, 1960 [1988] 61-77, esp. 61-63, 70-71, 73) suggesting therefore some kind of superiority of Pelopidas over his fellow-Thebans; cf. also the use of verbal forms in the singular: καταλαβών - ἔλυσε - διέκοψε; also earlier in the *Life* P. explicitly views Pelopidas as the leading figure of the coup: ὑφίσταται δὲ τὴν πρᾶξιν Πελοπίδας πρῶτος, 8.2.

Since the entire section is intended to make Pelopidas' undertaking emerge as more daring than that of Thrasyboulos, it is likely that P.'s reminder—see earlier mention of it at 8.3—of the twelve Theban exiles alludes to the much higher number of the Athenian exiles led by Thrasyboulos back to Athens; on the size of his group, see Krentz, 1982: 70n. 4.

εἰ δεῖ μεταφορᾷ... εἰπεῖν: apparently in order to alert the reader to the persuasive power of his argument, but also to soften the weight of the ensuing figure, P. announces that he is going to use a metaphor. The author of the *On the Sublime* 32.3 attributes such softening statements to Aristotle and Theophrastos; we may have here then an example of P.'s indebtedness to Aristotelian precepts regarding the measured use of metaphors; for similar expressions, cf. *Quaest. conv.* 692C, 747D, *Praec. ger. reip.* 803A, *Plat. quaest.* 1000F. For an example of Theophrastos' use of metaphors, see Plut. *De tuenda sanit.* 135E.

Ch. 14: Kleombrotos' invasion of Boiotia (379/8) and Sphodrias' raid on Peiraeus (378)

The events following the expulsion of the Spartan garrison from the Kadmeia and the election of boeotarchs by the Theban assembly until the Sphodrias incident and the outbreak of the Boiotian war are either omitted (cf. the dispatch of the Theban legation to Sparta offering peace terms reported by Isokrates c. 373, *Plat.* 29; views vary regarding the authenticity of the Athenian orator's claim, see Rice, 1971: 93-94, Cawkwell, 1973: 58-59, Hack, 1975: 68-77; Buck, 1994: 87) or merely hinted at by P., since they are only indirectly related to Pelopidas' career.

14.1 'Επεὶ τοίνυν... ἐμβαλόντων: after the digression on the greatness of the Thebans' achievement (13.4-7) P. resumes the thread of the events starting with the repercussions of Kleombrotos' punitive expedition to Boiotia in 379/8 (on the chronology of the expedition, see Cawkwell, 1973: 56, Kallet-Marx, 1985: 134-136; Burnett, 1962: 17; on Kleombrotos' campaign, see *Ages.* 24.3, Xen. *Hell.* 5.4.14-18, Diod. 15.27.3; see further, Munn, 1993: 139-145, Buck, 1994: 87-89). Kleombrotos advanced as far as Kynoskephalai and retired to Thespiai (Xen. *Hell.* 5.4.15). For an explanation of Agesilaos' refusal to undertake the command against Thebes, see Xen. *Hell.* 5.4.13, and a brief mention by P. in *Ages.* 24.3; see Rice, 1971: 95-96, id. 1975: 95-130, esp. 103-109, Hack, 1975: 77-84. On the transitional function of τοίνυν, see Denniston, 574-575.

οἱ 'Αθηναῖοι... Θηβαίοις: being alarmed at Kleombrotos' expedition in 379/8 and no doubt fearing that the Spartans would also retaliate against them if they continued their support of the Thebans, the Athenians decided to distance themselves from the Theban cause; although no official or unauthorized Athenian action taken during the siege of the Kadmeia is mentioned by P., there is evidence, albeit conflicting, of Athenian troops sent to help the Theban exiles in liberating Thebes; it is likely that this was an unofficial aid given to the Theban exiles by a group of pro-Boiotian Athenians (Buck, 199: 86, and note on 12.5) followed probably by a formal decree of help and the dispatch of troops (cf. the mention of Athenian peltasts led by Chabrias, Xen. *Hell.* 5.4.14; see Kallet-Marx, 1985: 142-144, Buck, 1994: 86-87). P. reports explicitly that the Athenians renounced their alliance with the Thebans (τήν τε συμμαχίαν ἀπείπαντο), without specifying at what stage this alliance was made (Bengtson, 1975, No. 254), and resumed it after the Sphodrias incident (πάλιν... συνεμάχουν, 15.1). Cawkwell rejects the existence of such an alliance and suggests that P. was probably using the term συμμαχία loosely (1973: 58; cf Accame, who calls it a reconciliation not a συμμαχία, [1941: 24]), while Fuscagni views the conclusion of an alliance between Thebes and Athens before Sphodrias' raid on Peiraeus unacceptable (1975: 33n. 9); for a different view, see Busolt, 1873-1875: 681-683, Westlake, 1939: 16, Burnett, 1962: 15-17, Hack, 1975: 110-112, Cargill, 1981: 58.

τῶν βοιωτιαζόντων... ἐζημίωσαν: while there is no earlier mention by P. of any pro-Boiotian Athenian activity except for their support of the Theban exiles in 382, in *Pel.* 6-7 (see note to 6.3) his reference to the trial and condemnation of pro-Boiotians after Kleombrotos' invasion of Boiotia suggests that he may have been aware of some sort of Athenian intervention in the liberation of Thebes. P.'s vague use of the term βοιωτιάζοντες is worth noting; by choosing to omit any reference to the Athenian assistance (see DeVoto, 1989: 114, Buck, 1992: 105-106) he is obliged to refer to the pro-Boiotians in question in general terms. Xenophon, however, is more informative about the generals' involvement and reports that of the two generals who were privy (συνηπιστάσθην, on which see Kallet-Marx, 1985: 141) to Melon's revolt against Leontidas and his group, one was condemned to death and the other was exiled (*Hell.* 5.4.19). Both P. and Xenophon agree that the motive for their condemnation was fear of war with Sparta; see Cloché, 1934: 56-57, Accame, 1941: 22-23, Cawkwell, 1973: 57-59, Kallet-Marx, 1985: 127-151, esp. 142-145.

βοιωτιαζόντων: "those siding with the Boiotians"; the expression is here confined to the anti-Lakonian group; the neologism βοιωτίζειν, instead of βοιωτιάζειν is found in *De gen.* 575D.

14.2 ὁ Πελοπίδας μετὰ Γοργίδου βοιωταρχῶν: at 13.1 P. says that Melon, Charon and Pelopidas were elected to the Boeotarchy on the eve of the liberation of the Kadmeia (379/8), while in *Ages.* 24.6 he only mentions Pelopidas and Melon as boeotarchs for that same period (see also note on 13.1). The lack of mention of Gorgidas' name in either instance is not an evidence of self-contradiction in P. for the simple reason that as a biographer he is not committed to giving a detailed picture of the state of political affairs in Thebes. Pelopidas and Gorgidas are mentioned here to the exclusion of their other colleagues, since they are the instigators of Sphodrias' raid on Peiraeus. In *Ages.* 24.6, however, it is Pelopidas and Melon that are mentioned as connected with the incident; this discrepancy may be due either to a slip of P.'s memory or to looseness of expression; cf. τῶν περὶ Πελοπίδαν καὶ Μέλωνα βοιωταρχῶν, *Ages.* 24.6.

ἐπιβουλεύοντες... μηχανῶνται: while it is generally assumed that P. depended on Kallisthenes for the entire Sphodrias episode (on which see Georgiadou, in *Boeotia Antiqua*) certain striking parallels in phraseology between the accounts of P. and Xenophon (*Hell.* 5.4.20-21) indicate that the latter's account of the incident may have been used in conjunction with Kallisthenes' *Hellenika*. Xenophon does not supply the names of the instigators of the scheme, but this is entirely in line with his methods; cf. Buckler, 1980a: 263-268, Fernández, 1984: 57-58, also [**2a**]. In contrast to Diodoros who attributes to Kleombrotos the initiative of Sphodrias (15.29.5), both P. (cf. also *Ages.* 24.6) and Xenophon make the Thebans responsible for Sphodrias' attack. Diodoros' testimony, however, cannot be flatly rejected on the basis of two converging sources. It is possible that Kleombrotos was maneuvered somehow into this situation by Agesilaos, his rival; note that both P. in *Ages.* 25.2 and Xenophon in *Hell.* 5.4.25 mention that Sphodrias belonged to the faction of Kleombrotos, and was therefore a member of the opposing royal circle; moreover, it is not accidental that it was Agesilaos who acquitted him, thus granting him the greatest favor possible; see Cartledge, 1987: 158-159, Lévy, 1990: 125-157, esp. 145. Kleombrotos' probable implication in the raid does not, of course, exclude the simultaneous involvement of the Thebans; the latter, suspecting that there was no serious possibility of Sphodrias capturing Peiraeus took advantage of his presence at Thespiai in order to precipitate an alliance with Athens (cf. 15.1); for a different view, see Munn, 1993: 145-146n. 23. On the possibility that neither the Thebans nor Kleombrotos were the instigators of Sphodrias' raid, see Marshall, 1905: 12, Cargill, 1981: 59n. 24. The example of Phoibidas and his success were, Marshall proposes, quite sufficient to induce a second Spartan commander to emulate his feat. For a survey of scholarly opinions regarding the testimony of the three sources, see MacDonald, 1972: 38-44 and Buck, 1994: 91-93, 153nn. 35 and 36. Regarding the date of Sphodrias' raid Cargill suggests that there is no reason to reject Diodoros' implication (15.28-29) that Sphodrias' assault on the Peiraeus came after the establishment of the Second Athenian League (1981: 58-59). Furthermore, Cawkwell speculates (1973: 47-60, esp. 51 and 55) that it was in fact the establishment of the League that prompted the raid; for a contrary view, see Marshall, 1905: 14, Ryder, 1965: 55, Rice, 1975: 95-130, esp. 109-110n. 34.

14.3 Σφοδρίας... μεστός: Kallisthenes, on the evidence of Harpokration, describes Sphodrias in very similar terms: εὐήθη τε εἶναι λίαν καὶ κοῦφον πρὸς τὰς ἐλπίδας (*FGrH* 124F9) in the second book of his *Hellenika*. This verbal similarity has prompted several scholars to assume that P. (cf. also *Ages.* 24.4) depended on Kallisthenes for the entire episode (cf. Queck, 1876: 28, Stern, 1887: 60-61, Bock, 1910: 61, Fuscagni, 1975: 36, Prandi, 1985: 37-39, 72, Buckler, 1980b: 75-76, Pédech, 1984: 34-35) and not on Ephoros (note Diodoros' characterization of Sphodrias: φύσει δ' ὄντος μετεώρου καὶ προπετοῦς, 15.29.5). With the portrayal of Sphodrias' character P. reintroduces the theme of bravery (see Proem, 1-2.8) which underlies both the Theban and the Roman biographies. In the Spartan's case it is accompanied by weak judgment, vain hopes and overriding ambition. Forrest (1968: 127) observes certain similarities between Sphodrias' and Phoibidas' acts and notes that Sphodrias was a 'would-be Phoibidas', and that the same pattern of 'brutality and aggressiveness' lay behind the actions of both men; see also Dillery, 1995: 232-235.

Σφοδρίας: a Spartan general; he was harmost at Thespiai in 378 (Xen. *Hell.* 5.4.20), and fell at the battle of Leuktra in 371 (6.4.14); the account of his trial and subsequent acquittal are reported at length by Xenophon (5.4.24-33) and P. in *Ages.* 25. Diodoros refers to him as Sphodriades (15.29.5), but this is not an exceptional case of his misspelling of names (cf. Parke, 1927: 160n. 3); see further, Fiehn, *RE* s.v. "Sphodrias, 1", 3.2, 1929: cols. 1749-1750, Poralla, 1985, No. 680.

ὑπόκουφος δὲ τὴν γνώμην: "light" or "weak with respect to his judgement", hence "with no sound judgement"; ὑπόκουφος is a rare word used by P. three times: *Artax.* 5.4, *Reg. et imp. apoph.* 205a; see Wyttenbach, s.v. ὑπόκουφος.

ἀπελείφθη... βοηθεῖν: Xenophon reports that Kleombrotos left one-third of each of the allied Peloponnesian contingents at Thespiai under the command of Sphodrias with enough money to hire mercenary forces (*Hell.* 5.4.15).

14.4 ὑποπέμουσιν: cf. πέμπουσιν... κρύφα, at 7.3; note also the use of historical present in both cases.

[ἰδίᾳ] <Διέμπορον>: the deletion of ἰδίᾳ is suggested by Sintenis. Keil emends ἔμπορον (codd.) to <Διέ>μπορον, who could be a descendant of the 431 BC boeotarch named in Thukydides, 2.2.1 (cf. Stern, 1887: 39, Ziegler, 1934: 238-9, Buckler, 1980a: 42). Kallett-Marx (1985: 151) argues against Διέμπορος and proposes Ἔμπορος; he connects it erroneously with Ἐμπόριος (*SEG* III.333, line 51), which he reads as Ἔμπορος. Sintenis' deletion of ἰδίᾳ (which could be a scribal gloss to make obvious the pertinence of ὑποπέμπουσιν), and the codices' reading of ἔμπορον are probably closer to P.'s frame of mind. He avoids consistently throughout the *Life* to provide the reader with names of minor figures (cf. *Pel.* 7.3, 8.6, 10.7) or sites (see note on 12.1); he apparently considers it unnecessary for the flow of events he is describing. It is, therefore, only remotely conceivable that he meant to include here the name of the messenger sent to Sphodrias by Pelopidas and Gorgidas. Furthermore, when he relates the same incident in *Ages.* 24.6, he is again far from specifying the names of the messengers (ὑπέπεμψαν γὰρ ἀνθρώπους λακωνίζειν προσποιουμένους). The idea that P. refers suggestively to the profession of the messenger rather than to his identity is more attractive, and finds support in the activity of merchants as carriers of information (cf. Xen. *Anab.* 5.6.21, *Hell.* 3.4.1, Lyk. *contra Leocr.* 14-15, Lys. 22.14), on which, see Starr, 1974: 19-28, esp. 22-24, F. Russell, *Information Gathering and Intelligence in the Greek World, ca 800-322 B.C.*, Ph.D. diss. University of California at Los Angeles 1994, 161-162.

μὴ φυλαττομένοις τοῖς Ἀθηναίοις: "when the Athenians were off their guard"; Xenophon, on the other hand, reports that Peiraeus was without gates (ἀπύλωτος, *Hell.* 5.4.20; cf. also 3.5.16) at the time of Sphodrias' raid, suggesting therefore that it was exposed to assault by land. On the fortification of Peiraeus, see Foucart, 1887: 129-144, R. Garland, *The Piraeus: from the Fifth to the First Century BC* (London 1987) 14-22, 37, 40-45, esp. 41-42.

14.6 νυκτὸς... Θεσπιάς: for a more detailed description of Sphodrias' failed raid, see Xenophon, *Hell.* 5.4.20-21. On the dating of the raid, see Kallet-Marx, 1985: 134-137. Another unsuccessful surprise attack on Peiraeus was attempted by the Syrian Erginos, a lieutenant of Aratos, in 242 BC (*Arat.* 33.2-4).

τῶν στρατιωτῶν... γενόμενος: neither Xenophon nor Diodoros comments on the cowardice of Sphodrias' soldiers, while all three report that he was detected by the Athenians (*Hell.* 5.4.21, Diod. 15.29.6).

συνταράξας <πόλεμον> οὐ φαῦλον οὐδὲ ῥᾴδιον αὐτοῖς ⟨τοῖς⟩ Σπαρτιάταις πολεμεῖν: Xylander's emendation of πολεμεῖν to πόλεμον following ⟨τοῖς⟩ Σπαρτιάταις is preferable; the emphasis is more on the negative adjectives (οὐ φαῦλον οὐδὲ ῥᾴδιον) than on the war itself (cf. συνταράξαι πόλεμον, *Arist.* 20.2).

Ch. 15: The formation of the Second Athenian Confederacy (378); Agesilaos' first campaign to Boiotia and the beginning of the Boiotian War (378-377)

15.1 Ἐκ τούτου... ἔχοντας: while Diodoros (15.28) refers explicitly to the foundation of the Second Athenian League and P. alludes to its creation, Xenophon is stunningly silent on this event. There is a chronological contradiction between our two sources: P. places it after the raid of Sphodrias, whereas in Diodoros' account the sequence of events is reversed. The presence of a Spartan embassy in Athens at the time of the raid of Sphodrias (Xen. *Hell.* 5.4.22-23) may, indeed, suggest that the Athenians had already begun to organize the League (*IG* II^2 23, ll.34-35, Tod, 1962, Nos.118-119), a view which is challenged by Buck (1994: 93). There is no consensus among scholars on the accuracy of either sequence; for formative accounts on the formation of the new defensive alliance and its dating, see Accame, 1941: 27-47, Burnett, 1962: 1-17, Ryder, 1965: 54-57, Rice, 1971: 112-116, Cawkwell, 1973: 47-60, Hack, 1975: 92-94, Sealey, 1976: 410-414, Cargill, 1981, Kallet-Marx, 1985: 127-151, Buck, 1994: 89-93.

ἐκ τούτου: refers to the raid of Sphodrias, and in particular to his subsequent trial and unexpected acquittal (cf. *Ages.* 24.9-26.1, Xen. *Hell.* 5.4.24-33, Diod. 15.29.6-7; see Cartledge, 1987: 136-138), which are not related by P. in the *Pelopidas.*

προθυμότατα Ἀθηναῖοι: to avoid hiatus read προθυμότατ' Ἀθηναῖοι.

Ἀθηναῖοι τοῖς Θηβαίοις συνεμάχουν: on the alliance, see *IG* II^2 40, *SEG* XVI.43, *IG* II^2 43, *SEG* XVI.44, Cargill, 1981: 51-60, esp. 60, Burnett, 1962: 1-17, Bengtson, 1975, No. 255. Diodoros reports that the Thebans were admitted to the συνέδριον of the Confederacy (15.28.5). Athens, of course, had every reason to encourage Thebes to join, because she feared that Thebes might become dangerously powerful in Boiotia, and she hoped to inhibit Theban assertiveness by making Thebes a member of the League Synedrion. Thebes, on the other hand, would not have to stand alone against Sparta and the Peloponnesian League, and, by ensuring Athenian support, would be able to start reorganizing the Boiotian League; see Burnett, 1962: 10-12, Hack, 1975: 95-96, Sealey, 1976: 411, Buckler, 1980a: 17-18.

15.2 οἱ δὲ Θηβαῖοι... ἀγώνων: P. is distressingly vague when he speaks about the battles, or rather the skirmishes, fought in Boiotia by the Thebans and the Spartans during the period 378-377; he attributes the successes exclusively to the Thebans (cf. also *Ages.* 26.2); he even specifies that these battles were fought without any allies (καθ' αὑτούς). The possibility that the Athenians were involved in these encounters cannot be ruled out (see Munn, 1993: 168-169). Diodoros reports that as soon as Agesilaos marched to Boiotia with 18,000 soldiers and 1,500 horsemen the Athenians went to the assistance of Thebes with 5,000 soldiers and 200 cavalry (15.32.2) under the command of Demeas and Chabrias (cf. Munn, 1993: 156n. 41). We are also informed that the Athenian general Chabrias won praise from both the Thebans (15.33.4) and the Athenians (Demosth. 20.76; cf. Nep. *Chabr.* 1, Polyain. 2.1.2) for saving the Thebans by the skill of his generalship. Xenophon does not speak of Athenian aid (*Hell.* 5.4.36-46; J.K. Anderson, "The Statue of Chabrias", *AJA* 67 [1963] 411-413, Buckler, 1972: 466-474) until Agesilaos' second expedition in 377 (5.4.54). Despite P.'s and Xenophon's reticence in that matter, there is no reason to assume that the

Athenians would not have honored their alliance with Thebes and sent soldiers to oppose Agesilaos' invasions of Boiotia (cf. also Hack, 1975: 97-98n. 1, Munn, 1987: 117-118). Nevertheless, P. (or his source) is not willing to acknowledge their presence and the military prowess of Chabrias, so that the Thebans emerge as victors and Pelopidas as chiefly responsible for these victories (15.4-5). Following P.'s account (*Ages.* 26.2) Agesilaos' first expedition to Boiotia in the summer of 378 ended with losses suffered on both sides, while Diodoros (15.33.1) refers to it as a "Lacedaemonian victory which did not entail any risk". Xenophon reports that Agesilaos returned to Sparta having done considerable damage to Theban territory (*Hell.* 5.4.41); on Agesilaos' first invasion of Boiotia, see Hack, 1975: 97-99, De Voto, 1987: 75-82, Munn, 106-121; id. 1993: 152-161, Buck, 1994: 93-95.

μαχόμενοι μάχας: cognate accusative nouns are frequently used in P. (cf. *Pel.* 4.4, 20.6, 30.5, 34.7). The μάχας not only qualifies, but also intensifies the present participle.

15.3 διὸ καί φασιν... διδάξας: Agesilaos is criticized by the Spartans for not bringing the Thebans to a decisive battle. The anecdote is often found in P. in connection with Lykourgos' *rhetra* which prohibited frequent campaigns against the same foes (*Lyk.* 13.10, *Ages.* 26.3, *Reg. et imp. apoph.* 189E-F, *Apoph. Lac.* 213F, 227C-D; cf. Polyain. 1.16.2, Stob. 4.13.59). P. uses also the anecdote as a progress report of the Thebans' military training (cf. previous mention of their training in the palaestra, 7.4-5, and later, after their victory at Tegyra, 17.11-13).

'Αγησίλαος... τετρωμένος: the wound may refer to the ailment described by Xenophon (*Hell.* 5.4.58).

'Ανταλκίδαν: a Spartan general; his name is linked with the King's Peace of 387/6; see Judeich, *RE*, s.v. "Antalkidas", 1.2, 1894: cols. 2344-2345, Ryder, 1965: 27-36, Rice, 1971: 10-23, Buckler, 1977a: 139-145, Poralla, 1985, No. 97.

'Αγησίλαος: son of Archidamos II; he succeeded his half-brother Agis II on the Eurypontid throne; see Poralla, 1985, No. 9, C.D. Hamilton, 1983: 119-127, Cartledge, 1987; also Xenophon's encomium of Agesilaos, and P.'s and Nepos' biographies.

διδασκάλια: "teacher's fee", similar to δίδακτρα. The plural seems to be confined to P.; cf. also Pelling, 1988: 141, note on 10.6; see Wyttenbach, s.v. διδασκάλιον.

15.4 ὧν μεγίστην δόξαν εἶχεν ὁ Πελοπίδας: P. resumes his praises of Pelopidas, but without specifying the role he had in these military engagements; nor is there any evidence of Pelopidas' activities at that time other than P.'s. Polyainos (2.1.2, 17.1) reports that Gorgidas commanded the Boiotian troops on this occasion and his maneuver, performed in coordination with Chabrias, took Agesilaos by surprise; see Buckler, 1972: 466-474, 1979: 53-56.

15.5 ἀφ' ἧς γὰρ εἵλοντο... ἔπραττεν: P. records that Pelopidas was constantly in office from the time he was first elected as leader in arms in 378, when the Boeotarchy was re-established (see note on 13.1), until his death in 364; he also specifies that during this period Pelopidas was being elected every year to high office either leading the Sacred Band, as for instance at the Battle of Leuktra in 371 (*Pel.* 20.3, Paus. 9.13.6-7) or, most often (τὰ πλεῖστα), serving as boeotarch until his death; it also appears that one office was not held to the exclusion of the other, although he makes it sound that way. Pelopidas must have been a boeotarch when he commanded both infantry and cavalry at the Battle of Tegyra (16.2; cf. Buckler, 1979: 55-56, 62n. 37). At 34.7 he also informs us that Pelopidas died during his thirteenth Boeotarchy. Diodoros, on the other hand, is apparently exaggerating when he declares in the death notice of Pelopidas that he continued to hold the office of boeotarch every year (βοιωταρχῶν πάντα τὸν χρόνον διετέλεσε, 15.81.4) until his death. Both P.'s mention of Pelopidas' thirteen Boeotarchies and Pausanias' testimony on the list of the seven boeotarchs at the Battle of Leuktra, in which Pelopidas is not included, have been accepted by scholars. Disagreement arises when it comes to assigning the other year during which Pelopidas could not have held the office of boeotarch; much of the dissent is related to the difficulty in assigning a date to a Theban decree granting proxeny to the Carthaginian Νώβας: *IG* VII 2407; Köhler, 1889: 639, Beloch, 1923, 3:2, 242, 252-253, G. Glotz, "Un Carthaginois à Thèbes en 365 avant J. -C.", in *Mélanges offerts à M. Nicolas Jorga* (Paris 1933) 331-339, esp. 332-333, Roesch, 1984: 45-60, esp. 47-51; for a discussion of Pelopidas' Boeotarchies, see

Bersanetti, 1949: 89-93, Cawkwell, 1972: 275-276, Fuscagni, 1972: 415-433, and note on 34.7.

15.6 'Εγένοντο... ἀνεῖλεν: P. recounts briefly some of the military engagements which took place in the years between Agesilaos' first invasion and the battle of Tegyra in 375. He mentions Spartan defeats at Plataia (it is unclear whether Charon's participation in the cavalry battle at Plataia mentioned by P. at 25.8 belongs to this period or to that before the battle of Leuktra) and Thespiai (cf. Diod. 15.33.6, Xen. *Hell.* 5.4.42-45), during which the Thebans killed Phoibidas, the Spartan harmost; on the possibility of two Theban raids on Thespiai, see Buckler, 1979: 53-56. The Thebans also expelled the Spartan garrison from Tanagra and Pelopidas killed the Spartan harmost, Panthoidas; cf. Xen. *Hell.* 5.4.49, Isokr. *Plat.* 9; cf. *IG* VII 1903, 1904.

τρεψάμενος... ἀνεῖλεν: sc. ὁ Πελοπίδας; the subject is disquietingly distant from the participle and the verb.

Πανθοΐδαν: Spartan harmost at Tanagra in 377, where he was killed by Pelopidas; he was probably nauarch in 403/2, on which see Beloch, 1879: 117-130, esp. 123, H. Schaefer, *RE* s.v. "Panthoidas", 18.3, 1949: cols. 776-777, Poralla, 1985, No. 585.

15.7 ἀλλ' οὗτοι μὲν οἱ ἀγῶνες... προῆγον: P. observes that the Thebans developed confidence and boldness during these skirmishes; cf. also Xenophon's comment on the rekindled spirits of the Thebans (*Hell.* 5.4.46).

οὐ... ἐδουλοῦντο τὴν γνώμην: "did not break their spirits"; for a similar metaphorical use, cf. Thuk. 4.34.

Chs. 16-17: The battle of Tegyra (375)

P. is our chief source for the account of the battle of Tegyra (cf. also *Ages.* 27.4). Xenophon, who confines Theban activity in 375 to two brief notices (5.4.62-63, 6.1.1), makes no mention of the battle, yet he seems to have known about the Spartan defeat, when he alludes to it at 6.4.10. This is not in the least surprising considering Xenophon's tendency to resist inclusion of material which was humiliating to the Spartans. Nepos, in his sketchy biography of Pelopidas, passes over Tegyra and proceeds to Pelopidas' part at Leuktra. Finally, Diodoros offers some useful information about the strength of the Lacedaemonian and Theban forces during this encounter although he does not name Tegyra at this point (15.37.1-2), but later, in Pelopidas' death-notice (15.81.2).

For a discussion and bibliography of P.'s sources for the battle of Tegyra, see [**2a.b**], and Georgiadou, in *Boeotia Antiqua*; scholarly opinion on the location of the battlefield was split until recently between Pyrgos and Polygyra, with the prevalent view now supporting the latter location; see Ulrichs, 1840: 196-197, K. Bursian, *Geographie von Griechenland*, vol. 1 (Leipzig 1862) 211, W. Vischer, *Erinnerungen und Eindrücke aus Griechenland*,[2] (Basel 1875) 583, H. Bulle, "Orchomenos", *AbhMünchen* 24 (1909) 116-124, R. Étienne and D. Knoepfler, *Hyettos de Béotie et la Chronologie des Archontes Fédéraux entre 250 et 171 avant J.-C.*, *BCH Suppl. III* (Athens, Paris 1976) 237, Buck, 1979: 8, Pritchett, 1982: 104-109, Lauffer, 1986: 151-153, Knauss, 1987: 71-77, 183-196, J.M. Fossey, *Topography and Population of Ancient Boiotia*, vol. 1 (Chicago 1988) 367-373, Buckler, 1992: 4804-4805, id. 1995: 43-58. For the date of the battle, see Beloch, 1922: 154-155, Judeich, 1927: 180n. 2, Buckler, 1971: 356-357, id. 1995: 43-58, Pritchett, *ib.*, 103. On its reconstruction, see Hack, 1975: 113-117, Pritchett, *ib.*, 115-122, De Voto, 1992: 9-10, and, most recently, Buckler, 1995: 43-58, whose topographical observations on Tegyra and the battle itself I follow in chs. 16-17.

16.1 Ὁ δὲ περὶ Τεγύρας... ἀπολιπών: Diodoros offers a similar evaluation of the significance of the battle of Tegyra (15.37.1-2, 81.2) and the unique role of Pelopidas. It would be tempting to attribute the similarity to both writers' use of the same source for this episode, namely Ephoros, were it not for Ephoros' own heavy indebtedness to Kallisthenes for the last three decades of his history. P.'s mention of Ephoros, Kallisthenes and Polybios as his sources for the size of the Spartan *mora* (17.4) makes each of the three historians a potential source for the entire episode; see discussion in [**2c**], note on 17.4, and Georgiadou, in *Boeotia Antiqua*.

τρόπον τινὰ τοῦ Λευκτρικοῦ προαγὼν γενόμενος: P. characterizes the engagement at Tegyra as something of a prelude to Leuktra, because he considered both as the most illustrious of contests (cf. *synkr.* 31(1).6).

προαγὼν: "preliminary contest". P. favors the transferred use of theatrical terms; cf. *Pel.* 34.1; on his theatrical metaphors, see O'Donnell, 1975, esp. 91-96 and note on 34.1.

οὔτε πρὸς κατόρθωμα... ἀμφισβήτησιν: P. probably alludes here to other instances where Pelopidas' leading role is less than fully certain, such as the battle of Leuktra and his first invasion of the Peloponnese (*synkr.* 32(2).2), apparently controversial matters among ancient sources; see also [**2e**].

τοῖς συστρατήγοις: the other commanders, left unnamed by P., who were boeotarchs like Pelopidas (see also note on 16.2).

οὔτε τῆς ἥττης... ἀπολιπών: the Spartans could not offer any excuse for their defeat since they outnumbered the Thebans (cf. 17.3, 11).

16.2 τῇ γὰρ 'Ορχομενίων πόλει... ἐστράτευσεν: Xenophon tells us that the Thebans, taking advantage of the Spartans' expedition against Timotheos, which they themselves had engineered, took the opportunity to recover neighboring cities of Boiotia and bring them under Theban control (*Hell.* 5.4.62-63). Since Orchomenos was an important Spartan stronghold in Boiotia the Thebans had tried for a long time to subjugate it and make it join the Boiotian Confederacy (cf. Xen. 3.5.6, Diod. 15.57.1, 79.3-6, Plut. *Lys.* 28.2; see further, Buckler, 1971: 356-357, id. 1977b: 77-78, Hack, 1975: 113-114, Salmon, 1978: 86-92). For this reason they awaited the opportunity to catch Orchomenos without adequate defense. The city finally capitulated in 371.

δύο... μόρας: mentioned again at 17.3. The *mora* is reported as the largest division of the Spartan army from at least 403 (Xen. *Hell.* 2.4.31); it was commanded by a polemarch (*Hell.* 4.5.7, 5.4.46) and had a strength of about 600 officers and men (Xen. *Reip. Lac.* 11.4, *Hell.* 6.4.12). After the battle of Leuktra the Spartan army was probably reorganized, since Xenophon refers to its units as *lochoi* (7.1.30, 4.20, 5.10). Thukydides, who regularly calls the largest units in the Spartan army *lochoi* (4.8.9), never uses the term *mora*, which is also unknown to Herodotos; on the *mora*, see Toynbee, 1969: 373-396, Anderson, 1970: 225-240, Lazenby, 1985: 5-10.

ἔχων μεθ' ἑαυτοῦ... πολλούς: when Pelopidas commanded the Sacred Band and a contingent of cavalry at Tegyra he was both lochagos of the Band and a boeotarch (Diod. 15.81.2); see Buckler, 1979: 56, 62n. 37. On the Sacred Band, see note on 18-19; on the Boiotian cavalry, see Salmon, 1978: 182-185.

16.3-4 ἀπῆγεν... ἐποίει: Pelopidas led his army back again along the surrounding foothills of Tegyra. The use of πάλιν clearly indicates that he passed through the territory of Tegyra the first time, when he was marching northwestward against the city of Orchomenos (16.2). A retreat through the plain lying between Orchomenos and Tegyra was made impossible by the flooding of the river Melas, which, P. says, was swallowed up in swamps and pools immediately below its source (cf. *Su.* 20.6, where he notes that Melas is held to be the only one in Greece which abounds in water and is navigable at its source; see observations of the river by Knauss, 1987: 184-186). On the circular route taken by Pelopidas along the rim of Lake Kopais in order to avoid the flooded plain, see Buckler, 1995: 43-58; a different route is suggested by Pritchett, 1982: 109.

16.5 P. disrupts the unity of the narrative with a digression on the temple of Apollo Tegyraeus thus displaying his erudition in religious matters—drawing probably from local records, as he himself suggests at 16.8 (ὁ πάτριος λόγος)—and his apparently first-hand knowledge of the topography of the site of Tegyra (Buckler, 1992: 4804-4805). P., Buckler notes, must have passed by the site on his way to Hyampolis (*Quaest. conv.* 660D; see Teodorsson ad loc., vol. 2, p. 18), a city in eastern Phokis about twenty kilometers from Chaironeia. P. indulges in the use of digressions, especially before launching into the description of battles; cf. 20.4-22, 31.5-6, *Marc.* 14.7-15, 17. The significance of this digression in the *Pelopidas* lies in the fact that the Boiotian biographer seizes the opportunity here to remark that great heroes have come from Boiotia, such as Herakles and Dionysos, who were made gods for their ἀρετή (virtue, bravery); this is connected with the theme of bravery which is central to this Pair of *Lives*, and also anticipates Pelopidas' great individual triumph at 17. On the value of digressions in developing ideas which P. considers important in a *Life*, see Sansone, 1980: 63-74, Georgiadou, 1992b: 147-164.

16.5-6 Μικρὸν δ'... ὀνομάζομεν: P. makes a similar statement in *De def. orac.* 412B-C (cf. also 414A) about the flourishing of the oracle of Tegyra during the Persian wars, with more details on the prophecy of Greek victory; he then adds other information on the sanctuary at the time of the Peloponnesian war. The whole passage indicates that P. was familiar with the history of the oracle, as is to be expected of him, with his interest both in Boiotian matters and oracles, and thus makes it most unlikely that his principal source was Kallisthenes. On the sanctuary of Apollo Tegyraeus, see A. Schachter, 1981: 75, P. Bonnechère, "Les Oracles de Béotie", *Kernos* 3 (1990) 53-65, esp. 56-57, Rescigno, 1995: 279. On the cult of Apollo Tegyraeus, see Schachter, "A Boeotian Cult Type", *BICS* 14 (1967) 1-16, esp. 6; on the remains of the temple, see Lauffer, 1986: 151-153, Buckler, 1995: 43-58. On P.'s interest in oracles, see Levin, 1989: 1599-1649.

ὑπὸ τὰ ἕλη νεώς ἐστιν: unless ὑπό + accus. bears the meaning of proximity, of which I have not found any convincing evidence so far (cf. Kühner-Gerth, vol. 1:2, 525), it should read ὑπέρ, following Wilamowitz's emendation. The temple of Apollo of Tegyra was definitely built on a higher level than the swamps; for a different usage of ὑπό + accus., see note to *Pel.* 30.3.

δύο ῥήγνυνται πηγαὶ... ὀνομάζομεν: cf. *De def. orac.* 412B. These two springs have been located at Polygyra (see Lauffer, 1986: 151-153, id. 1989: 652) and verified by J. Knauss ("Munich Kopais Expedition, Topographical Surveys of October 1986 and May 1987: Progress Report", *Teiresias* 17, 1987, 1-5, esp. 4) and Buckler (1995: 43-58). In personal observation of 24 of July, 1996, Buckler observed only traces of one spring at Polygyra. Earlier he had found evidence of a second spring fifteen minutes east of Polygyra. Recent work in the area seems to have obscured the identity of the latter spring. Buckler suggests that if P. describes the battle from Pelopidas' point of view then the first spring must be Φοίνιξ and the second one Ἐλαία.

<ἐνταῦθα μυθολογοῦσι γενέσθαι τὸν θεόν>: Ziegler transposes the sentence and places it after ὀνομάζομεν, with a change of the word order to avoid hiatus (ἐνταῦθα μυθολογοῦσι τὸν θεὸν γενέσθαι). Since Kallisthenes also associates Tegyra with the birthplace of Apollo (Steph. Byz. s.v. Τεγύρα; *FGrH* 124F11) it has been conjectured (Queck, 1876: 28, Stern, 1887: 60, Fuscagni, 1975: 36, Prandi, 1985: 40-42, 72, Sordi, 1989: 123-125, Buckler, 1980b: 75-76, Pédech,

1984: 36) that the biographer's digression on the oracle of Apollo Tegyraeus derives from him; for a more reserved view, see [**2c**], note above and on 17.4, also Georgiadou, in *Boeotia Antiqua*. For the various claims regarding Apollo's birthplace, see H. Gallet de Santerre, *Délos Primitive et Archaïque* (Paris 1958) 239n. 91.

οὐ φυτῶν... λοχευθείσης: as in the Delian story of the birth of Apollo and Artemis. The standard Hellenic story had Leto bear Apollo on the island of Delos, between two trees, a palmtree (Φοῖνιξ) and an olive ('Ελαία); he later killed the snake Python at Delphi, prior to setting up his oracle there. So the sanctuary at Tegyra appropriated both stories for itself, changing the trees into fountains.

16.7 τὸ Πτῷον ἐγγύς: on Ptoion, a mountain south-east of Lake Kopais, there was an oracular sanctuary of the Ptoan Apollo (cf. Hdt. 8.133-135, Paus. 9.23.6). For its history, see Ducat, 1971: 439-450; for the identification of the site (Perdikovrysi), see Lauffer, *RE* s.v. "Ptoion" 23.2, 1959: cols. 1528-1538, Borelli, 1965: 535-538; Ducat, *ib.*, esp. 3-5, P. Roesch, "Ptoion", in *The Princeton Encyclopedia of Classical Sites* (Princeton 1976) 741-742, F. Croissant, "Les Kouroi du Ptoion", *RA* (1977) 87-94; for a detailed survey of excavations, topographical observations, inscriptions, discussion of the cult of the Ptoan Apollo and literary references, see Schachter, 1981: 52-73, Roesch, 1982: 225-243. The use of ἐγγύς here by P. is discussed by Lauffer, *ib.* cols. 1510-1513.

ὅθεν αὐτὴν ἀναπτοηθῆναι... λέγουσι: the dedication of a statue of Leto and the boar is recorded on a statue base signed by Teisikrates of Sikyon (4th-3rd century BC); see L. Bizard, "Fouilles du Ptoïon (1903): II. Inscriptions", *BCH* 44 (1920) 227-262, esp. 242-245. If Bizard's restoration is correct, it would give earlier currency to the tradition referred to here by P. about Leto, which otherwise appears only in later literary sources (see Schachter, 1981: 60). Leto was surprised (ἀναπτοηθῆναι) by the sudden appearance of a wild boar when she was on the point of giving birth; cf. also *schol. Lyk. Alex.* 265.

τὰ περὶ Πύθωνα καὶ Τιτυὸν: Python is the dragon whom Apollo killed with his arrows, when he decided to found a sanctuary at the foot of Parnassos, not far from Delphi. He could pronounce oracles, which is why Apollo had to eliminate this rival before setting up his oracle at

Delphi; cf. Hom. *Hym. Apol.* 370-374, Call. *Hym.* 2.100-104, Eur. *Iph. Taur.* 1245-1254, Paus. 10.6.5, Ael. *V.H.* 3.1, Hyg. *Fab.* 140, Ovid, *Met.* 1.438-447; also Fontenrose, 1959. P. denies the battle between Apollo and Python (*De def. orac.* 417F) and attributes the tale to the poets and prose-writers. Tityos was a giant, the son of Zeus and Elara. When Leto gave birth to Artemis and Apollo, Hera unleashed Tityos against her, but Tityos was struck by one of Zeus' thunderbolts; see Hom. *Od.* 11.576-581, Apollod. *Bibl.* 1.4.1, Apoll. Rhod. *Arg.* 1.761, Paus. 10.4.5, 10.29.3; Str. 9.3.14, Lucr. 3.984-994, Verg. *Aen.* 6.595-600, *App. Verg.*, *Aetna* 80, Hyg. *Fab.* 55, Ovid, *Met.* 4.457-458; Rescigno, 1995: 286-287.

τὰ γὰρ πλεῖστα τῶν τεκμηρίων παραλείπω: the proofs which P. omits in the *Life* are recounted in the *De def. orac.* 412C, on which, see Rescigno, 1995: 279-280.

16.8 οὐ γὰρ ἐν τοῖς ἐκ μεταβολῆς... τηλικούτων: P., who had become a permanent priest at Delphi (Flacelière, 1943: 72-111) probably under Domitian, devoted much of his work glorifying the shrine in his Pythian dialogues and many of his *Lives* (see Brenk, 1977: 236-255, id. 1987: 248-349, esp. 330-336, Buckler, 1992: 4808-4811). His partiality towards the Delphic oracle and his manipulation of sources in favor of the importance of the shrine is frequently attested (Fontenrose, 1978: 268-454; for further bibliography, see Brenk, 1987: 333-335). It is to be expected then that he would advocate Apollo's superiority over Herakles and Dionysos. P. states that it is incompatible with his native tradition that Apollo should have ever been a demon who later turned into a god; he is not averse, however, to accepting this interpretation of Dionysos and Herakles. In the *De E ap. Delph.* 388E-389B Ammonios gives a long philosophical eulogy of the god Apollo as eternal being, symbolized by the E (Brenk, 1977: 104). For the death and apotheosis of Herakles, cf. Soph. *Trach.* 1191 (also Easterling, 1982: 17-18), Eur. *Herakl.* 910-916, Diod. 4.38.1, Luc. *Herm.* 7, Apollod. *Bibl.* 2.7.7, Ovid, *Met.* 9.136-137, *Her.* 9, Hyg. *Fab.* 36, Sen. *Herc. Oet.* 485-490, 1483-1485; also Boardman, 1984, col. 2.1, 294.

Ἡρακλέα καὶ Διόνυσον: as the national divinity of Thebes Herakles was frequently represented on Boiotian coins in various forms, among which the most distinctive was his buckler; see Lacroix, 1958: 5-

30, P. Guillon, *Le Bouclier d'Héraclès*, Publication des Annales de la Faculté des Lettres (Aix en Provence 1963), Head, 1974: 10 and plates I-VI; on the cult of the Theban Herakles, see Cloché, 1952: 28-29, Demand, 1982: 49-52, Schachter, 1986: 14-30. Dionysos, like Herakles, was traditionally held to have been born in Thebes. On his cult in Thebes, see Ziehen, *RE* s.v. "Thebai, Kulte", 5.A2, 1934: cols. 1509-1511, Demand, *ib*., 54-55, Schachter, 1981: 185-192, Symeonoglou, 1985: 56-58, 72-73, Valgiglio, 1988: 233-247. Dionysos and Herakles are often associated; cf. Plut. *Lys*. 27.5-6, Luc. *VH* 1.7.

17.1 δ' οὖν: resumptive; it leads back to the main topic of the battle of Tegyra after the digression on the temple and oracle of Apollo Tegyraeus (16.5-8); see Denniston, 463-464.

ἐξ ἐναντίας αὐτοῖς... ἀναζευγνύντες: at the same time that the Lacedaemonians were returning to Orchomenos from Lokris (16.2) the Thebans, led by Pelopidas, were marching back from Orchomenos to Thebes through Tegyra (16.3). P. draws attention to the element of surprise when the two armies confronted each other thus allowing Pelopidas to emerge as a competent military leader combining both preparedness and skill when dealing with emergency situations; for an explanation of the element of surprise on topographical grounds, see Buckler, 1995: 43-58.

ἐξ ἐναντίας: "face to face"; ἐναντίος is commonly used by P. to denote the position of an army facing their enemy at the time of the encounter; see Georgiadou, 1990: 81-82.

ἀναζευγνύντες: "marching back", "returning home".

17.2 καί τις εἶπε... ἐκεῖνοι;: cf. *Reg. et imp. apoph*. 194D. A similarly playful stichomythy is attributed to Leonidas, *Apoph. Lac*. 225B; also *ib*. 234B.

17.3 καὶ τὴν μὲν ἵππον... συνήγαγεν: Pelopidas ordered the rapid deployment of his cavalry in front of the hoplites to lead the charge, while at the same time he drew up his three hundred hoplites of the Sacred Band in close formation (συνήγαγεν) hoping to cut his way through the enemy (διακόψειν τοὺς πολεμίους). His masterly coordination of cavalry and infantry seems to have been the key factor for the Theban victory at Tegyra. The Sacred Band is first seen at this battle acting as an integral unit and it continued to do so until its destruction at the battle of Chaeronea (cf. 19.3-4 and note ad loc.); see De Voto, 1992: 9, Buckler, 1995: 43-58.

προεμβαλοῦσαν: cf. Diodoros' similar wording in his sketchy description of the battle, 15.81.2. Pritchett's understanding of the role of the cavalry as cutting through the Spartan phalanx (1982: 120) is refuted by Buckler, (1995: 43-58). For a study of Theban tactics at this time, see id. 1985: 134-143.

τὴν μὲν ἵππον: similar metonymies are not infrequent in the *Life*; cf. 32.7, and 35.5 (τυραννίς instead of τύραννοι).

17.4 ἦσαν δὲ δύο μόραι... Πολύβιός ἐστι: other than being the only extensive account of the battle of Tegyra preserved by the ancient sources, the importance of this passage lies in the unique information P. gives about the historians he may have consulted in the composition of the *Pelopidas,* although it should be noted that his reference to them could be of a purely informative character regarding the varying size of the Spartan *mora* over the years rather than an indication that he used them as a source. Considering, however, P.'s emphasis on Pelopidas' great individual triumph, the introduction of the variant figures for the size of the *morai* may be attributed to his desire to augment the glory of the achievement: the bigger the *morai*, the greater the achievement. He names three authorities for the strength of the Spartan *mora*, Ephoros, Kallisthenes and Polybios, without committing himself openly to any. P.'s reference to Kallisthenes has been adduced as an additional proof that the latter has been his principal source not only for the episode at Tegyra, but for the entire *Life*; see also notes on 14.3, 16.5, 20-23 for other indications of his reliance on this historian.

Following the three historians, the total number of the Lacedaemonians' two *morai* would be 1,000, 1,400 and 1,800 men respectively. P.'s reference to the authority of Ephoros on the *mora* finds support in Diodoros' account of the same battle (15.37.1), who reports that the ratio of the opposed forces was 1:2. In P.'s account the Theban army consisted of the 300 hoplites of the Sacred Band together with τῶν ἱππέων οὐ πολλούς (16.2); despite the lack of precision his note suggests that the total strength of the Thebans would hardly exceed 500, which is the number given by Diodoros (15.37.1). Furthermore, P.'s evaluative comments on the Thebans' achievement (cf. 17.3, 11) do not, in any way, suggest, as has been long assumed (cf. Stern, 1887: 61, Shrimpton, 1970: 41-42, Fuscagni, 1975: 36, Prandi, 1985: 40-42, Sordi, 1989: 123-125), the overwhelming superiority of the Lacedaemonians to the Thebans, which the ratios of Kallisthenes and Polybios indicate (1:3 and 1:4, respectively). Noticeable correspondences in language between P.'s and Diodoros' accounts should not be overlooked (οὐδέποτε, 17.11; Diod. οὐδέποτε, 15.37.1; cf. also similar laudatory remarks on the uniqueness of the Theban victory). Furthermore, given Ephoros' interest in the struggle for hegemony in Greece and in themes, such as the political and moral decline of Sparta (Sacks, 1990: 42-44), as demonstrated by her fourth century military defeats (cf. also P.'s similar interest, 13.4-7, 17.11-13), one is tempted to suggest that the primary source for the narrative of this battle is Ephoros. It is pointless to speculate further on P.'s sources here and it is safe to assume that he was aware of Kallisthenes' account, as well, if not of Polybios' also, depending on whether or not the latter dealt with this episode by way of digression. For a more detailed discussion of P.'s sources for the Tegyra episode, see [**2c.b**], and Georgiadou, in *Boeotia Antiqua*. On the size of the *mora*, see Michell, 1964: 245-247, Toynbee, 1969: 373-396, Anderson, 1970: 225-240, Lazenby, 1985: 5-10.

17.5 θαρροῦντες οἱ πολέμαρχοι τῶν Σπαρτιατῶν: they were confident of victory because of their numerical superiority (cf. 17.3).

Γοργολέων καὶ Θεόπομπος: Spartan polemarchs are only mentioned by P.; see Poralla, 1985, Nos. 191 and 364, respectively.

ὥρμησαν: for its use in military contexts, cf. *Reg. et imp. apoph.* 202E, *Apoph. Lac.* 222F, *Publ.* 19.5, *Per.* 10.1, *Pomp.* 35.4; also Wyttenbach, s.v. ὁρμάω.

17.6-7 γενομένης δέ πως... βουλομένοις: the attack was aimed at the positions where the commanders of both armies were posted. When the Spartan polemarchs fell the army was seized with fear and, hoping that the Thebans simply wanted to force their way through the phalanx to the opposite side and escape (ὡς διεκπεσεῖν... καὶ διεκδῦναι), created a lane (διέσχε) through which the Thebans would retreat. "The reason for this maneuver", Buckler notes (1995: 43-58), "was to allow the hoplites to strike those fleeing on their unprotected side and to close up again to assail them in their rearmost ranks". For the use of διέχω and διεκπίπτω in military contexts, cf. *Ages.* 18.8, *Rom.* 19.3, *Aem.* 20.9, *Ant.* 66.5; for a similar use of διεκδύομαι, cf. *Tim.* 10.5.

συρράξαντες: the only instance in P. of the use of the verb συρρήγνυμι in the active.

17.8 ἐπεὶ δὲ... δίωξις: Pelopidas disregarded the opportunity for a breakthrough that was offered to him and persisted in the combat; however, he did not carry on the pursuit for a long time because they were near Orchomenos, from which an army of Orchomenians and the Spartan replacement garrison might come at any time. Despite the Theban victory, the Spartans continued to hold Orchomenos.

τὴν δεδομένην ὁ Πελοπίδας ⟨χαίρειν ἐάσας⟩: a possible reconstruction of the vexed passage may be: τὴν δεδομένην ὁ Πελοπίδας ⟨δίοδον παριδών,⟩; the participle δεδομένην becomes substantially clearer with a word like δίοδος, for instance or perhaps the word πάροδος, (*Nik.* 26.1). The phrase διδόναι δίοδον is found in a military context in P. (cf. *Pomp.* 34.2, *Apoph. Lac.* 215F) and is frequently used by Thukydides, Xenophon, Polybios, Diodoros and Polyainos. The participle παριδών is suggested by Flacelière and Chambry, 1966: 154. The anonymous suggestion ⟨χαίρειν ἐάσας⟩ was probably made to furnish a homoioteleuton with -ιδας.

17.10 στήσαντες τρόπαιον: cf. also Diod. 15.81.2. For a survey of studies on the history of the trophy, see Pritchett, 1974: 246-275, Lonis, 1979: 129-146, 268-269.

17.11-12 Although the Theban victory at Tegyra demonstrated the vulnerability of Spartan military supremacy, it did not change the situation in Boiotia immediately; the Spartans continued to have Orchomenos and other Boiotian cities under their control preventing the Thebans from consolidating their power in Boiotia. Some striking verbal similarities between P.'s and Diodoros' laudatory comments on the uniqueness of the Theban victory (cf. 15.37.1-2, 81.2) may suggest that they both draw here directly or indirectly upon the same source; see **[2c, b]**, and Georgiadou, in *Boeotia Antiqua*. There are echoes of the laudatory piece on the liberation of the Kadmeia (13.4-7) in P.'s praise of the Theban success at Tegyra.

οὐδ' ἴσοι... συμβαλόντες: P. probably alludes to the previous unconsequential encounters of the Spartans and the Thebans (15.6-8) and ascribes the exceptional importance of this battle to the fact that it was the first defeat of the Spartans by a considerably smaller army in a pitched battle (cf. *Ages*. 27.3). The same pattern of negative clauses, reiterated at 17.13, is also used in the account of the liberation of Kadmeia (13.7). On P.'s cumulative use of negatives, see Ambrosini, 1991: 19-20.

17.13 ὁ Εὐρώτας: common metonymy for Sparta; cf. also 30.3, *synkr*. 32(2).2, *IG* VII 2462.

ὁ μεταξὺ Βαβύκας καὶ Κνακιῶνος τόπος: Babyka and Kankion were the geographical delimitations of Sparta according to the *rhetra* (*Lyk*. 6.2-3). P. quoting Aristotle (fr. 536, Rose) says that Babyka (cited also as Βαβύας and Βαβύη by Hesychios, s.v. Βαβύας) was a bridge and Knakion a river and adds that in his time they were given different names, ⟨Χείμαρρος⟩ (an emendation by L. Ulrichs, "Über die Lycurgischen Rhetren", *RhM* n.s. 6 [1848] 194-242, esp. 215-216, based on Hesychios) and Οἰνοῦς. Tzetzes calls the river Κνηκείων (*schol. Lykophr*. 550). Following the *rhetra* of Lykourgos the Spartans would hold their assemblies between these two physical boundaries; see Oberhummer, *RE* s.v. "Babyka", 2.2, 1896: col. 2667, and Geiger, *RE* s.v. "Knakion", 11.1, 1956: cols. 907-908.

ἀλλὰ παρ' οἷς... φεύγοντες: P. attributes something very similar to Plato in *Quom. adolesc. poet. aud*. 29E; cf. also *Ages*. 2.1, Isokr. *Dem*. 43.

Chs. 18-19: The Theban Sacred Band

The earlier discussion of the harmonious cooperation of Pelopidas and Epameinondas (4.2-8) is expanded here in a lengthy digression which revolves around the close relation of lover and beloved, the idea of harmony in the community and the importance of the Sacred Band functioning as a unit. This digression provides also a suitable transition between Tegyra and Leuktra, especially given Pelopidas' support of Epameinondas (23.5).

18.1 Τὸν δ' ἱερὸν λόχον... τριακοσίων: Gorgidas (on whom see note to 12.2) is credited with forming the Sacred Band (cf. also Polyain. 2.5.1). Athenaios, in a discussion of homosexual love, attributes the formation of this unit to Epameinondas on the authority of Hieronymos (602a [F34, Wehrli]; cf. also Dio Chrys. *Orat.* 22.2-3, who is probably drawing on Hieronymos here). The attribution of its formation to Epameinondas, Shrimpton suggests (1970: 23), may very well originate from a tradition which ascribed to him every successful Theban venture. The creation of the Sacred Band falls after the liberation of the Kadmeia, in the Spring of 378. Similar elite units are reported to have been operating in several city-states, such as Syracuse (Diod. 11.7.62), Sparta (Xen. *Reip. Lac.* 4.1-6), Argos (Thuk. 5.67.2), Arkadia (Xen. *Hell.* 7.4.22); see further, Salmon, 1953: 349-350, Pritchett, 1974: 221-224, De Voto, 1992: 5-6.

ὁ ἐκ πόλεως λόχος: "the battalion of the city"; by πόλις the acropolis Kadmeia is meant.

τὰς γὰρ ἀκροπόλεις... ὠνόμαζον: P. notes the semantic evolution of the word πόλις initially designating the fortress of a city. By the fourth century the two words were used interchangeably. In the *Life*, whenever P. refers to the Theban acropolis he simply names it (6.1, 13.1) or calls it ἄκρα (5.3) or ἀκρόπολις (13.1); in the *De genio* he clearly distinguishes the 'city' from the 'Kadmeia' (587B); on this distinction, see Schachter, 1981: 166n. 2.

ἐπιεικῶς: "generally", "usually"; cf. *Arist.* 26.4, *Cat.Ma.* 2.6, *Mar.* 21.8.

18.2 ἔνιοι δέ φασιν... τὸ σύστημα τοῦτο: P. does not name his sources. It is unclear whether he does so because he wants to distance himself from the account, or simply because he cannot vouch for its truth. Dio Chrysostom, a contemporary of P., makes a similar remark (*Orat.* 22.2), but indicates that the band's composition of lovers and beloved belongs to Epameinondas; see also note on 18.1.

καὶ Παμμένους... τάττειν: the same anecdote is quoted in *Quaest. conv.* 618D and in *Amat.* 761B; in the *Amatorius* P. refers to the same Homeric line as here without quoting it, and then goes on to add a citation of another line (13.131). This suggests that P. adapts his anecdotes to the context.

Pammenes, Epameinondas' protegé, was an illustrious Theban general and statesman, probably the best to emerge between Epameinondas' death (362) and the battle of Chaeronea (338). Around the time of the latter's second invasion of the Peloponnese (369) he was chosen to be host for the young Macedonian prince Philip (26.6); after Epameinondas' death he took command of the Sacred Band; on his military and political career, cf. Dem. *contra Aristocr.* 183, Diod. 15.94.2-3, 16.34.1-2, 16.34.28, Paus. 8.27.2, 8.27.28, Polyain. 5.16, 7.33.2; see further, Lenschau, *RE* s.v. "Pammenes", 18.3, 1949: cols. 298-299, Κουμανούδης, 1979, No. 1571, Buckler, 1980a: 108, 118, 134, id. "Pammenes, die Perser und der heilige Krieg", in H. Beister - J. Buckler, eds., *BOIOTIKA: Vorträge vom 5. Internationalen Böotien-Kolloquium zu Ehren von Professor Dr. Siegfried Lauffer* (Munich 1989) 155-162.

τακτικόν: "tactician".

18.3-4 τὸ δ' ἐξ ἐρωτικῆς φιλίας... παρόντων: the exploitation of homosexual desire for military purposes is mentioned by Plato (*Symp.* 178d-179a; cf. Xen. *Kyn.* 12.20). On the idea of bravery as a result of αἰδώς, see Arist. *EE* 1230a30. For examples of supreme devotion between ἐρασταί and ἐρώμενοι at the price of life, see Xen. *Hell.* 4.8.39, *Symp.* 8.35, Plut. *Amat.* 761D, Paus. 9.26.7-8, Athen. 602c-d; see further, Marrou, 1965: 61-64, K.J. Dover, *Greek Homosexuality* (Cambridge, Mass. 1989) 191-192.

ὡς ἐκεῖνος... αἰσχυνθῇ: a more elaborate version of this anecdote is found in Aelian (*N.A.* 4.1), who in his discussion of Cretan lovers records that the ἐραστής requested from the enemy to be struck in the front, in the breast (κατὰ τῶν στέρνων ἀντίαν παῖσον), in order that the ἐρώμενος might not judge him guilty of cowardice and refrain from laying out his dead body. Both P. and Aelian may derive the anecdote from the same source.

18.5-6 'Αριστοτέλης... προσεῖπε: P. citing Aristotle as his source (fr. 97, Rose) relates that the ἐρασταί and the ἐρώμενοι offered vows to Eros at the tomb of Iolaos, Herakles' beloved (also *Amat.* 761D-E). These vows, unknown to us, seem to have played a great role in the cohesion of the Sacred Band; for lists of military vows, see Pritchett, 1979: 230-239. The fragment is assigned to Aristotle's 'Ερωτικός by Ross (1955), Rose (1967: 96) and Laurenti (1987: 548).

εἰκὸς οὖν... προσεῖπε: P. sees the love of the ἐρασταί and the ἐρώμενοι of the Sacred Band through Platonist eyes and assumes that it was so called because of Plato' description of the lover as "divine friend" (cf. *Phaedr.* 255b, *Symp.* 179a). He suggests thereby that their love did not seek physical expression, but aimed at the good of the beloved object; cf. his description of Pelopidas' and Epameinondas' divine desire of seeing their country glorious (cf. 4.4 and note ad loc.); also Wardman, 1974: 60. On P.'s cautious attitude toward male homosexual love, see Stadter, 1995: 221-236, esp. 231-233.

18.7 λέγεται δὲ διαμεῖναι... ἀήττητον: Alexander was the first to break the ranks of the Sacred Band about fifty years after its creation; for a reconstruction of Alexander's frontal assault on the Thebans, see Buckler, 1990: 75-80.

τῆς ἐν Χαιρωνείᾳ μάχης: P. often refers to the defeat of the Thebans by the Macedonians in his native Chaeronea in 338 and the annihilation of the Sacred Band by Alexander's cavalry (*Alex.* 9.2-4, *Dem.* 19-20, *Cam.* 19.8, *Mul. virt.* 259D-E). On the battle, see also Diod. 16.86, Paus. 9.10.1, Polyain. 4.2.2, 8.40, Just. 9.3, Kromayer, 1903, vol. 1, 127-195, Hammond, 1938: 186-218, esp. 201-216, Pritchett, 1958: 307-311, Cawkwell, 1978: 144-149, Hammond and Griffith 1979, vol. 2: 596-603, Buckler, 1990: 75-80, De Voto, 1992: 16-17.

ὡς δὲ μετὰ τὴν μάχην... ἀναμεμειγμένους: although P. does not fail to mention Philip's admiration for the valor of the Sacred Band, his omission of the famous Lion of Chaeronea, erected by the Thebans as a memorial to the three-hundred and marking their common grave, which he certainly knew as a native of Chaeronea, is puzzling. P. knew of the *polyandreion* which Philip had erected for the Macedonians who died at the battle, *Alex.* 9.2. The Lion identified by Pausanias (9.40.10) and Strabo is still a remarkable monument overlooking the battlefield; see Hammond, 1938: 186-218, esp. 216-218, Cawkwell, 1978: 148-149, De Voto, 1992: 18, Buckler, 1992: 4801-4802n. 43. Philip is reported by Justin (9.4.6) to have permitted a burial of the Sacred Band separate from the rest.

ἐναντίους: the adjective denotes the position of the Sacred Band facing the enemy at the time of the encounter; see Buckler, 1990: 75, and Georgiadou, 1990: 81-82.

ταῖς σαρίσαις: P. records that the three hundred lay mixed up together among their weapons opposite the sarissai they had encountered in the struggle; he nowhere specifies, however, whether the sarissa-bearing troops were footsoldiers or cavalrymen. Both Rahe (1981: 84-87) and Buckler (1990: 75-80) argue convincingly that the Sacred Band was annihilated by Alexander's infantry; for a different view, see M.M. Markle, "The Macedonian sarissa, spear and related armor", *AJA* 81 (1977) 323-339, id. "Use of sarissa by Philip and Alexander of Macedon", *AJA* 82 (1978) 483-497, De Voto, 1992: 16-17; see further on the Macedonian sarissa, Andronikos, 1970: 91-107, Lonis, 1985: 356-357.

ἐν τοῖς [στενοῖς] ὅπλοις: Schaefer and Sintenis, followed by Ziegler, delete στενοῖς; Claviger and Coraes see στέρνοις behind it, and emend accordingly. The expression ἐν τοῖς στέρνοις is commonly used with reference to wounds inflicted on the breast. Although it does not necessarily entail a frontal assault, in this case it does, especially in view of the code of honor existing among lovers and beloved, mentioned earlier (18.4); for wounds inflicted on the back are regarded as disgraceful by the ἐρώμενος; cf. also Ael. *N.A.* 4.1, on which see note to 18.4. Claviger's and Coraes' emendation strengthens the argument of Rahe and Buckler about the frontal encounter of the Sacred Band with Alexander's infantry; cf. similar use of ἐξ ἐναντίας at 17.1. See also Georgiadou, 1990: 81-82.

εἰπεῖν... ὑπονοοῦντες: Philip denounces those who criticized the Sacred Band for fighting against him and deemed their death shameful. The notion of a shameful death (πάσχειν αἰσχρόν) relates to the programmatic statement offered by P. in ch. 1.

19.1 P. rejects the poetic tradition which ascribes to Laios the practice of homosexuality among Thebans (*Parall.* 313E, *Amat.* 750B; cf. Ael. *V.H.* 13.5, *hyp.* III and VI Thom. Magist. Aesch. *Sept.*) and gives it an official stamp by attributing it to the Theban legislators. His description of a Theban custom in *Amat.* 761B according to which the lover offered the beloved a set of armor upon his official coming-of-age probably reflects some aspects of Theban law regarding homosexual relationships. P.'s claim may also find support in Aristotle (*Pol.* 1274a32-b4), who in his story about the lovers Diokles and Philolaos reports that the latter served as a lawgiver in Thebes. It is likely then that in this capacity Philolaos dealt also with laws pertaining to the practice of homosexuality. On Laius' passion for Chrysippos, Pelops' son, see also *hyp.* Aristoph. gram. Eur. *Phoen.*, *schol.* Eur. *Phoen.* 21, Athen. 602f, Apollod. 3.5.5, Dio Chrys. *Orat.* 10.24, Ael. *H.A.* 6.15, *V.H.* 13.5, Cic. *Tusc.* 4.33.71, Hyg. *fab.* 85.

P. makes the link between homosexuality and military valor (cf. *Amat.* 761B and his earlier reference to Herakles and Iolaos, 18.5, who provided powerful role models for young Thebans; also Xen. *Symp.* 8.32) and regards the Boiotians, the Spartans and the Cretans as susceptible to love but also as being most warlike (*Amat.* 761D; cf. Xen. *Reip. Lac.* 2.12, *Symp.* 8.32); see further, Demand, 1982: 94-95.

ἀνυγραίνειν: "soften", "mollify", used metaphorically by P.; for the similar effect of wine on ἦθος, cf. *Sept. sap.* 156D, *Quaest. conv.* 620D, 712B. P. notes elsewhere, however, that the aulos helps to increase the fighting spirit of the Spartans (*De virt. mor.* 452B); ἀνυγραίνειν is used in a negative sense at *De sera num.* 566A.

πολὺν μὲν ἀνεμείξαντο... ἄγοντες: the Thebans' high esteem for the αὐλός is perceived by P. as a deliberate attempt by Theban lawgivers to soften the impetuous nature of their youth. The general theory about the ethical effect of music in quelling and harmonizing the irrational passions of the soul derives from Plato (cf. *Resp.* 400a-c, 410c-d, 411a-412a, 424c-d, Plut. *De cohib. ira* 167B-C, Ps. Plut. *De mus.* 1140C, E, 1145E) which P. seems to mould to make it fit this particular musical instrument. Plato, however, who is in turn indebted to the Pythagorean tradition for his theory of the educational role of music, is scornful of the αὐλός (*Resp.* 399d), but this does not seem to affect P.'s views here. In the same vein, Aristotle banished the instruction of the αὐλός from the educational curriculum for its morally bad effects (οὐκ ἔστιν ὁ αὐλὸς ἠθικὸν ἀλλὰ μᾶλλον ὀργιαστκόν, *Pol.* 1341a21-23), but conceded that αὐλός-playing had purificatory effects. P.'s linking, however, of αὐλός-playing to homosexuality is either late, or is his own; for the association of αὐλός with peace, see *Marc.* 22.6; on the reputation of the Thebans as αὐλῆται, see Roberts, 1895: 32-37, A. Baines, *Aulos*, vol. 1 (London 1954) 263-264, J.P.H.M. Smits, *Plutarchus en de Griekse Muziek* (Bilthoven 1970) 64-68, 89-101, Demand, 1982: 86-89, P. Roesch, "L'aulos et les aulètes en Béotie", in H. Beister and J. Buckler, eds., *BOIOTIKA: Vorträge vom 5. Internationalen Böotien-Kolloquium zu Ehren von Professor Dr. Siegfried Lauffer* (Munich 1989) 203-214;

generally on αὐλός, see K. Schlesinger, *The Greek Aulos* (London 1939), Baines, 1954, G. Comotti, *Music in Greek and Roman Culture* (Baltimore and London 1989) 71-72.

19.2 To give further credence to his earlier argument about the balanced education of the young Thebans achieved through the harmonious blending of music and physical training P. draws attention to the goddess Harmonia, wife of Kadmos. She was particularly honored by the Thebans, because through her, P. notes, all the citizens come to the most melodious and orderly form of society. "The contrast with Rome", Pelling observes (1989: 200n. 1) at this point, "where everything had to serve the interests of continual warfare (cf. *Marc.* 1.4-5), is very clear".

From Hesiod on, Aphrodite is associated with Ares at Thebes as the mother of Harmonia (cf. Hes. *Theog.* 933-937, 975, Aesch. *Sept.* 135-144, Eur. *Phoen.* 7, Paus. 9.5.2, Nonnos, *Dion.* 2.665; the qualities of Πειθώ and Χάριτες as divinities are associated with Aphrodite also in P.'s *Conj. praec.* 138C; on the divine parentage and character of Harmonia, see Plut. *De Is. et Os.* 370C-371A). Schachter (1981: 40) regards this linking as an allegorical representation of the formation of the community by a fusion of its destructive and generative, or its male and female, elements. He also indicates (*ib.*, 230) that there is no evidence for an actual cult of Harmonia in Thebes (cf. also Paus. 9.12.3). P.'s reduction here of Ares, Aphrodite and Harmonia to the status of psychological faculties for the glorification of Thebes is apparent; see further Hardie, 1992: 4743-4787, esp. 4779.

Two silver coins of Thebes, dated c.446-426 BC, depict on their reverse, a female figure seated on a low rock, and another seated on a chair and holding a helmet. They have been tentatively named Thebe and Harmonia respectively; see Cloché, 1952: 65, Vian, 1963: 118-121, 134, 141-143, 147-151, Schachter, *ib.*, 168n. 1, Demand, 1982: 53.

19.3-5 P. concludes his extensive digression on the Sacred Band by recapitulating the brilliance of the Sacred Band's exploit at Tegyra. He points to Pelopidas' superb innovation in using the corps as an integral unit directing its strength upon a common object, as opposed to the earlier formation, introduced by Gorgidas, according to which the Sacred Band was distributed among the front ranks of the hoplites' entire phalanx. It is difficult to tell to what extent P.'s eulogistic remarks on

Pelopidas' military skill and the Sacred Band are his own or, as Sordi suggests (1989: 125-126), derive from his source.

19.3 τὰ πρῶτα ζυγά: "the front ranks"; on this technical term, see Pritchett, 1985: 87.

19.5 ὥσπερ γὰρ οἱ ἵπποι... προθυμοτάτους: the idea of collaboration and its merits for the interest of the community, which is a recurring motif in both the *Pelopidas* and the *Marcellus*, is here reintroduced in an elaborate simile; cf. *Praec. ger. reip.* 817D; also [**2e**], and note on 4.2-4.

συνεκκαίει τὸν θυμὸν: similar metaphorical use of the verb at *Caes.* 44.8.

Chs. 20-22: Preliminaries to the Battle of Leuktra: The King's Peace of 371; Pelopidas' dream and the legend of the daughters of Skedasos

By autumn 375 the conflict between the Lacedaemonians and the Thebans was suspended and a shortlived treaty, initiated by Artaxerxes, was concluded by the principal belligerents, Athens, Sparta and, most likely, Thebes (Xen. *Hell.* 6.2.1, Diod. 15.38.2-4, Nep. *Tim.* 2.2). The terms of the Peace laying down the principle of autonomy reaffirmed those of the King's Peace of 387/6. An additional term of the treaty demanded the evacuation of foreign garrisons from all cities; for a discussion of the Peace of 375, see Accame, 1941: 143-151, Cawkwell, 1963: 84-95, Ryder, 1965: 58-63, 124-126, Buckler, 1971: 353-361, Rice, 1971: 145-150, Hack, 1975: 126-141, C.D. Hamilton, 1991: 190-195, Buck, 1994: 101-103, Jehne, 1994: 57-64, Bengtson, 1975, No. 265.

20.1 Ἐπεὶ δὲ Λακεδαιμόνιοι... χιλίους: P. omits all key events that led to the Peace of 371. The Thebans, having taken advantage of the withdrawal of Spartan garrisons from Boiotia, started compelling Plataia, Thespiai, Tanagra and other Boiotian cities to rejoin the League (Xen. *Hell.* 6.3.1, Isokr. *Plataikos*, Diod. 15.46.4-6, Paus. 9.1.4-8). The Athenians, being annoyed by the Thebans' aggressive policy with Plataia and Thespiai and their interest in Phokis (Xen. *Hell.* 6.3.1), decided to invite them to join them with the other Greeks to negotiate peace in 371 (*Hell.* 6.3.2). The treaty provided that all harmosts be withdrawn from the cities, that all land and naval forces be disbanded, and that all cities be left autonomous. A fourth provision required that if any city violated the terms of this treaty, anyone who wished might come to the aid of the wronged cities, but no state would have any legal obligation to do so (Xen. *Hell.* 6.3.18; Bengtson, 1975, No. 269). The Spartans signed on behalf of themselves and their allies, while the Athenians and their allies swore separately; the Thebans signed first in their own name, but on the following day, having apparently reassessed the binding stipulations of the autonomy-clause, Epameinondas requested that "Boiotians" be substituted for "Thebans"; by doing so, he asserted the right of the Thebans to lead the newly unified Boiotia. Agesilaos, who was the presiding officer of the conference, rejected the Theban request and excluded Thebes from the peace-treaty (Xen. *Hell.* 6.3.2-20, Diod. 15.50.4, Plut. *Ages.* 27.5-28.4, Paus. 9.13.2, Nep. *Epam.* 6.4).

Theban intransigence on the question of Boiotian autonomy meant the end of the Theban alliance with Athens and the reopening of war between Thebes and Sparta. Diodoros (15.51.3) refers to an ultimatum offered to the Thebans to leave the Boiotian cities independent, while Xenophon only hints at it (ἄγειν ἐπὶ τοὺς Θηβαίους, εἰ μὴ αὐτονόμους ἀφίοιεν τὰς πόλεις, *Hell.* 6.4.3). The last opportunity for the Spartans to resolve their dispute with the Thebans by way of diplomacy and dissolve, therefore, the Boiotian Confederacy failed and Kleombrotos was ordered to march on Boiotia. It is unclear whether he was already stationed at Phokis or whether he was sent out for a second time in 371 (Rice, 1971: 190-192n. 101, Seager, 1974: 53, Buckler, 1980a: 54-55, C.D. Hamilton 1991: 204, Buck, 1994: 161n. 66).

We have only one piece of evidence of Pelopidas' military activity in the period between the Battle of Tegyra and Leuktra (375-371): Polyainos (2.38.1) associates him with the siege of the Phokian city of Elateia; his failure to take the city may have cost him, according to Bersanetti (1949: 56), the loss of the Boeotarchy in 371. On the Peace of 371, see Cloché, 1934: 80-93, Accame, 1941: 152-158, Sordi, 1951: 34-64, Mosley, 1962: 41-46, id. 1972a: 312-318, Ryder, 1963: 237-241, id. 1965: 64-67, 127-133, Rice, 1971: 162-169, Cawkwell, 1972: 264-265, Seager, 1974: 50-53, Hack, 1975: 174-187, Tuplin, 1977: 51-56, id. 1993: 148-149, Dusanic, 1979: 319-347, Buckler, 1980a: 48-54, C.D. Hamilton, 1991: 199-202, id. 1992: 4219-4220, Hornblower,

1991: 219, Tuplin, 1993: 101-110, Buck, 1994: 111-113, Jehne, 1994: 65-79, A.G. Keen, 1996: 113-125.

ὁ δὲ κίνδυνος... κατεῖχεν: despite their earlier victory at Tegyra the Thebans were right to be afraid of encountering the Spartans, who were famous for their soldierly qualities (cf. 17.12-13, 23.4), and their allies in a pitched battle and endangering their status quo in Boiotia, especially now that they only had the support of other Boiotians (Xen. *Hell.* 6.4.4, Diod. 15.51.1).

20.2 ἐξιὼν μὲν... σῴζουσιν: also at *Reg. et imp. apoph.* 194D; for analogous scenes, cf. *C.Gracch.* 15.2-3, *Caes.* 63.10. This one seems to have been modelled after the famous Homeric episode in which Andromache makes a passionate plea to Hektor to stay away from the battle (*Il.* 6.399-465). Pelopidas' answer, which probably reflects P.'s own ideal of a general's attitude rather than his own (note, for instance, his inconsiderate attitude at 27 and 32) conforms also to the views expressed in the proem, especially at 2.1 and 2.5, where P. stresses the enormous responsibilities of the general vis-à-vis his soldiers.

[ἀεὶ] χρή: ἀεί is deleted by Reiske; cf. *Reg. et Imp. Apoph.* 194D, where δεῖν is used instead of χρή.

20.3 Pausanias, who does not mention Pelopidas in connexion with the battle at Leuktra, gives more details about the dispute among the boeotarchs before the battle (9.13.6-7; cf. Diod. 15.53.3). He provides the names of the two boeotarchs who were in favor of Epameinondas' decision to do battle at once (Μάλγης and Ξενοκράτης), and of those who were initially reluctant (Δαμοκλείδας, Δαμόφιλος and Σιμάγγελος); he also names the seventh boeotarch, who added his vote to the side of Epameinondas (Βραχυλλίδης). Sordi (1989: 125-126) agreeing with Fuscagni (1975: 45-46 and n. 53) argues that the prominent role given to Pelopidas by P. in the Leuktra episode (20-23), namely the significance of his vote, his prophetic dream, which Pausanias ascribes to Epameinondas, and his part as a leader of the Sacred Band, which is omitted by Pausanias, should be attributed merely to his source, Kallisthenes. While P.'s indebtedness to his source(s) cannot be disputed, it is wrong to view his Theban *Life* as a servile imitation and reproduction of any of his alleged sources; see further **[2c]**, and Georgiadou, in *Boeotia Antiqua*. Furthermore, P. or his source(s) does not substitute Pelopidas for Pausanias' seventh boeotarch, as Sordi claims (ib. 126), nor does he suggest that Pelopidas voted as a boeotarch, as Fuscagni takes the text to mean (ib. 45); P. simply notes that "Pelopidas was first to side with (προσέθετο γνώμην) Epameinondas"—he does not say προσέθετο ψῆφον; cf. γενόμενος σύμψηφος, 24.3—"when the latter voted to give the enemy battle immediately, since he was not appointed a boeotarch". Pelopidas' support of Epameinondas' decision confirms the earlier discussion of the two statesmen's genuine friendship and flawless collaboration (ch. 4) and may be P.'s own fabrication; see Fortina, 1958: 28n. 55.

20.4 περὶ τὰ Λεῦκτρα... ἀντεστρατοπέδευον: P. suspends the actual account of the battle and introduces a lengthy digression on the local tradition about the daughters of Skedasos, who were buried in the Leuktrian plain, and Pelopidas' visionary dream. It is a familiar technique followed earlier (16.5-8), in his account of the battle at Tegyra; cf. also the digression before the battle at Plataia in *Arist.* 11.3-9. On the different sources about the events that led up to the encounter at Leuktra, see Xen. *Hell.* 6.4. 3-12, Diod. 15.51.4-54, Paus. 9.13.3-8; see further, Beister, 1970: 13-59, Hack, 1975: 194-200, Buckler, 1980a: 54-63, Buck, 1994: 113-114. For a study of the road which Kleombrotos took to reach Leuktra and the two armies' movements before the battle, see Pritchett, 1965: 52-57, Beister, 1970: 35-51, C. Mazzucchi, "Cleombroto a Cheronea: i precedenti della battaglia di Leuttra", *RIL* 107 (1973) 671-675, Tuplin, 1987: 72-107, and, most recently, Buckler (1996: 127-139), who, following the sounder testimony of Xenophon and Pausanias, convincingly argues against the historical validity of Diodoros' account.

περὶ τὰ Λεῦκτρα: Leuktra was probably the name of a village located close to the battlefield and did not simply designate the Leuktrian plain where the battle took place; see note on 20.7.

ὄψιν εἶδε... αὐτόν: Pelopidas' annoyance at the dream, which is recounted at 21.1, recalls Agesilaos' equanimity (οὐ μὴν διετάραξεν αὐτόν) after having had a similar dream before his expedition against the Persians in 396 (*Ages.* 6.8).

20.5-6 The story about Skedasos' daughters is narrated in a greater detail in the *Amat. narrat.* 773B-774D. According to the fuller account the names of the Leuktridai are Ἱππώ and Μιλητία or Θεανώ and Εὐξίππη (cf. Μολπία and Ἱππώ in Paus. 9.13.5), which suggests that there were different versions of the same legend. A fragment of Aelian (77, Hercher) seems to suggest the existence of an even more detailed and perhaps independent version (cf. Westlake, 1939: 13n. 4). Xenophon's brief account of the same legend (*Hell.* 6.4.7) has Skedasos' daughters commit suicide after the rape (cf. also Paus. 9.13.5), and not killed by the Lacedaemonians, as P. informs us in the *Amat. narrat.* Diodorus' account (15.54.2-5) differs from the two previous with regard to the persons involved in that incident, but agrees with Xenophon's version as to the daughters' suicide. In P.'s account Pelopidas stands out as the central, enlightened, figure, who is, however, heavily disturbed by the dream, while Diodoros has nothing to say about the dream, and mentions Epameinondas, and not Pelopidas, in connection with the legend of Skedasos (15.54.2-3; cf. also Paus. 9.13.6); on the discrepancies and the different traditions these authors draw upon, see Georgiadou, in *Boeotia Antiqua*; for references to the Skedasos tale in later literature, see Schachter, 1986: 122; on its variations, see Fontenrose, 1978: 145-148; see also note on 21.1.

ἀρὰς κατὰ τῶν Σπαρτιατῶν ἀρασάμενος: cf. Λακεδαιμονίων ἀρὰς ποιουμένου, *Amat. narrat.* 773E; on the use of adnominal genitive in this case, see G. Giangrande, "Linguaggio e struttura nelle *Amatoriae Narrationes*", in G. d'Ippolito and I. Gallo, eds., *Strutture Formali dei Moralia di Plutarco*, Atti del III Convegno Plutarcheo, Palermo 3-5 maggio 1989 (Naples 1991) 287-288.

20.7 ἀλλ' ἀμφιγνοούντων τὸν τόπον: Xenophon, who is our earliest known source for the Skedasos story, includes an important information about the virgins' monument. He says that it stood at the spot where the Lacedaemonians were destined to be defeated by the Thebans and also notes that the Thebans decorated this monument before the battle (6.4.7). Neither P., in the *Life*, nor Diodoros mentions anything about this monument, while the former goes out of his way to inform us that the Lacedaemonians were not even certain about the exact location where the Skedasos incident took place. In the *Amat. narrat.* 773B P. locates the Skedasos incident in Thespiai and in 774D he relates that Pelopidas first had the exact location of Skedasos' tomb reconfirmed and then proceeded to the battle. Given this noticeable disagreement between Xenophon's and P.'s accounts it seems fair to assume that both writers are using here different sources.

ἐπεὶ καὶ τῆς Λακωνικῆς... ὁμώνυμος: the only evidence indicating that Leuktra was a hamlet (κώμιον) appears in P.'s *Narr. amat.* 773B (Wrede, *RE* s.v. "Leuktra", 12, 1925: col. 2306) and in Strabo (8.4.4). The *Etymologicon Magnum* uses the word τόπος to designate Leuktra, while Suda calls it χωρίον. Leuktron, as a place-name, is found written on the Linear B tablets as re-u-ko-to-ro (L.R. Palmer, *Mycenaeans and Minoans* [New York 1965] 87) and is presumed to be an early and common name. P. himself informs us here that there used to be more than one place with the same name: a πολίχνιον in Lakonia and a τόπος near Megalopolis in Arkadia, but his free use of the word τόπος as designating both a general location and a little town rather obfuscates the matter. Xenophon (6.4.4) leads us to infer that the place called Leuktra was on a hill, for he speaks of the Boiotians stationed on the opposite hill to Kleombrotos' camp at Leuktra (cf. Diod. 15.53.2; Pritchett, 1965: 50). Leuktra was apparently a small village located to the S.E. of Thespiai and the name of the plain itself suggests the existence of a village in the vicinity which most probably lent its name to it. The absence, however, of any archaeological evidence pointing to the existence of a Leuktra village before the battle forces us to regard this possibility with some cautiousness. On the topography of the battlefield, about which P. gives no information, see H.N. Ulrichs, vol. 2, 1840-1863: 103-104, Frazer, 1898: 50-51, Wolter in Kromayer, vol. 4, 1924-1931: 290-299, Pritchett, 1965: 49-56, Beister, 1970: 44-46, Buckler, 1980a: 59-61.

21.1 Analogous pairs to the Leuktridai, who, after their death, have the power to guarantee victory—in this case they seem to guarantee the 'dispersion' (cf. Σκέδασος) of the hostile army—in battle to their homeland, are found, as Schachter notes (1986: 122), elsewhere in Boiotia; these are the daughters of Antipoinos of Thebes (id. 1981: 35, s.v. "Antipoinides"), who die to guarantee victory over the Orchomenians, and the Koronides (1986: 117 s.v. "Koronides"), the daughters of Orion at Orchomenos, who die to save their homeland from famine; for an explanation of the ritual slaughter of young maidens in terms of male aggression and sexuality, see Burkert, 1979: 72-77, id. 1983: 58-67; cf. Wilkins, 1990: 177-194, esp. 178-180, id. 1993: 104; for a different view, see Kearns, 1989: 58-63; on mythical human sacrifices and their symbolism, Henrichs, 1980: 195-235, Dowden, 1989: 17, 35-37, 41-47, 199-204, Hughes, 1991, esp. 73-130, Vernant, 1991: 244-257, esp. 256-257, Larson, 1995: 135-138; on the legend of the Leuktrian maidens as part of the Theban folklore tradition, see Vian, 1963: 202-215, esp. 215; for a comprehensive survey of mythical instances of human (self-) sacrifice, see P.E. von Lasaulx, "Die Sühnopfer der Griechen und Römer und ihr Verhältnis zu dem einen auf Golgotha", *Studien des classischen Alterthums* (Regensburg 1854).

ἔδοξε... ὁρᾶν: a common expression in relating a dream or a vision; cf. *Per.* 3.3, *Alk.* 39.2, *Arist.* 11.5, *Alex.* 24.5, 24.9, 68.2. In many instances δοκέω is left out; cf. *Alex.* 24.8, 41, 6, *Luc.* 23.3, *Crass.* 12.4.

21.2-6 Pelopidas relates the dream to the Boiotian seers and officers and a debate follows in which opinions are split as to the course of action they should follow. It is difficult to determine whether or not P. manufactured this discussion in order to throw light on the different beliefs regarding the power of dreams and their interpretation, as well as the existence of avenging demons. His interest in a variety of superstitious practices, such as credulity in dreams, penitential exercises, human sacrifice, fear of punishment in the afterlife, is well attested. More specifically, the themes in question (namely the role of dreams and belief in human sacrifices) he developed at an earlier period of his life in treatises, such as the *De superstitione* and the *De defectu oraculorum*: in the former he satirizes, among other things, the abhorrent abominations of the Gauls and the Scythians, who sacrifice their children in order to propitiate divinities (171B-171C); in the latter, in Kleombrotos' speech about the demons, he links the demons to human sacrifices and rejects the idea that the gods either demanded or welcomed any such apotropaic rites (417C-D). In the *Pelopidas*, P. subtly alludes to the overreliance on dreams (cf. *De superst.* 165E-166C); elsewhere he indicates their dubious interpretation, (cf. *De gen.* 587A-C) and relates two cases where the dreams seem to be encouraging the recipients to patently immoral behavior, Pelopidas' and Agesilaos'; in both instances the heroes object to human sacrifice (cf. also *Them.* 13.2). Although P. does not commit himself here to belief or to disbelief in the necessity of a human sacrifice (in fact, as Brenk observes, he seldom himself gives full support to the demonological views offered by his speakers, 1973: 5), the very fact that he dedicates so much space to this demonological issue and has Pelopidas emerge in the end as an enlightened figure by choosing to sacrifice a foal instead of a maiden (ch. 22) clearly suggests that he endorses the more humane path taken by the Theban (for a contrary view, see R.M. Jones, 1916 [1980]: 38-39). For a detailed discussion of the meaning and structural function of the dreams (about 53 in number) in P.'s *Lives*, see Brenk, 1975: 336-349, id. 1977: 9-27, 214-235, id. 1987: 322-327, Valgiglio, 1988: 79-88; for discussions of P.'s demonology, see Latzarus, 1920 [1987]: 98-120, Soury, 1942, Babut, 1969: 388-440, Brenk, 1977: 49-183, id. 1987: 275-294, Dillon, 1977: 216-224, Hardie, 1992: 4764-4766; on θεός and δαίμων in Greek literature, see François, 1957.

21.3 τῶν μὲν παλαιῶν... τῶν δ' ὕστερον: P. clearly distinguishes between heroic and historical times citing as instances from ancient times the examples of Menoeceus and Makaria. There is no compelling reason to eliminate the possibility that P. found the arguments in favor of human sacrifice in his sources for the battle of Leuktra, as Hughes suggests (1991: 117); the entire debate may, however, have been invented by P. himself in order to demonstrate the triumph of reason over superstition.

Μενοικέα τὸν Κρέοντος: Teiresias revealed that Thebes could not survive the assault of the Seven unless atonement were made for the killing of the dragon by Kadmos, the founder of Thebes. The victim had to be a descendant of the Spartoi, who sprang from the jaws of the dragon. Menoeceus killed himself while standing on the walls, in order that his blood would drip outside the city. Menoeceus' story is best known from Eur. *Phoen.* 910-1012; cf. also Stat. *Theb.* 10.651-655, 756-782; see further D.W.T.C. Vessey, "Menoeceus in the *Thebaid* of Statius", *CPh* 66 (1971) 236-243; see further, Vian, 1963: 208-212, 214-215, R. Rebuffat, "Le sacrifice du fils de Créon", *REA* 74.1 (1972) 14-31, O'Connor-Visser, 1987: 73-98.

Μακαρίαν τὴν Ἡρακλέους: the soothsayers revealed to the Athenian king Demophon that their enemies, the Argives, could only be defeated if he sacrificed a girl of noble birth to Persephone. Makaria, one of Herakles' children, offered herself for sacrifice, on which see Eur.

Herakl. 474-607, *schol. Plat. Hp. Mi.* 293a, and discussion by M. Schmidt, "Makaria", *Antike Kunst* 13 (1970) 71-73, R. Guerrini, "La Morte di Macaria", *SIFC* 45 (1973) 46-59, O'Connor-Visser, 1987: 19-49.

Φερεκύδην... φρουρουμένην: Pherekydes the Wise (mid-sixth century BC) should probably be identified with Pherekydes the Theologian of Syros reported by P. to have been afflicted with a terrible disease (*Su.* 36.5); on the identification, see Hughes, 1991: 118; for a dissenting view, E.N. Tigerstedt, "Pherecydes in Sparta", *Eranos* 51 (1953) 8-13, id. 1965-1978, vol.1: 62, Flacelière and Chambry, 1966: 252. The non-sacrificial term ἀναιρεῖν used here by P. raises doubts about Pherekydes' actual sacrifice, which the context implies; for other versions of his death, cf. Diog. Laert. 1.118, Diod. 10.3.4, Porph. *Vit. Pyth.* 55; see further, H.S. Schibli, *Pherekydes of Syros* (Oxford 1990) 6-10; on sacrificial language, see Casabona, 1966. P., who is our only source for the preservation of Pherekydes' skin, may be confusing this story with a similar one known about Epimenides (Paus. 2.21.3), on which see Hughes, *ib.*

Λεωνίδαν... Ἑλλάδος: Leonidas' heroic stance at Thermopylai is presented here metaphorically (τρόπον τινά) as a sacrifice. No such imagery is used by Herodotos (7.205-209, 223-224; see further, Hughes, 1991: 117, 236n. 142). The oracle referred to by P. is preserved by Herodotos (7.220).

ἔτι δὲ τοὺς ὑπὸ Θεμιστοκλέους... ναυμαχίας: the seer is said to have urged Themistokles to sacrifice three Persian prisoners to Dionysos the Raw-Eater before the battle of Salamis (*Them.* 13.2-5). Themistokles' reaction is similar to that of Pelopidas and Agesilaos, though, in the end, he was compelled by the πολλοί to give in; a more detailed account of the youths' sacrifice is given at *Arist.* 9.2, which in the *Themistokles*, P. attributes to Phanias (fr. 25, Wehrli). Against the historicity of this sacrifice, see Frost, 1980: 150, Henrichs, 1980: 195-235, esp. 208-224, Hughes, 1991: 111-115; on the battle of Salamis, see N.G.L. Hammond, "The Battle of Salamis", *JHS* 76 (1956) 32-54, G. Roux, "Eschyle, Hérodote, Diodore, Plutarque racontent la bataille de Salamine", *BCH* 98 (1974) 51-94, Frost 1980: 133-169.

Ὠμηστῇ Διονύσῳ: "Dionysos the Raw-Eater"; on the epithet Ὠμηστής, most attestations of which occur in P., see Henrichs, 1980: 221-224. On the multiple titles of Dionysos including that of Ὠμηστής, cf. *Anton.* 24.4-5 and Pelling's note (1988: 180).

21.4 τοῦτο δ' Ἀγησίλαον... γενομένην: in spring 396, when Agesilaos was about to embark on his expedition to Asia, he spent the night at Aulis. In his sleep a voice told him that he should make the same sacrifice that Agamemnon made before departing for Troy. But Agesilaos, refusing to imitate Agamemnon's insensitive behavior, instructed his seer to sacrifice a hind; cf. Xen. *Hell.* 3.4.3-4, Plut. *Ages.* 6.6-9, Paus. 3.9.4-5; see further, Brenk, 1977: 55n. 7, C.D. Hamilton, 1992: 4214-4215, id. 1994: 208-210, Hughes, 1991: 109-111. The association of Agesilaos' expedition with that of Agamemnon's may have been part of an anecdotal tradition in which historical events and characters were shaped to evoke certain well-known mythical episodes or figures (cf. Hektor and Andromache at 20.2), or were directly compared to such (cf. *De laud. ips.* 542B, where Menekleidas reproaches Epameinondas for emulating Agamemnon and the Theban statesman retorts by demonstrating the superiority of his victory at Leuktra over Agamemnon's at Troy; also Nep. *Epam.* 5.5-6). On the other hand, it may go back to Xenophon, who knew Agesilaos; the latter no doubt thought of himself as another Agamemnon, since there is no other reason for him to have made the detour to Aulis. P.'s attitude toward Agesilaos is ambiguous, as Brenk observes (1977: 55n. 7), because he leaves the impression that by ignoring the message of the dream Agesilaos brought his expedition to an inglorious ending.

ἀπομαλθακισθείς: "having showed weakness", "having been tender-hearted"; Agesilaos' attitude is contrasted with Agamemnon's ἀπάθεια, *Ages.* 6.8, in which case ἀπάθεια seems to make more sense than ἀμαθία given by Y; cf. also R. Flacelière and É. Chambry, *Sertorius/Eumène, Agésilas/Pompée*, vol. 8 (Paris 1973) 102n. 1.

21.5 οἱ δὲ τοὐναντίον... θυσίαν: having enumerated the ancient examples brought forward by the partisans of human sacrifices (21.3-4), P. goes on to explain the more enlightened view, which he himself seems to have shared. Westlake's opinion that the latter view reflects Epicurean influence (1939: 13; cf. also W. Schmid ["Götter und

Menschen in der Theologie Epikurs", *RhM* 94 (1951) 97-156]) is challenged by Brenk (1977: 56n. 8). P.'s suggestion that "such a lawless and barbarous sacrifice was not acceptable to any of the superior beings above us for it was the father of all gods and men who governed the world, not the fabled typhons and giants" appears to be consistent with the Platonic view expressed in *De Is. et Os.* 369B. For a recent discussion of P.'s attitude toward divination, see Opsomer, 1996: 165-194.

οὐ γὰρ τοὺς Τυφῶνας... ἄρχειν: similar views are expressed in *De superst.* 171D-E. On the fight of the gods and the giants, see Vian, 1952. Typhon, often confused with the Giants, (Hor. *C.* 3.4.53), was born by Earth to Tartaros after the defeat of the Titans. He is invested with demonic power by the Pythagoreans (*De Is. et Os.* 363A) and is associated with the idea of harm or destruction in nature (*ib.* 369A); on the myth of Typhon in epic poetry, see A. Ballabriga, "Le dernier adversaire de Zeus: Le mythe de Typhon dans l'épopée archaïque", *Revue de l'Histoire des Religions* 207 (1990) 3-30. P. makes the Titans out to be demons as well (*De def. orac.* 421C). Both the Typhon and the Giants are connected with the demons (*De def. orac.* 421C; see Brenk, 1977: 58-59, Fontenrose, 1959: 77-93; cf. also Hom. *Il.* 2.783-785, *Hymn Apol.* 306-309, Hes. *Theog.* 820-880, Apollod. 1.6.1, Nonnus, *Dion.* 1.154-162, 481-534, Verg. *Aen.* 9.715-16).

21.6 δαίμονας... ἀδυνάτων: the point here is that if demons, which rejoice in human blood, exist, then they must be ineffectual and are not to be feared. This passage has received much attention from scholars. Erbse argues that it indicates a development in P.'s thought toward an increasing belief in the power of evil demons (1952: 295-314), while Brenk demonstrates convincingly that P. does not promote the belief in evil demons in his later writings, such as the *Pelopidas*; on the contrary he resists this belief (1973: 1-11, esp. 2-3, id. 1977: 52-59). The idea of demons as being intermediate between gods and mortals appears prominently in P.'s demonological treatises (*De def. orac.* 415A, 416C-D, *De Is. et Os.* 361A). P. recognizes also different degrees of excellence within the demons (*De def. orac.* 417B); on his disagreement with Epicurean views about the location of the demons, see Flacelière, 1959: 202-204, Hershbell, 1992b: 3379. For an extensive bibliography on P.'s demonology, see Brenk, 1977, id. 1987: 275-294. On the development of the notion δαίμων, see H. Nowak, *Zur Entwicklungsgeschichte des Begriffes Daimon*, Diss. Bonn 1960.

δαίμονας... πιστεύειν ⟨εἶναι⟩: Ziegler's addition of εἶναι here finds support from the use of this verb in *Brutus* 37.6 and *De def. orac.* 418F and 420C.

22 The solution to the debate is given when a substitute sacrificial victim, a filly, fitting the color demanded by Skedasos in Pelopidas' dream forced her way to death. Pelopidas yields to the more humane path advocated by those who objected to human sacrifices. P. shows here how the enlightened reasoning (vindicated by the Theban victory at Leuktra, 23.5) prevailed upon the superstitious one, which endorsed human sacrifice. Moreover, the theme of superstition provides a further link between the two *Lives*—Roman religiosity is a recurrent theme in the *Marcellus* (cf. 5.5-7, 6.11-12, 7.4, 8, 10.6, 12.2, 28.2-3, 29.8-11)—and allows the drawing of instructive contrastive comparisons between them; no doubt, what is supposed to emerge in *Pelopidas* 21-22, is that "human sacrifice is barely tenable for an enlightened Greek" (Pelling, 1989: 199-232, esp. 200-201), whereas this is not at all the case with the Romans, whose superstitious beliefs seem to have played a prominent role in their lives; cf. the practice of *devotio* by the Romans (Liv. 8.6.9-16, 9.1-13, 10.11-14, 10.28.12-17, Cic. *Pro Sest.* 48); for a discussion of the practice of human sacrifices among the Greeks and the Romans, see S. Eitrem, *Opferritus und Voropfer der Griechen und Römer* (Kristiania 1915; repr. New York 1977), F. Schwenn, *Die Menschenopfer bei den Griechen und Römer*, Religionsgeschichtliche Versuche und Vorarbeiten 15, 3 (Giessen 1915; Berlin 1966). For a discussion of P.'s attitude toward superstition, see A. Pérez Jiménez, "Δεισιδαιμονία: el miedo a los dioses en Plutarco", *Studia Hellenistica* 32 (1996) 195-225, and bibliography therein.

The substitution of an animal for a human victim is common in Greek myth, with the most famous example being the deer replacing Iphigeneia in Euripides' *Iphigeneia in Aulis*; see Henrichs, 1980: 195-235, Burkert, 1985 passim, O'Connor-Visser, 1987: 99-147, Hughes, 1991: 82-86, Loraux, 1991: 34-37.

22.3 ὦ δαιμόνιε: on this type of apostrophe, see E. Brunius-Nilsson, *Δαιμόνιε, An Inquiry into a Mode of Apostrophe in Old Greek Literature*, Diss. Uppsala 1955.

22.4 λόγον... διδόντες: Coraes' emendation to διαδιδόντες conveys the idea intended here by P., namely an account spreading, circulating through the whole army; it is a common expression in P. (cf. *Thes.* 6.1, *Alk.* 31.3, *Arist.* 7.1, *Philop.* 19.5, *Su.* 20.2).

Ch. 23: The Battle of Leuktra

The Battle of Leuktra has been the object of systematic study by several scholars mainly because of the discrepancies among our major sources regarding the events that led up to Leuktra, the topography of the battlefield and the course of the battle (Xen. *Hell.* 6.4.3-15, Plut. *Pel.* 20-23, Diod. 15.52-56, Paus. 9.13.3-12). P.'s account of the battle was generally dismissed by earlier historians as a rhetorical elaboration, slanted toward Pelopidas (Delbrück, 1908 [1975]: 168) and as impossible and inaccurate (Wolter, vol. 4, 1924-1931: 306). It is only after Anderson's study (1970: 192-220) followed by Cawkwell (1972: 260-263), and, more recently, by Buckler (1980b: 75-93) and Tuplin (1987: 72-107, esp. 84-93; cf. also 1986: 37-66, esp. 50) that P.'s testimony has been given greater credence. Nepos mentions the battle briefly (*Pel.* 4.2, *Epam.* 8.3), while less significant references to it include Plut. *Ages.* 28.5-6, *Reg. et imp. apoph.* 193A-B, *Apoph. Lac.* 214B, *Aet. Rom. et Gr.* 282E, *De glor. Athen.* 346b, App. *Syr.* 11.41, Front. 1.11.6, 16, 1.12.5, 4.2.6 and Polyain. 2.3.8.

The different accounts of the battle vary in length and amount of detail and depend upon each writer's scope, intentions and available sources. Only Xenophon's account is contemporary and, despite the fact that it carries greater authority for us than the other extant accounts, it is regarded as incomplete: Xenophon stresses the preceding cavalry combat and the fact that the Lacedaemonians were defeated in this encounter (Delbrück, *ib.*); his account is more a list of reasons for the Spartan defeat than a narrative of what actually happened (Anderson, 1970: 198); furthermore, it is drawn from the Spartan side only and is marked by the deliberate omission of important information on the battle; writing with a pro-Spartan bias he fails to give credit to the generalship of Epameinondas, suppresses Pelopidas' contribution to the Theban victory and attributes the Spartan defeat to divine retribution for the seizure of the Kadmeia and the Theban victory to τύχη, but also to rational calculation; see further, Cawkwell, 1972: 255-257, Buckler, 1980a: 263-268, Lazenby, 1985: 156, Tuplin, 1986: 37-66, esp. 49-51, 1987: 87, id. 1993: 134-138, 215, Dillery, 1995: 179-194, 221-225. P., mainly concerned with the description of the battle from the Theban side, describes it through the maneuvers of Epameinondas and Pelopidas, and makes, therefore, a valuable complement to Xenophon's testimony. Diodoros' account is generally viewed as unreliable, since he tends to displace events in his account of the battle: he places, for instance, Jason's intervention and Archidamos' arrival before the battle instead of after it, and puts Archidamos and Kleombrotos on the Spartan wings. Finally, Pausanias has very little to say about the actual battle.

The date of the battle has been assigned with some precision to the middle of the second week of August 371; see Beister, 1970: 59-71, Pédech, 1972: 1-6, Buckler, 1980a: 233, Tuplin, 1987: 77-84, Sergent, 1991: 137-143.

P.'s sources for the battle of Leuktra: Kallisthenes has generally been acknowledged as P.'s source for the description of the battle of Leuktra, mainly on the basis of the exclusion of Ephoros and Xenophon as the other two possible sources known to us (Stern, 1887: 59-61, Peper, 1912: 35-38, Fuscagni, 1975: 37, 45, Buckler, 1980b: 75-76, Prandi, 1985: 46-47, Sordi, 1989: 125-129; cf. Tuplin, 1987: 84). Some significant discrepancies between the accounts of P. and Ephoros/Diodoros or P. and Xenophon seem to justify this view. Ephoros/Diodoros makes Epameinondas the focal point of the account of the battle (15.54-56) and says nothing about Pelopidas' contribution to the victory at this point, while P. describes the incident, as we would expect, chiefly from Pelopidas' standpoint. Moreover, certain omens which foretold the impending defeat of the Lacedaemonians are attributed by Cicero to Kallisthenes (*De div.* 1.74-76 [*FGrH* 124F22a], 2.54 [124 F22b]); two of these portents—none of which, however, is mentioned in the *Pelopidas* (cf. also *Ages.* 28.6)—resemble those related by P. in *De Pyth. orac.* 397E-F (see Schröder, 1990: 168-171). While such arguments tend to support the claim that P. depended on Kallisthenes, his use of this historian, as noted before, cannot be determined with complete certainty; see [**2c.c**], and Georgiadou, in *Boeotia Antiqua*.

23.1-3 Ἐν δὲ τῇ μάχῃ... ἐπιβαλών: P. does not give a complete picture of the battle, but focuses on three phases of the clash of Epameinondas' left wing with Kleombrotos' right wing: 1) Epameinondas drew his phalanx obliquely towards the left with the intention of attacking Kleombrotos' right wing; 2) Kleombrotos realizing that the Thebans were massing on their left wing opposite his own position began to change the formation of the Spartans; 3) Pelopidas heading the Sacred Band attacked the Spartan line before Kleombrotos could complete his evolutions; for a reconstruction of the battle from P.'s account, see Buckler, 1980b: 76-86, and counter-arguments by Tuplin, 1987: 84-93; see also Hack, 1975: 199-205.

P. builds up a crescendo of force and excitement of the battle-scene in the first sentence (ἐν δὲ τῇ μάχῃ... ἐπιβαλών) with an impressive accumulation of verbs, participles and infinitives, all of which denote intense climactic activity, movement and speed. When the climax of the battle-tension is being reached with the confusion which arose amongst the Lacedaemonians, P. pauses abruptly to comment on Spartan military expertise (23.4); only after the tension has dropped significantly for a little is the conclusive scene of the battle described (23.5). This battle scene has its parallel in *Marc.* 15.1-16.3, where the Romans attack Syracuse; the dramatic tension is similarly built up in two long sentences (σχάσαντος δὲ τὰς μηχανὰς... συντριβόμεναι and πολλάκις δὲ... ἀνείσης); a similar drop of suspense is produced by the description of the sambyca (15. 5); the tension builds up again in 15.6-7 leading to the first retreat of the Romans; a second and final release of the tense atmosphere takes place (15. 8-9), leading this time to the Romans' failure to seize Syracuse in their first attempt and their defeat (16.3).

23.1 τὴν φάλαγγα λοξὴν: the phrase occurs only once in P.'s writings; see Wyttenbach, s.v. λοξός. Both Diodoros (15.55.2) and P. speak of an "oblique formation" of the phalanx. According to Anderson (1970: 324n. 60) the two writers do not seem to have had the same conception of what this meant: Diodoros' use of the term corresponds to the technical meaning given to λοξὴ φάλαγξ by the tacticians (Arr. *Tact.*, 20.3, 26.3, Onas. 21.8, Asklep. 10.1, 21, 11.1, 5, 7), while P. uses it to indicate the diagonal march to the left of the whole line as it advanced. Buckler, on the other hand, argues for the similar use of the term by both Diodoros and P. as a "slanting phalanx", in which one wing is advanced and the other refused" (1985: 138-139; cf. Cawkwell, 1972: 261n. 3). Tuplin (1987: 84n. 44) takes λοξήν adverbially with ἕλκοντος and translates "Ep. was drawing the phalanx obliquely to the left". It is questionable whether P. would have used λοξήν adverbially, since he could have written πλαγίως or ἐγκαρσίως. The "oblique phalanx" was among the great tactical innovations of Epameinondas' plan of attack at Leuktra; see G. Busolt, "Spartas Heer und Leuktra", *Hermes* 40 (1905) 387-449, esp. 434-438, Lévêque and Vidal-Naquet, 1960: 294-308, esp. 294-297, Kromayer and Veith, 1963: 93-95, Pritchett, 1965: 58, Cawkwell, 1972: 254-278, esp. 261-262, Buckler, 1980b: 88-89, Lazenby, 1985: 161-162. Hanson (1988: 190-207) argues unconvincingly against novelties in tactics by Epameinondas, while A.M. Devine ("EMBOLON: A Study in Tactical Terminology", *Phoenix* 37 [1983] 201-217) paying little attention to P.'s testimony, disputes the traditional view that Epameinondas' dispositions at Leuktra were deep columns; see Buckler for counter-arguments (1985: 134-143).

ἐπὶ τὸ εὐώνυμον ἕλκοντος: the slanting phalanx was in addition to making a move to the left (cf. Cawkwell, 1972: 261, Tuplin, 1987: 84n. 44). It seems that Epameinondas led his phalanx on a leftward diagonal march so that the Spartan right wing could be drawn away (ἐξώσῃ, on which see Krentz, 1985: 55-59, Pritchett, 1985: 65-68) from the left, where their allied forces were stationed, and thus defeat the Spartan right without the left being able to interfere (J. Buckler, "Epameinondas and the EMBOLON", *Phoenix* 39 [1985] 138; Lazenby, 1985: 157).

προσπεσὼν ἀθρόως κατὰ κέρας: Epameinondas planned to overpower Kleombrotos by a charge of his massed wing; ἀθρόως: in a military context it is attested only here in P.'s writings; κατὰ κέρας: "in column", "on the wing", "on the flank"; see Tuplin, 1987: 85n. 45; for a listing of occurrences of κατὰ κέρας, see F. Reuss, "Zu Xenophons anabasis", *Neue Jahrbücher für Philologie und Pädagogik* 127 (1883) 817-831, esp. 818.

23.2 οἱ μὲν πολέμιοι... τὸν Ἐπαμεινώνδαν: when the Spartans realized that their own right wing was in danger of being outflanked by the Thebans who were forming up on the left, opposite their own position, they began to make changes in their own formation. Kleombrotos started by withdrawing elements from the left of his right wing and moving them behind the line in order to encircle Epameinondas' column (Buckler, 1980b: 85-88, Lazenby, 1985: 158-159, Tuplin, 1987: 85-93).

ἀνέπτυσσον: "fold back", "unfold"; it describes, like περιῆγον, a movement in progress. For a description of Kleombrotos' maneuver, see Buckler, 1980b: 83-85; cf. Tuplin, 1987: 85, 89-93. On the different meanings of this verb in military contexts, see Anderson, 1970: 324-325n. 61, Delbrück, 1975: 216, Ferrill, 1985: 158; on P.'s use of the verb, see Wyttenbach, s.v. ἀναπτύσσειν.

περιῆγον: "lead round". Kleombrotos was trying to march his men around the men on his extreme right wing in order to outflank Epameinondas' wing. For an interpretation of this maneuver, see Buckler, 1980b: 85-86, Ferrill, 1985: 158, Tuplin, 1987: 85. The imperfect in ἀνέπτυσσον and περιῆγον indicates that these actions were not completed; since, however, it implies nothing as to the actual length of the maneuver (cf. Kühner-Gerth, vol. 2.1, p. 142 [ch. 383.1]), it is impossible to determine how advanced Kleombrotos' maneuvers were when Pelopidas attacked the Spartan line. Tuplin (1987: 85) suggests that "the two successive processes, ἀναπτύσσειν and περιάγειν, were to produce a continuous battle-line of the same depth as the original one but shifted laterally so as to extend much further to the right than the Theban forces opposite it".

ὡς κυκλωσόμενοι: "to wheel in a circle", "to encircle". Buckler (1980b: 91n. 20; cf. Pritchett, 1985: 74-75, Tuplin, 1987: 89-90) takes the verb κυκλόω here to mean "an operation in which the line pivots on one man or unit and moves in a circular motion through an unbroken arc".

περιβαλοῦντες: "to surround"; its only occurrence with a military meaning in P. is here. Buckler (1980b: 80; cf. Tuplin, 1987: 90) notes that the verb does not specify the tactical disposition by which Epameinondas was to be enveloped.

23.3 ὁ δὲ Πελοπίδας... τὴν τάξιν: Pelopidas, stationed in the front ranks of Epameinondas' wing, was the first to engage Kleombrotos (cf. also Diod. 15.81.2, Nep. *Pel.* 2.2). P. describes Pelopidas' charge with the Sacred Band in two stages: he first ran forward from his position (προεξέδραμε; cf. *Cor.* 9.7; see Buckler, 1980b: 78) and reached the Spartan line before Kleombrotos could complete the deployment of his force. For a discussion of Pelopidas' position in the front of the Theban phalanx, forming the "cutting edge" of Epameinondas' column and his crucial role in the battle, see Buckler, 1980b: 76-80, 89, Lazenby, 1985: 157, 159.

φθάνει: "to be beforehand with". P. makes a brusque transition from the past tense to a historical present to express the antecedent and the subsequent of the two evolutions. The use of the present tense here, which coincides with the culmination of the Theban successful combined attack, shows an enduring result and imparts dramatic immediacy to the event. On the use of φθάνω by P., see Buckler, 1980b: 78. Pelopidas managed to catch the Spartans "before Kleombrotos stretched out his wing" (πρὶν ἀνατεῖναι τὸν Κλεόμβροτον τὸ κέρας).

ἀνατεῖναι: "to stretch up", "stretch out", "reach up", "reach out". In P. it usually refers to hands or things held in the hand (*Arist.* 7.8, *Rom.* 18.8, *Alex.* 30.11, *Per.* 31.4). Buckler (1980b: 81-82; cf. Anderson, 1970: 326n. 61, Tuplin, 1987: 87-88) suggests that it is here used descriptively rather than in a tactical sense and proposes that in P.'s view "Kleombrotos intended to bring his wing into a position from which it could reach forward towards the enemy in order to fall upon Epameinondas' flank".

συναγ⟨αγ⟩εῖν: "to bring together", "to collect"; in military contexts the term, commonly used by Herodotos and Thukydides, generally signifies "to assemble" the soldiers (cf. *Brut.* 22.3, 28.4, *Reg. et imp. apoph.* 206A, *Apoph. Lac.* 209B).

συγκλεῖσαι: "to shut in", "to close"; rarely found in a military context in P. (cf. *Crass.* 25.10); here it probably means "to close up" the line (Buckler, 1980b: 82-83); cf. Xen. *Hell.* 6.4.14.

23.4-5 The Spartan model, earlier posited as a model worthy of imitation (7.5), is now superseded by the Theban (cf. also *IG* VII 2462, Diod. 15.50.5). P.'s intention of magnifying the Thebans through contrast with the Spartans is so clearly expressed that one gets the impression that the whole account of the battle is only peripheral to his main objective, i.e. to underline the decaying glory and expertise of the Spartan army, and hence the end of their supremacy in Greece. His comments on the Spartans' skill in the art of war appear to echo

Xenophon's praise of their military organization: Λακεδαιμονίους δὲ μόνους τῷ ὄντι τεχνίτας τῶν πολεμικῶν, *Reip. Lac.* 13.5; see further, Anderson, 1970: 104-105, Pritchett, 1974: 209-213.

23.5 τότε δ' ἡ τοῦ Ἐπαμεινώνδου... γενέσθαι: P. moves on to the final stage of the battle focusing anew on the successful attacking movements of the architects of the Theban victory, Epameinondas and Pelopidas. He does not mention here Kleombrotos' death, but only in *Ages.* 28. 5. P.'s report on the number of the Lacedaemonians who fell at the battle (*Ages.* 28.5) agrees with that of Xenophon (*Hell.* 6.4.15) and Pausanias (9.13.12) and differs from Diodoros' exaggerated figures (15.56.4). P. makes no mention of Boiotian losses, on which see Diod. 15.56.4 and Paus. 9.13.12.

23.6 διὸ τῷ Ἐπαμεινώνδᾳ... κατορθώματος: both P. and Diodoros (15.81.2) recognize Pelopidas' major contribution to the victory at Leuktra, while Nepos is no less cognizant (*Pel.* 4.2). Xenophon mentions nothing about the Sacred Band and Pelopidas' role at Leuktra (or Tegyra). Only later, in his account of the Greek embassy at Sousa, does he refer to the Theban victory at Leuktra, and even then the connexion between the victory and Pelopidas is expressed in rather vague terms (7.1.35). Pausanias (9.13.6), who is generally believed to have drawn either on P.'s lost *Epameinondas* or on the same source as P. for the account of these events (Peper, 1912: 16-21, Momigliano, 1934: 196, Beister, 1970: 81, Geiger, 1985: 62), mentions nothing about Pelopidas' presence at Leuktra. Tuplin (1984: 346-358), however, raises serious doubts about Pausanias' dependence on P.'s lost *Life* in this section; for further discussion, see Georgiadou, in *Boeotia Antiqua*. On the Spartan defeat at Leuktra and its consequences, see Hack, 1975: 203-224, Buckler, 1980a: 65-69, E. David, "Revolutionary Agitation in Sparta after Leuctra", *Athenaeum* 58.3-4 (1980) 299-309, Cawkwell, "The Decline of Sparta", *CQ* 33 (1983) 385-400, esp. 397-399, Cartledge, 1987: 382-384, C.D. Hamilton, 1991: 211-214, Hornblower, 1991: 219-222, Davies, 1993: 200-201, Tuplin, 1993: 149-151.

P. makes no mention here of the trophy set up by the Thebans on the battlefield as a memorial of their victory (cf. Diod. 15.81.2; see bibliography in Tuplin, 1987: 98n. 87 and [**2c**] s.v. "Autopsy and Oral Tradition"), but only in *Non posse* 1098B (τὸ ἐν Λεύκτροις τρόπαιον), and there he associates it merely with Epameinondas. This trophy was replaced with a bronze one, apparently the first permanent trophy commemorating a Greek victory over Greeks (Cic. De *invent.* 2.23, Frazer, 1898, 5: 51-52). It may not have met with P.'s approval—he criticizes trophies of stone or bronze in *Aet. Rom. et Graec.* 273C-D)—and so he omitted it. He also eschews any reference to the agonistic festival established in honor of Zeus Basileus at Lebadeia (Diod. 15.53.4), which he must have known about, although it had ceased to be celebrated by his day; see Schachter, 1994: 111-118. Perhaps he included these in his lost *Life of Epameinondas.*

ἴσον ἠνέγκατο... κατορθώματος: P. expresses reservations about this in the *synkrisis* (32.(2).2), on which see [**2e**].

Ch. 24: Epameinondas' and Pelopidas' first Invasion of the Peloponnese (370-369)

After the battle of Leuktra, which marked the end of Spartan hegemony in Greece and the beginning of the Theban, the Thebans were concerned with consolidating their position in Boiotia and across central Greece before entertaining any ambitions beyond this region (Xen. *Hell.* 6.5.22-23, *Ages.* 2.24, Diod. 15.57.1). Expansion to the north was hindered by their powerful northern ally, Jason of Pherai, the tagos of Thessaly (*Hell.* 6.4.20-32); it was only after his murder in late summer 370 that the Thebans could extend their influence northward. The Athenians meanwhile, apparently displeased with the Theban victory, summoned a new peace conference in late 371 (Accame, 1941: 159-167, Sordi, 1951: 34-64, Ryder, 1965: 70-78, Seager, 1974: 36-63, esp. 54-55, Buckler, 1978b: 85-96, esp. 91, 1980a: 68-69, C.D. Hamilton, 1991: 215-217, Urban, 1991: 173-174, Jehne, 1994: 74-79), in which they tried to reassert the principles of the King's Peace of 387 (Xen. *Hell.* 6.5.1-3, Bengtson, 1975, No. 270 [Wickersham and Verbrugghe, 1973, No. 34]). A new stipulation in the treaty made assistance to threatened states compulsory for all signatories. In the meantime, political unrest in the Peloponnese in 371/370 provided the Thebans with a good basis for an extension of power in that area; see Seibert, 1979, vol. 1: 119-121.

24.1 The first Theban invasion of the Peloponnese (370-369) had been the result of a plea from the states of Argos, the Arkadian League and Elis to assist them against the Spartans. In their search for allies they appealed first to Athens, and when the Athenians refused (Dem. 16.12, 19, Diod. 15.62.3), probably owing to Elis' recalcitrance at the peace conference in 371 (cf. Xen. *Hell.* 6.5.3), they turned to Thebes, who voted to accept the alliance (*Hell.* 6.5.17-22). Though it is nowhere mentioned explicitly, it may be deduced from Nepos' reference to Menekleidas (*Epam.* 5.2-3; cf. Plut. *Pel.* 25.5-15) that he had been one of those who opposed Epameinondas on the policy of intervening in the Peloponnese. After the Eleians secured the assistance of the Boiotians by financing their operation with ten talents (*Hell.* 6.5.19), the latter, under the command of Epameinondas and Pelopidas (Diod. 15.62.4, Nep. *Epam.* 7.3, Plut. *Ages.* 31) marched south. For the foreign policy of Athens in this period, see Cloché, 1934: 97-99, Accame, 1941: 159-167; on the first Theban invasion of the Peloponnese, see Wiseman, 1969: 177-199, esp. 178-180, Roy, 1971: 569-599, esp. 572, 588, 590-591, Buckler, 1980a: 72-90, C.D. Hamilton, 1991: 220-231; on P.'s possible sources for this, see [**2c**], esp. s.vv. "Theopompos and Xenophon". On the Arkadian Confederacy, see Larsen, 1968: 180-195.

ἀμφότεροι βοιωταρχοῦντες: sc. as opposed to the battle of Leuktra, where Epameinondas was a boeotarch, but not Pelopidas (23.6). Sordi (1989: 128-130) argues against the Boeotarchy of Pelopidas during the expedition to the Peloponnes, which she dates to 371/0, and regards it as an invention of P.'s alleged source, Kallisthenes.

Λακεδαιμονίων... σύμπασαν: following the battle of Leuktra and the Peace Conference at Athens political turmoil broke out in several Peloponnesian cities, which reflected widespread discontent with Spartan oligarchic leadership. By late 370 several cities had banded together to compose the Arkadian League (*Hell.* 6.5.11), which was built on democratic principles (H. Schaefer, "Πόλις μυρίανδρος", *Historia* 10 [1961] 292-317, esp. 310-314, J.A.O. Larsen, 1968: 183-188). Both the Eleians and the Argives welcomed the alliance with the Arkadians (Diod. 15.62.3). P. describes the unification of the Argives, Arkadians and Eleians against the Spartans and their detachment from the Lacedaemonian Confederacy as a personal achievement of Epameinondas and Pelopidas (βοιωταρχοῦντες... ἐνέβαλον... προσήγοντο); it is clear, however, that it is the Spartans' defeat at Leuktra and their subsequent weakening of their power which triggered unrest among the democratic circles and united the Peloponnesians against them.

24.2 καίτοι χειμῶνος... παραδιδόντας: the first Theban expedition to the Peloponnese took place in the winter of 370-369 (24.3; see Roy, 1971: 569-599, esp. 590-591, Buckler, 1980a: 233-234; dated to 371/370 by Sordi, 1989: 129) and did not end until four months later; when they returned it was still winter (25.2; cf. *Ages*. 32.13, Xen. *Hell.* 6.5.50). The term of office for all boeotarchs ended at the winter solstice, the first day of Boukatios, which was the first month of the Boiotian civil year, and the penalty for not surrendering the office to the successor on that day was death (Nep. *Epam.* 7.5). According to P. the boeotarchs kept their office four months beyond that date (25.2), while Appian (*Syr.* 41.214) says six; see Wiseman, 1969, 177-199, esp. 178-180, Roesch, 1972: 61-70, esp. 62-64, id. 1982: 287-290, and note on 25.2.

24.3 οἱ δ' ἄλλοι... στράτευμα: while P. specifies that the other boeotarchs, except for Epameinondas and Pelopidas, were reluctant to invade Lakonia because it was winter and for fear of the law, Xenophon attributes their hesitation to the difficulty of the operation and fear of the Spartan garrisons (*Hell.* 6.5.24). Diodoros simply says that the boeotarchs willingly relinquished authority to the two Thebans in recognition of their experience and courage (15.62.5). Despite their abdication of command, however, the other boeotarchs were still held responsible for any infringement of the constitution (Buckler, 1980a: 27; cf. *Reg. et imp. apoph.* 194A-B).

Πελοπίδας δὲ... σύμψηφος: on P.'s tendency to highlight Pelopidas' support for Epameinondas' decision to invade Lakonia, see [**2e**], and note on 4. Pausanias does not mention Pelopidas in connection with the first invasion, but makes Epameinondas solely responsible for both it and the foundation of the modern Messene (9.14.4-5, 15.6). For a discussion of Pausanias' sources regarding the period of Theban hegemony, see Fuscagni, 1975: 31-55, esp. 45-46, Tuplin, 1984: 346-358, Sordi, 1989: 128; also [**1c**], and Georgiadou, in *Boeotia Antiqua*.

24.4 ἦγεν... στρατιᾶς: a series of singular verbs and participles by which P. foregrounds Pelopidas in events which either involved both men or in which Epameinondas played the leading part (cf. Plut. *synkr. Pel./Marc.* 32(2).2). P. describes in greater detail the Theban invasion of Lakonia and its consequences in *Ages.* 31-32 (cf. also Xen. *Hell.*

6.5.27-32, 50, *Ages*. 2.24, Diod. 15.65 (see A. Bauer, "Der zweimalige Angriff des Epameinondas auf Sparta", *HZ* 65 [1890] 240-274, esp. 242-243). Though he attaches too much importance to the fact that the two Thebans crossed the Eurotas—the first time in 600 years (*Ages*. 31.3; 500 years in *Reg. et imp. apoph.* 194B) that a foreign army had invaded Lakonia—, he avoids commenting at this point on their failure to capture Sparta (cf. *synkr.* 32(2).2). This would no doubt give an anticlimactic tone in his account of their successive successes. On the crossing of the Eurotas, see Buckler, 1980a: 81-83.

ἡγούμενος ἑπτὰ μυριάδων... μέρος: Fuscagni (1975: 45-46) suggests that P.'s description of Pelopidas and Epameinondas as leaders of the numerous allied Greek forces reflects the pro-Macedonian tendencies of his source Kallisthenes, in which Philip assumed the role of the commander of a panhellenic army (cf. also Prandi, 1985: 71, and [**2c**]). Diodoros estimates the number of the allied forces as more than 50,000 at 15.62.5, while in the eulogy of Pelopidas at 15.81.2 he makes him a commander of 70,000 men (cf. Plut. *Ages*. 31.1-2). Queck (1876: 24) ascribes this inconsistency to the historian's use of different sources. Xenophon lists the Greek contigents which joined the Thebans in the expedition (*Hell*. 6.5.23, *Ages*. 2.24).

24.5 ἀλλ' ἡ δόξα... ἐκείνοις: a weakness of the alliance between the Boiotian Confederacy and the Peloponnesian states of Argos, the Arcadian League, and Elis was that its members did not defer to any one state as leader of the expedition, even if Thebes did in fact play that role (cf. Xen. *Hell*. 7.1.22); see Roy, 1971: 569-599, esp. 597-599, Buckler, 1980a: 73.

24.7 κἂν ὥσπερ οἱ πλέοντες... ἔχουσι: a maritime metaphor with a Platonic touch; cf. *Polit*. 272e, 296e-297a, 302a-b, *Resp*. 488a-489a, *Leg*. 709b, *Crit*. 109c. P. frequently employs seafaring metaphors: cf. *Phok*. 3.4, *Per*. 33.6, *Arat*. 38.5, *Aem*. 18.3, *Mar*. 11.1, *Caes*. 28.5, 34.3, *De coh. ira* 454A, *De vit. pud.* 536C-D, *Quaest. conv*. 619D, *An sen. resp*. 787D-E, *Praec. ger. reip*. 801C, E-F, 807B-C, 812C; see further, F. Fuhrmann, *Les Images de Plutarque* (Paris 1964) 49-50, 69-70n. 3, 234-237.

24.9 καὶ πᾶσαν μὲν Ἀρκαδίαν... συνέστησαν: P. attributes to Pelopidas and Epameinondas the unification of Arkadia into one power, while Diodoros (15.59.1) makes Lykomedes of Tegea responsible for it. Whereas the formation of the Arkadian League must have been undoubtedly strengthened by the Peloponnesian expedition of the Boiotians and their allies, the connecting of Epameinondas and Pelopidas with the new Confederacy seems to originate with a tradition which credited important achievements either to both men or to Epameinondas only. Pausanias links the founding of the capital of the Arkadian Megalopolis—for which ancient sources offer various dates from 371/0 to 368/7 (see Roy, 1971: 569-599, esp. 591)—with Epameinondas (9.14.4); Diodoros does not make this connection; see J.A.O. Larsen, 1968: 186 and n. 1, E. Lanzilotta, "La fondazione di Megalopoli", *RSA* 5 (1975) 25-46, Buckler, 1980a: 70-72, 107-108, McKechnie, 1989: 47-48, Sordi, 1989: 129-130.

τὴν δὲ Μεσσηνίαν... συνοικίσαντες: perhaps the most important stage of Epameinondas' campaign to the Peloponnese was the re-establishment of the independence of Messenia and the construction of a new city along the slopes of Mt. Ithome in the plain of Pamisos; by liberating a city which for centuries had been subject to the Spartans, and populating it with helots and Messenian people summoned from Italy, Sicily, Libya and from wherever else they had settled (Paus. 4.26.5), the Thebans were striking a strong blow at Sparta's economic and military system. Xenophon omits all reference to both this event and Epameinondas' role in it, whom he only introduces on the eve of his third invasion of the Peloponnese; he alludes, however, to Messenia's independence in 7.1.27, 1.36; cf. Paus. 9.14.5, 15.6, Plut. *Pel.* 30.7, *Ages.* 34.1, *De laud. ips.* 540E, Isokr. 5.49, 6.28, Lyk. *contra Leocr.* 62, Nep. *Epam.* 8.5, Ael. *V.H.* 13.42. Diodoros ascribes Epameinondas' restoration of Messenian independence and the foundation of a new and independent capital to his desire for everlasting fame (15.66.1) rather than to his strategic prescience of the effects this would have on Sparta; on the different sources regarding the founding of Messenia, see Roebuck, 1941: 31-34, L. Pearson, *The Lost Histories of Alexander the Great*, in Philological Monographs publ. by the American Philological Association, vol. 20, 1960, 30n. 43, Oliva, 1971: 195-196, Dipersia, 1974: 54-61, Buckler, 1980a: 84-87, 268-269, Asheri, 1983: 27-42, esp. 36-39, Sordi, 1989: 128-130. For a discussion of the liberation of Messenia by the Thebans as a theme of fourth century Theban propaganda, see W. Den Boer, "Political Propaganda in Greek Chronology", *Historia* 5 (1956) 162-177, esp. 169-170.

24.10 ἀπιόντες... διὰ Κεγχρεῶν: P. is more informative on reasons for the Theban withdrawal from Lakonia in *Ages.* 32.13-14; cf. Xen. *Hell.* 6.5.50-51.

'Αθηναίους... πορείαν: the liberation of Messene by the Thebans made the Spartans turn to Athens for help, since, according to the Common Peace, the Athenians had sworn to help any participant of the peace that suffered attack; on the Spartan embassy, cf. Xen. *Hell.* 6.5.33-39, Kallisthenes, *FGrH* 124F8, Arist. *NE* 4.3.25, Isokr. 5.44, 159, 7.7, 69, 8.105, Aeschin. 2.164, Dem. 16.12, 18.98, 19.75, Deinarch. 1.76. The Athenians voted aid for the Spartans and sent Iphikrates to Corinth (Xen. *Hell.* 6.5.49-52, Diod. 15.63.2). Xenophon does not speak of a Theban victory over the Athenians, but finds fault with Iphikrates' strategy to leave unguarded the pass through Kenchreae, which let the Thebans return to Boiotia unobstructed (*Hell.* 6.5.51); see Sealey, 1956: 178-203, esp. 193, Buckler, 1980a: 87-89.

Ch. 25: Pelopidas' and Epameinondas' Trial

25.1 'Επὶ δὲ τούτοις... αὐτοῖς: P. again takes the opportunity to comment on envy (cf. 4.3-4), here as inherent among the fellow-citizens of oustanding individuals (cf. *Marc.* 22.1). The envy which is aroused by the eminence and success of another individual and is directed at the honors he has received is probably the most common variety described in the *Lives* (*Per.* 32.6, *Them.* 22.1, 5, *Caes.* 69.1, *Cat.Ma.* 16.4, *Dion* 11.7, *Fab.* 23.4, 25.2, *Praec. ger. reip.* 805C, 811B, *De inv. et od.*, *De laud. ips.*). P. criticizes the Thebans for their envy of the valor and good fortune of Pelopidas and Epameinondas and their unfitting reception of the two men by subtly differentiating them (ὁ δὲ συγγενὴς καὶ πολιτικὸς φθόνος; cf. Diod. 15.72.2) from the other Greeks (οἱ μὲν ἄλλοι). For a treatment of φθόνος in P., see note on 4.2-4, Wardman, 1974: 69-78, Lührmann, 1978: 363-366, Walcot, 1978: 52-76, Barigazzi, "Un esercizio di scuola: distinzione fra invidia e odio", in *Studi su Plutarco*, Università degli Studi di Firenze, Dipartimento di Scienze dell'Antichità (Florence 1994) 57-73; and more generally, E. Milobenski, *Der Neid in der griechischen Philosophie* (Wiesbaden 1964) esp. 157-165.

25.2-3 θανάτου γὰρ... ἀπελύθησαν: upon their return from the first invasion of Lakonia in Spring 369 Pelopidas and Epaminondas were brought to trial on the charge of having added four months to their terms as boeotarchs. According to federal law, the boeotarchs were required, under penalty of death, to lay down their authority by the first of Boukatios, which was the first month after the winter solstice (24.2). P. seems to have known of one tradition of the trial of Pelopidas and Epameinondas, which he relates differently in different contexts in his *Moralia* (*De laud. ips.* 540D, *Reg. et imp. apoph.* 194A-C, *Praec. ger. reip.* 817F). In each case he adds or suppresses details to suit his narrative purposes (Buckler, 1978a: 36-43). The most representative instance of his twisting of the facts, however, is his account of the trial in *De laud. ips.* 540D: P. says that the Thebans came near condemning Pelopidas; however, he gained his acquittal by weeping and throwing himself on the mercy of the court. Epameinondas, on the contrary, dwellt on his achievements, and told the judges that he was unwilling to die if they would admit that he had founded Messene, ravaged Lakonia and organized Arkadia, all against the will of the Boiotians. The judges did not cast their votes, but left the court laughing. It is obvious that the version of the boeotarchs' trial in the *Pelopidas* supports the characterization of Pelopidas as a man of virtue, while the version of the *De laud. ips.* 540D shows a sharp contrast between the fortitude of Epameinondas and the pusillanimous behavior of his friend Pelopidas. Furthermore, the choice of details in the *Pelopidas* is at least partly determined by the need to maintain a balance with the portrayal of the Theban's Roman counterpart. On P.'s tendency to avoid any comments on Pelopidas' personality that would damage him in the comparison with Marcellus, see Georgiadou, 1992a: 4222-4257, esp. 4233-4235. For ancient sources on the trial, see Paus. 9.14.5-7, Nep. *Epam.* 7.3-8, App. *Syr.* 41.215-218, Ael. *V.H.* 13.42; for discussion, see Cary, 1924: 182-184, Westlake, 1939: 11-22, esp. 13-14, Bersanetti, 1949: 43-101, esp. 59-62, Cuff, 1954: 259-264, Wiseman, 1969: 177-199, esp. 189-190, Beister, 1970: 75-105, Cawkwell, 1972: 254-278, esp. 276-278, Buckler, 1980a: 138-142. Epameinondas' trial was used as a *corpus vile* for argumentation in the schools, to judge from Cicero, *De Invent.* 1.55-56, 69-70.

ὃν Βουκάτιον ὀνομάζουσι: the first month of the Boiotian calendar year (*IG* VII 3172, *SEG* XIX 335); it may have been named after a festival Boukatia (Schachter, 1981: 137). The *Pelopidas* is the principle text which provides information on the first month of the Boiotian year (cf. also 24.2), and determines its relation to the solar year (Roesch, 1982: 33-36; for a list of the Boiotian months, *ib.* 33-48), and for discussion, see Samuel, 1972: 66-70.

ἀμφότεροι δ' ἀπελύθησαν: P. simply states that both Pelopidas and Epameinondas were acquitted; he omits, however, all unflattering information regarding Pelopidas' conduct at the trial, which he includes in *De laud. ips.* 540D; see Buckler, 1978a: 36-43. Furthermore, there is no reference to the other boeotarchs being brought to trial in the *Life*; this is mentioned in *Reg. et imp. apoph.* 194A.

25.4 τὸ δὲ συκοφάντημα... ποιούμενος: P.'s description of Epameinondas' dignified attitude to the calumnies of his fellow-citizens echoes Nepos' characterization in *Epam.* 7.1. Epameinondas' πρᾳότης is one of the qualities which both Philopoemen (*Philop.* 3.1) and Philip II (26.8) tried to imitate unsuccessfully; it can be parallelled to Marcellus' forbearance, when he was accused by the Syracusans for being responsible for their sufferings (23.8-9). On P.'s use of πρᾳότης, see [**2e**], H. Martin, 1960: 65-73, Aalders, 1982: 46.

Πελοπίδας... αἰτίας: P. offers inadequate justification of Pelopidas' vengeful conduct in the subsequent trial (25.6). Buckler (1978a: 41) rightly suggests that what must have prompted P. to insert the second trial was the desire to show Pelopidas at his best by using the defects of Menekleidas as a foil for the noble traits of his hero; by doing this P. made up for his silence on Pelopidas' cowardly deportment during the first trial (*De laud. ips.* 540D; see earlier note on 25.2-3).

φύσει θυμοειδέστερος... ὤν: fiery temper and inability to control it were the traits above all in Pelopidas' character which brought about his death (32.9, *synkr.* 33(3).5); see further [**2e**].

25.5-6 Μενεκλείδας... πολὺν χρόνον: P. seems to have known of a second trial of Epameinondas to which he refers briefly here and apparently at 28.1. It seems to have taken place in 369 upon Epameinondas' return from the second invasion of the Peloponnese and, according to Diodoros—the only other source which alludes to this incident (15.72.1-2)—was provoked by the truce that he had made with the Spartan polemarch on Mt. Oneion (Xen. *Hell.* 7.1.17). Although no ancient source links Menekleidas with this episode, it is safe to assume that he was responsible for Epameinondas' prosecution. Nepos (*Epam.* 5) lays special emphasis on Menekleidas' animosity towards the Theban and his vituperative remarks on both his military policy and his personal life (note P.'s mention of Menekleidas' long-continuing attacks on both Thebans, 25.5). On the historicity of the second trial, see discussion by Bersanetti, 1949: 43-101, esp. 93-101, Weisman, 1969: 177-199, esp. 186-192, Beister, 1970: 105-110 (cf. Roesch's review, 1971: 176-177), Buckler, 1980a: 142-145.

Μενεκλείδας... κρείττονας: it is nowhere explicitly attested that Menekleidas was opposed to Epameinondas' invasion of the Peloponnese, but it may be deduced from Nepos' references to him (Nep. *Epam.* 5.) that he was among those Thebans who were not in favor of it (see note on 24.1). P.'s emphasis on the former's oratorical prowess (cf. Nep. *Epam.* 5.2) helps to justify the impact his slanderous speeches had on the people and Epameinondas' deposition from office. On Menekleidas, see Cawkwell, 1972: 254-278, esp. 266-267, Buckler, 1980a: 142-150. On his slanderous inclinations and envious deportment, see *Praec. ger. reip.* 805C, 811C, Nep. *Epam.* 5.2.

συνελθόντων: the reading συνεισελθόντων is given by C, and is commonly used by P.; cf. *Dem.* 7.1, Dion 54.1, *Pomp.* 29.3, *Phok.* 22.4.

ἐχρῆτο τῇ φύσει πρὸς τὸ συκοφαντεῖν: P.'s language suggests that Menekleidas was a professional prosecutor; see Bonner and Smith, 1945: 11-23, esp. 20n. 79.

μετὰ ⟨τὴν⟩ δίκην ἐκείνην: sc. the first trial, after Epameinondas' first invasion of the Peloponnese (25.3).

'Επαμεινώνδαν... βοιωταρχίας: "he prevented Epameinondas from holding the Boeotarchy"; this could only be after Epameinondas' second invasion of the Peloponnesos (369/8), when he was charged with treason (cf. Diod. 15.72.2). He must have been acquitted, otherwise it is hard to justify a punishment as lenient as deposition from office. In his attempt to be re-elected boeotarch in 368, Epameinondas was apparently defeated (cf. δι' ὀργήν τινα πρὸς τὸν 'Επαμεινώνδαν, 28.1), which explains why he was a private citizen in 368, when he was sent off to Thessaly to liberate Pelopidas and Ismenias. Menekleidas' successful attempt to exclude Epameinondas from the Boeotarchy apparently belongs to this year. Buckler suggests that Epameinondas' deposition from office referred to by Diodoros (15.72.2) "was probably nothing more than the usual deposition of a magistrate preparatory to his trial, the *apocheirotonia*" (1980a: 144).

κατεπολιτεύσατο: "reduced (sc. him) by policy"; the verb signifies political rather than legal tactics to overcome an opponent (Buckler, 1980a: 305n. 69); cf. also Plut. *Pomp*. 51.1, *Galba* 20.7.

25.7-8 καὶ κοινήν τινα... κακίους: "and since it is a solace to the envious to make those men, whom they themselves cannot excel, appear worse than others"; ὧν ἂν αὐτοί: the order of ὧν ἄν, very common in P., is given by QL[2]; ἂν ὧν is suggested by Stephanus and edd.; cf. *De tranq. an.* 466A.

πολὺς ἦν... ποιῆσαι: probably the ideal period for Menekleidas to persuade the people of the extraordinariness of Charon's achievements (on Charon, see notes on 7.4, 15.6) and to dedicate a painting to him was during the early part of 367, when Pelopidas was still a hostage in Thessaly. At that time, Menekleidas could have had his proposal accepted with little opposition. Pelopidas' indictment is assigned to 366, after his liberation by Epameinondas; see also note on 25.13.

πολὺς ἦν... αὔξων: "he vigorously enlarged upon his (sc. Charon's) exploits"; on this meaning of πολύς, see Kühner-Gerth, vol. 2.2, p. 58 (ch. 482.8).

τῆς δὲ πρὸς Πλαταιὰς... Χάρωνος: on Charon's successful cavalry attack on Plataia, see note on 15.6.

25.9-10 'Ανδροκύδης... κατέσχον: with the exception of outstanding national events, paintings of historical rather than mythological subjects, T. Hölscher observes (*Griechische Historienbilder des 5 und 4 Jahrhunderts v. Chr.*, Beiträge zur Archäologie, No. 6 [Würzburg 1973] 112-121), are rare in the Greek artistic tradition; see further, Borthwick, 1976: 148-151, esp. 150-151.

'Ανδροκύδης ὁ Κυζικηνός: a contemporary and emulator of Zeuxis and Parrhasios (first half of the 4th cent. BC); cf. Athen. 341a, Plut. *Quaest. conv.* 665D, 668C, Plin. *HN* 35.64; see also, Rossbach, *RE* s.v. "Androkydes, 3", 1.2, 1894: col. 2150, Hölscher, (*ib.*, see above,

113-115), A.J. Reinach, (*Recueil Milliet: textes grecs et latins relatifs à l'histoire de la peinture ancienne* [Paris 1985] 242-244) and bibliography in J.-M. Croisille, *Pline l'Ancien, Histoire Naturelle, Livre 35* (Paris 1985) 184n. 5.

25.11 ἔπειθεν: conative imperfect; "he was trying to convince them", but, judging from the rest of the passage, his efforts did not fare well. Ziegler's suggestion of the aorist form ἔπεισεν would not convey the idea of Menekleidas' unwavering efforts to convince the Thebans; the conative imperfect also balances Pelopidas' ἐγκωμιάζων... διετέλεσε... ἐξελέγχων, in 25.14.

25.12 ἦν δ' ἀβέλτερος... ἀγαπωμένης: an interjection of authorial opinion. P. probably alludes here to the preceding victories at Tegyra, Leuktra, and Epameinondas' and Pelopidas' invasion of the Peloponnese.

φιλοτιμία: on the meanings of φιλοτιμία in P. see Frazier, 1988b: 109-127, and bibliography in Scardigli, 1995: 9n. 60; for a definition by P., cf. *Praec. ger. reip.* 819F.

25.13 τοῦτο τὸ ψήφισμα... παρανόμων: Pelopidas challenged the proposed decree as illegal and brought a γραφὴ παρανόμων—a process of impeachment for unconstitutional proposals—against Menekleidas claiming that it was not a Theban custom to honor a single individual for a victory but for the whole country to have the honor (see Bonner and Smith, 1945: 11-23, esp. 19-20); generally on the γραφὴ παρανόμων, see Bonner and Smith, 1930, vol. 1: 170-173, 225, 264-267, Hignett, 1952: 210-213, A.R.W. Harrison, 1971, vol. 2: 172-176, MacDowell, 1978: 50-52. Buckler has proposed the year 366 as the date of Pelopidas' indictment (1980a: 146-147; see also note on 25.7-8). The historical present (γράφεται) is common in referring to decrees: *Them.* 10.4, *Per.* 17.1, 30.2, *Arist.* 22.1, *Pomp.* 30.1.

ἰσχυριζόμενος... σῴζειν: P. has Pelopidas advocate honoring the state rather than the individual—a view which he substantially shares (cf. *Praec. ger. reip.* 820B-F)—, although elsewhere he portrays Pelopidas' victories as clearly individualised achievements (13.7, 30.2-3, *synkr.* 31(1).6, 32(2).2). On the custom of connecting great events and achievements with the names of individuals rather than with the state, see Dem. contra *Aristocr.* 198, Aeschin. *contra Ctesiph.* 186.

τῇ πατρίδι κοινῶς: P.'s emphasis on the communal appropriation of victories (cf. also 4.3-4, 19.5, 20.3, [**2e**]) creates a sharp contrast with the Roman celebration of the triumph described at length in the parallel *Life* (8, 21-22), which magnified the role of the individual general.

25.15 καλὸν πέπρακται ⟨**⟩ †ὃ μὴ Μενεκλείδαν ζημιῶσαι:** a possible emendation could be: ⟨, ἕως ἔπεισε τὸν⟩ Μενεκλείδαν ζημιῶσαι ("until he persuaded [the Thebans] to penalize Menekleidas"), or perhaps with the use of a continuative participle: ⟨πείσας τὸν⟩: "and he persuaded".

ἃ μὴ δυνάμενος... πολιτείαν: Menekleidas lost the γραφὴ παρανόμων brought against him by Pelopidas and was charged a heavy fine; his failure to pay it apparently resulted in ἀτιμία, the loss of his civil rights, which is the usual punishment according to Athenian law (A.R.W. Harrison, 1971, vol. 2: 169-176). He was probably sent into exile, since P. mentions that he tried to effect a rebellion, perhaps the one mentioned by Diodoros (15.79.3), in 364; see Buckler, 1980a: 147-148, 306n. 74.

καὶ τοῦ βίου ⟨**⟩ ἀποθεώρησιν:** ⟨τοῦ ἀνδρὸς⟩ is proposed by Hermann; his emendation finds support by analogy with *Publ.* 10.9, *synkr. Per./Fab.* 28(1).1; for the phrase τοῦ βίου ἀποθεώρησις, see *Lyk.* 4.4, *synkr. Agis-Kleom./Gracch.* 41(1).1, *synkr. Ages./Pomp.* 84(4).11.

Ch. 26: Pelopidas' activities in the north: Thessaly and Macedonia (369)

In the summer of 369 the Argives, Arkadians and Eleians appealed again to the Thebans for help. The Thebans responded positively to this request and undertook a second expedition in the Peloponnese under Epameinondas (Xen. *Hell.* 7.1.15-22, Diod. 15.68.1-70.1, Paus. 9.15.4); this enabled Epameinondas to further his work of the previous invasion; see Buckler, 1980a: 90-102, 233-242, C.D. Hamilton, 1991: 231-234.

During the same period (according to Diodoros, 15.71.2, after Epameinondas' second expedition) the Thebans, now unhampered by the presence of Jason of Pherai, the tagos of Thessaly, started expanding their influence northwards using as a pretext an appeal by the Thessalian cities against Jason's successor, Alexander of Pherai (Diod. 15.67.3-4). On the dating of Pelopidas' first intervention in the north in 369, see Westlake, 1935: 129, Sordi, 1958: 204, Hammond and Griffith, 1979, vol. 2: 180-185, Buckler, *ib.* 245-247, Hammond, 1989: 87, Roy, 1971: 569-599, esp. 591, id. in *CAH*, 1994: 194-195. On Xenophon's omission of the Theban activities in the north, see Buckler, 1980a: 263-268, esp. 267, Tuplin, 1993: 149.

26.1 Ἐπεὶ δ' Ἀλεξάνδρου... νομίζων: Alexander of Pherai's seizure of power and his war with the Thessalian cities led to Pelopidas' first expedition to Thessaly. The accounts of P. and Diodoros regarding Pelopidas' activities in the north are noticeably similar, and yet one cannot fail to notice some significant divergences between the two. Diodoros points to the expediency of the Theban intervention (15.67.3; cf. also Nep. *Pel.* 5.1), whereas P. presents it as motivated by Pelopidas' eagerness for activity (26.1), administering justice and helping the wronged parties (26.4-5). It has been proposed (Sordi, 1958: 193-202, Fuscagni, 1975: 31-55, esp. 37-41, 47-52) that P.'s portrayal of both Pelopidas and Epameinondas as defenders of justice (26.5), promoters of autonomy (30.7), and liberators of the oppressed people in Greece (31.6) were motivated by the ideological coloring of his main source, Kallisthenes; that the external policy of the Thebans during the period of the Theban hegemony, invested, as it is described in the *Pelopidas*, with panhellenic ideals under the leadership of the two Thebans, was used later on as a model by Philip II in devising his involvement in Thessaly, in shaping his policy in central and southern Greece, and in dealing with Artaxerxes (cf. also Momigliano, 1934: 196). Kallisthenes, it has been argued, was the first historian to give a panhellenic tone to the external policy of the Thebans in order to adjust it to the demands of Macedonian policy in the period around 343. Despite the validity of certain aspects of this theory, P.'s own creativity, personal judgment and preconceptions in composing the *Life* are overlooked; see subsection **d** in [**2c**], and Georgiadou, in *Boeotia Antiqua*.

Ἀλεξάνδρου τοῦ Φερῶν τυράννου: he was the nephew of Jason of Pherai, who was tagos of Thessaly. After the latter's murder (Ephoros: *FGrH* 70F214 [Diod. 15.60.5]) his brothers Polyphron and Polydoros succeeded him to the ταγεία. Polydoros was probably the victim of fratricide. Polyphron was, in his turn, murdered by Alexander of Pherai (29.8), who posed as avenger of Polydoros and destroyer of the tyranny (Xen. *Hell.* 6.4.33-34, Diod. 15.61.2); see further, Kaerst, *RE* s.v. "Alexandros von Pherai", 1.2, 1894: cols. 1408-1409, Westlake, 1935: 126-132, Berve, 1967, vol. 1: 285-293, Buckler, 1980a: 67, 110-111, 245, Gehrke, 1985: 189-194, and note on 29.8.

ἐπρέσβευσαν... δύναμιν: P. does not provide a comprehensive narrative of Pelopidas' intervention in the Thessalian affairs, but contents himself with incidents that are directly related to his activities; he omits therefore any reference to the Thessalians' initial appeal for help to King Alexander of Macedonia which is related by Diodoros (15.61.3).

αὐτὸς ἑαυτὸν... νομίζων: P. avoids looking for any political motives for Pelopidas' intervention to Thessaly and Macedonia (cf. Diod. 15.67.3, Nep. *Pel.* 5.1, Isokr. *Phil.* 53). His statement that "wherever Epameinondas was there was no need of another general" may suggest that Pelopidas' abstention from Epameinondas' second expedition to the Peloponnese (on which, see Buckler, 1980a: 90-102) may be partly due to his rivalry with his friend, who was the acknowledged leader of the Peloponnesian alliance. Any further speculation on Pelopidas' motives might undermine the picture P. has built up of their amicable and enduring comradeship (4, 20.3, 24.3); see Buckler, *ib.*: 111-112.

26.2 ὡς οὖν ἐστράτευσεν... παρέλαβε: in his convincing reconstruction of Pelopidas' activities in Thessaly Buckler observes (1980a: 245-247) that when Pelopidas arrived at Larisa he took over the city from the Macedonian garrison. He notes that P. "often uses παραλαμβάνω to indicate a peaceful change of authority over a place or object" (246); he also suggests that since King Alexander II was feeling insecure with his political opponents in Macedonia, such as Ptolemy of Aloros, and could not afford to prolong his stay in Thessaly and keep his garrisons in the Thessalian cities any longer, he must have made an agreement with the Thebans to hold Larisa and Krannon until they arrived (see also note on 26.4). Buckler's solution accounts for both Diodoros' notice of the second Thessalian embassy (15.67.3) and Pelopidas' immediate take over of Larisa without using force. It has been argued (Fuscagni, 1975: 49-51; cf. Prandi, 1985: 71) that P.'s incomplete description of Pelopidas' expedition to Macedonia and Thessaly in 369 and his silence on Pelopidas' alleged military encounter with the Macedonian king in Larisa are to be attributed to his source, Kallisthenes; he, it is argued, purposefully left this incident out of his account lest it damage the image of the Macedonians as liberators of Greek cities, and expose their military inferiority vis-à-vis the Thebans. P.'s possible manipulation of his sources is again overlooked. His omission of the Macedonians' presence could reflect a desire to magnify Pelopidas' role in the management of Thessalian affairs; see [**2c.d**], and Georgiadou, in *Boeotia Antiqua*.

καὶ τὸν 'Αλέξανδρον... δεόμενον: in an effort to be on good terms with the Thebans, Alexander probably asked for a treaty hoping to renew the agreement which they had had with Jason (Xen. *Hell.* 6.1.10, 6.4.20-24) and to be recognized as the tagos of Thessaly.

διαλλάττειν... νόμιμον: P.'s obscuring, for moralistic purposes, of Pelopidas' real *modus operandi* and the depiction of him as the tyrant's "educator" is unconvincing; it is reminiscent of Plato's desire to change Dionysios' despotic regime to a lawful rulership, (*Ep.* VII. 327d, 331e, 332c, 334c-d). The Theban is clearly not acting through altruism, as is suggested here, but is rather concerned about Alexander's ambitious plans and his activity in Thessaly. Pelopidas doubtless sees him as an impediment to Boiotian ambitions beyond central Greece; see Georgiadou, 1992a: 4222-4257, esp. 4235-4238. Buckler comments (1980a: 299n. 10) that P.'s words (καὶ ποιεῖν ἐκ τυράννου πρᾷον ἄρχοντα τοῖς Θεσσαλοῖς καὶ νόμιμον) suggest that Pelopidas was prepared to recognize Alexander's authority as a constitutional ruler; cf. also B. Niese, "Chronologische und historische Beiträge zur griechischen Geschichte der Jahre 370-364 v. Chr.", *Hermes* 39 (1904) 84-132, esp. 116; for a contrary view, see Westlake, 1935: 133. The Thessalians, however, were not prepared to accept such a compromise with Alexander, and his claim to command the Thessalian Confederacy was denied.

Πρᾳότης is often associated in P. with legality and contrasted with cruelty, violence and tyranny; cf. *synkr. Lys./Su.* 40(2).1, *synkr. Dion/Brut.* 55(2).2, *Phok.* 29.5, *Pyrrh.* 23.3; see further, H. Martin, 1960: 65-73, esp. 69-70, Wardman, 1974: 49-57, Romilly, 1979.

26.3 ὡς δ' ἦν ἀνήκεστος... κατηγορεῖτο: Alexander "was being denounced" presumably by the Thessalians. P. stresses the tyrant's unscrupulous and gruesome tyranny (cf. also Xen. *Hell.* 6.4.35, Diod.

15.61.2) in order to contrast Pelopidas' democratic and peaceful means of arbitration (26.4). Further description of the tyrant's character (27.6, 28.3, 9, 29.5-11) gives P. ample opportunity for similar contrasts and for justifying his murder at the hands of his wife (35.4-12). Alexander's ὠμότης and ἀσέλγεια are elucidated at 28.9 and 29.5-8.

ἀνήκεστος καὶ θηριώδης: "incurably brutal"; hendiadys.

τραχυνομένου... δορυφόρων: seeing that there is no hope of a settlement acceptable to himself, Alexander withdrew from Larisa.

26.4 ὁ δὲ Πελοπίδας... ὁμόνοιαν: Pelopidas' activity in Thessaly after Alexander's withdrawal cannot be ascertained because of the compressed accounts of both P. and Diodoros. There seems to have been a certain amount of fighting against Alexander (Front. *Strateg.* 3.8.2 [Polyain. 2.4.1], Front. *Strateg.* 1.5.2 [Polyain. 2.4.2], Polyb. 38.6.2) probably belonging to Pelopidas' first campaign in Thessaly (Westlake, 1935: 132, Buckler, 1980a: 247, 299n. 11; cf. Sordi, 1958: 210n. 2). P. suggests that Pelopidas made Alexander accept a peace treaty (cf. *FGrH* 115F409, on which see [**2c**] s.v. "Theopompos and Xenophon") by which the tyrant agreed to leave the Thessalian cities in peace. On the concept of ὁμόνοια in a political context, see the literature quoted in G.J.D. Aalders H. Wzn., *Die Theorie der gemischten Verfassung im Altertum* (Amsterdam 1968) 23n. 74, id. 1975: 29n. 102; see also A. Moulakis, *Homonoia. Eintracht und Entwicklung eines politischen Bewussteins* (Munich 1973).

αὐτὸς εἰς Μακεδονίαν... γενησόμενον: while in Thessaly Pelopidas was invited to Macedonia to act as an arbitrator between the two claimants to the throne, king Alexander II and his rival Ptolemaios of Aloros. Pelopidas' invitation was apparently related to the agreement that Alexander and himself are presumed to have made, when the latter was asked to intervene in Thessaly (see note on 26.2). Pelopidas established a formal alliance with king Alexander (Diod. 15.67.4) with which Ptolemy seems to have complied temporarily relinquishing his claim to the throne; see Buckler, 1980a: 118-119; for a different view, Hatzopoulos, 1985: 247-257, esp. 252 and n. 55.

ὡς διαλλακτὴν... ἀδικεῖσθαι: emphasis is imparted by the accumulation of noble roles assigned to Pelopidas (on which, see Fuscagni, 1975: 31-55, esp. 49-50) akin to those mentioned at 26.5, 30.7, 31.6;

see also note on 26.1. P. avoids giving any evidence to substantiate his characterizations. He does not comment on the real political advantages which Thebes would gain from Pelopidas' interference in Macedonia. His mission appears as if it was determined by moral imperatives rather than by considerations of political advantage.

26.5-6 ὅμηρον ἔλαβε... δίαιταν εἶχεν: as a guarantee of his compliance with the alliance and the acceptance of his status as a subject-ally, king Alexander had to give hostages, including his brother Philip, to Pelopidas. Diodoros concurs with P. at this point (15.67.4), but at 16.2.2-3 (cf. Just. 7.5.1-3), probably drawing on Demophilos, Ephoros' son, he reports that Philip was handed over to the Illyrians, who passed him on to Epameinondas' father. P. has Philip as a child under the tutelage of Pammenes with Epameinondas an adult; Diodoros, in his second passage, has Philip as a fellow-pupil of Epameinondas. Views vary regarding Diodoros' sources in Bk. XVI, on which see Momigliano, 1932: 523-543, Hammond, 1937: 79-91, esp. 80, Sordi, 1969: XI-XXXIII, 5. On the various accounts of Philip's stay at Thebes, see Bersanetti, 1949: 43-101, esp. 71n. 5, Aymard, 1954: 15-36, Sordi, 1975: 56-64, Hammond and Griffith, vol. 2, 1979: 205-206, Hatzopoulos, 1983; on the disputed chronology of this event, see Aymard, ib., Hatzopoulos, 1985: 246-257. Mention of Philip's stay at Thebes is also made by Dem. *De fals. legat*. 135, Ael. *V.H.* 13.7, Justin 6.9.7, 7.5.1-3.

26.7-8 ἐκ δὲ τούτου... μετέσχε: Philip apparently remained in Thebes long enough to witness Epameinondas' success on the battlefield and be inspired by his tactical skill. For a comparative study of the two men's military tactics, see J.C. Mitchell, 1980; ἐκ δὲ τούτου: causal.

ζηλωτὴς... Ἐπαμεινώνδου: Epameinondas is depicted by P. as a role-model also for Philopoemen (*Philop*. 3.1-2) and Timoleon (*Tim*. 36.1).

γεγονέναι <τισὶν> ἔδοξεν... κατανοήσασ<ιν>: Coraes eliminates the hiatus and changes κατανοήσας to dative plural; on this emendation, see Ziegler, 1934: 211-250, esp. 243-4, Kronenberg, 1937: 303-314, esp. 308.

<τισὶν>... κατανοήσασι<ιν>: P. is apparently hinting here at historians who wrote about Philip's career and discussed, among other topics, his borrowing of military ideas from Theban tactics. Theopompos is a likely candidate, although he started his *Philippika* with the year 360, nine years after Pelopidas' first intervention in Thessaly (369) and the beginning of Philip's sojourn at Thebes. He must have had various opportunities in his work to include information about the period during which Philip II was a hostage in Thebes. Theopompos may not have drawn a comparison between the two (cf. 26.8), but his denunciation of Philip on both moral—"maniacal by nature", "lacking in self-control"—and political—"inclined to rush headlong into danger"—grounds (Connor, 1967: 133-154, Shrimpton, 1991: 6-7, 122-180) could have inspired P. in his own comparison; on Theopompos as a possible source for the *Life*, see [**2c**], Hatzopoulos, 1985: 247-257, esp. 249.

Ephoros, or his son Demophilos, who wrote along his father's lines (Sordi, 1969: XIX-XXX; 5), is another possibility. The unconditional admiration for the Macedonian King's character—he is described as having excelled in ἀγχινοίᾳ στρατηγικῇ καὶ ἀνδρείᾳ καὶ λαμπρότητι ψυχῆς (16.1.6), and as being equal to Epameinondas in φύσιν τε καὶ φιλοπονίαν (16.2.3)—may have prompted P. to counteract what he regarded as an unwarranted evaluation of Philip's ethos. Kallisthenes, as P.'s source here, is proposed by Sordi, 1975: 56-64, esp. 57-61.

ἴσως: "perhaps"; the uncertainty implied by ἴσως may express P.'s reservations about the historians' understanding of military tactics; cf. Polybios, who criticizes not only Ephoros for his factual shortcomings in his description of battles (12.25f.2-5), but also Theopompos (12.25f.6); see Meister, 1975: 67-80.

Ch. 27: Pelopidas' second mission to Thessaly and his imprisonment by Alexander of Pherai (368)

27.1 Μετὰ δὲ ταῦτα... ἀναγκαζόμενος: in 368 the Thessalians accused Alexander of Pherai again of creating political perturbation within the Thessalian cities. P.'s words ὡς διαταράττοντα τὰς πόλεις seem to indicate that the nature of the problem was not military, since Pelopidas set out this time for Thessaly with Ismenias without taking any troops; they were simply sent as ambassadors (Nep. *Pel.* 5.1; cf. Diod. 15.71.2-3, Polyb. 8.35.7, Paus. 9.15.1) to settle the renewed disputes by diplomacy; on the second mission to Thessaly, see Westlake, 1935:

139-141, Bersanetti, 1949: 70-72, Sordi, 1958: 199-202, 209-212, Buckler, 1980a: 119-124.

'Ισμηνίου: probably the son of the homonymous statesman, who was put to death by the Spartans in 382 (see note on 5.1). Ismenias, Pelopidas' friend, should be distinguished from Ismenias the aulos-player, on whom see Stadter, 1989: 56-57. He was sent as ambassador along with Pelopidas to Sousa in 367 (*Artax.* 22.8-12); see Swoboda, *RE* s.v. "*Ismenias 2*", 9, 1916: col. 2139, Κουμανούδης, 1979, No. 1007, Buckler, 1980a: 135.

αὐτοῖς... ἀναγκαζόμενος: once in Thessaly and without any Boiotian troops Pelopidas may have realized that Alexander of Pherai was more threatening than he had expected and so, as a defensive measure, he began to recruit in Thessaly.

27.2-3 In the meantime the situation in Macedonia had been complicated by the assassination of king Alexander II. Ptolemy of Aloros had the king murdered and ruled as regent for Perdikkas, Alexander's brother (Diod. 15.71.1, Aesch. *De fals. legat.*, 26-29, Athen. 629d [*FGrH* 135-136F11]). Complaints against Ptolemy brought Pelopidas into Macedonia. The situation there, however, was further aggravated by another pretender to the throne, Pausanias. Ptolemy enlisted the support of the Athenian Iphikrates, who drove Pausanias out of Macedonia (Aesch. 2.26-29, Nep. *Iphicr.* 3.2). When Pelopidas arrived in Macedonia, apparently fearing that Athenian intervention in the north could thwart Theban ambitions, he hired mercenaries with which he marched against Ptolemy. On Pelopidas' second involvement with Macedonian affairs, see Cloché, *Histoire de la Macédoine* (Paris 1960) 123-124, Wiseman, 1969: 177-199, esp. 185-186, Cawkwell, 1978: 26, Hammond and Griffith, vol. 2, 1979: 182-184, Buckler, 1980a: 121-123, Hatzopoulos, 1983: 22-23, Hammond, 1989: 87, Hornblower, 1991: 227-228, Roy, 1994: 194-195. On mercenaries in general, see Parke, 1933, A. Mockler, *Mercenaries* (London 1970), L.P. Marinovic, *Le Mercenariat grec au IVe siècle avant notre ère et la crise de la polis*, Annales Littéraires de l'Université de Besançon 372, Centre des recherches d'histoire ancienne, vol. 80 (Paris 1988).

27.4 ὡς δ' ἐγγὺς... τῶν ἑταίρων: Ptolemy came to terms with Pelopidas and surrendered his son Philoxenos and fifty of his companions as hostages, much as Alexander II had done in the previous year (see note on 26.5). P. simplifies the historical record in his description of the circumstances which forced Prolemaios to negotiate with Pelopidas. Hornblower (1991: 227) rightly observes that it is improbable that Ptolemy made an alliance with Pelopidas because he was overawed by the Theban's reputation, as P. asserts; it is more likely that Athenian territorial ambitions in Macedonia posed a far greater threat to the Macedonians than Thebes' northern policy so far; see Momigliano, 1934: 34, Buckler, 1980a: 121-123, Hammond, 1989: 87-88.

27.5 αὐτὸς δὲ βαρέως φέρων... Φάρσαλον: Pelopidas, being indignant at the desertion of his mercenaries—they had been bribed by Ptolemy, 27.4)—marched to Pharsalos, where their families were then staying, with the intention of avenging their disloyalty.

27.6-7 Having recruited some Thessalians, Pelopidas marched to Pharsalos with Ismenias and was unexpectedly confronted by Alexander of Pherai. Disregarding the tyrant's perfidy (26.3) and naively relying on the truce which Pelopidas had formerly made with him (26.2-4; cf. Paus. 9.15.1, Nep. *Pel.* 5.1, *FGrH* 115F409) and on their status as ambassadors (τὸ περὶ αὐτοὺς ἀξίωμα), the two Thebans walked into his hands, and were imprisoned (*Reg. et imp. apoph.* 194D). P. stresses persistently Alexander's savagery and unscrupulous conduct in order to mitigate the two Thebans' (and especially Pelopidas') rashness and overconfidence. His silence on the repercussions that Pelopidas' miscalculated and ineffectual performance at Pharsalos had on Theban operations in Thessaly is in line with his usual method of relating only events which can be exploited in Pelopidas' favour (on which, see [**2a, d, e**], and Georgiadou, 1992a: 4222-4257; cf. Bersanetti, 1949: 43-101, esp. 72n. 3.). It has been suggested (Westlake, 1935: 141-142; cf. Hornblower, 1991: 226, 228) that Alexander's act of violence should not so much be attributed to his savagery, as P. asserts, as to his confidence in his improved relations with Athens (*IG* II^2 116, 39-40, *SEG* 21, No. 243, *SEG* 32, No. 62, Bengtson, 1975, No. 293, Xen. *Hell.* 7.1.28, Dem. *contra Aristocr.* 120, Plut. *Pel.* 31.6, Diod. 15.71.3).

οἱ περὶ τὸν Πελοπίδαν: on this construction, see note on 6.2.

ἐβάδιζον... προσδοκήσαντες: Pelopidas' imprudence in trusting the tyrant and thereby damaging both Thebes and his own reputation is sharply criticised by Polybios (8.35.6). The historian's phraseology partly echoes P.'s own (cf. also 31.6), and it has been suggested (Sordi, 1958: 197-200) that the source of these comments may have been Kallisthenes, who was occasionally critical of Pelopidas' conduct.

ἐξώλη μὲν ὄντα καὶ μιαιφόνον: P.'s portrayal of the tyrant's brutality and perfidy may have derived from a digression on Alexander in Theopompos' *Philippika*, on which see [**2c**] s.vv. "Theopompos and Xenophon".

27.7 τηλικαύτην: the pronoun preceded by an article is found in ABD: cf. also *Arat.* 33.4, *Reg. et imp. apoph.* 180A, *Quaest. conv.* 682B, 316C. It commonly appears, however, without an article.

Ch. 28: Pelopidas' captivity at Pherai

28.1 Thebes responded immediately to Alexander's imprisonment of Pelopidas and Ismenias and dispatched with all speed a force of eight thousand hoplites and six hundred cavalry (Diod. 15.71.3); the expedition was led to Thessaly by the boeotarchs Kleomenes and Hypatos with Epameinondas serving in the ranks as a hoplite (Paus. 9.15.1, Diod. 15.71.3-72.1, Nep. *Epam.* 7.1). Alexander in the meantime had made an alliance with Athens, on which, see note on 27.6-7. The Athenians sent thirty ships and a thousand men under Autokles to protect the tyrant against Boiotian reprisals (Diod. 15.71.3-4); see further, Buckler, 1980a: 125, Hornblower, 1991: 228.

δι' ὀργήν τινα πρὸς τὸν Ἐπαμεινώνδαν: Epameinondas probably stood for the Boeotarchy in 368 and failed to be re-elected, which explains why he was a private citizen in that year; see note on 25.5-6, Buckler, 1980a: 144-145.

28.2-10 P. interrupts his account of the expedition against Alexander following his favorite method of suspending the narrative when dramatic tension is heightened (cf. 16.5-8, 20.5-22, 34); he pauses in order to highlight Pelopidas' and Thebe's victimization at the hands of Alexander and to provide thereby an undeniably creditable motive for their unrelenting hatred of him and their desire to avenge his savagery. Pelopidas' discussions with Thebe may well be the product of P.'s own imagination; he is our only source for this incident. The visual impact of the scene, the strong emotional tone and directness of the dialogue, and the poignant terms which P. uses to describe both victims' pitiful situation and Alexander's licentiousness serve to justify, at least in part, Pelopidas' uncontrollable anger at him and his desire to have a final reckoning with the tyrant at all costs (31.5, 32.9); see Bersanetti, 1949: 73-74, and Georgiadou, 1992a: 4222-4257, esp. 4238-4242.

28.3 ὡς νῦν μάλιστα... δίκην: P. hints repeatedly at Alexander's upcoming punishment; he first presents it as a firm belief of Pelopidas—he had every reason to believe that he would not be left by his fellow-citizens abandoned in Alexander's hands—and later, when Epameinondas was sent out against the tyrant (29.3), as an equally firm conviction of the Thessalians.

πρὸς δ' αὐτὸν... διαφύγῃ: P. is far from criticizing Pelopidas for his provocative attitude towards Alexander; on the contrary, he portrays him as fearless and high-spirited earning even the admiration of the tyrant himself. Later he seems to contradict himself: when he relates the second Theban expedition to release Pelopidas he lays particular stress on Epameinondas' cautious avoidance of irritating the tyrant and on his determination to place Pelopidas' safety above his personal glory (29.4-5). P.'s change of perspective is alarming, and yet it is in line with his ability to define character traits according to the different moral purposes and emphases he has set up in a *Life* or in a Pair, on which, see Larmour, 1992: 4154-4200, esp. 4157-4174.

28.4 τί δὲ... θεομισὴς γενόμενος: the same anecdote in *Reg. et imp. apoph.* 194D; it foreshadows the tyrant's murder by Thebe's brothers at her instigation (35.4-12); θεομισής: to die at the hands of one's own wife seems to go beyond the realm of human revenge and is thus attributed to the gods.

σὺ τάχιον ἀπολῇ: a Homeric reminiscence; cf. Achilles' words to Hektor in *Il.* 20.429.

28.5 Θήβη: on Jason's daughter, see Konon, *FGrH* 26F1(L) s.v. Τισίφονος ἢ Θήβη, Theopompos, 115 F337, Stähelin, *RE* s.v. "Thebe, 4", 5A2, 1927: col. 1595, Stadter, 1965: 102 and n. 254.

28.7 ἦν γὰρ τῷ 'Ιάσονι... φίλος: Pelopidas' friendship with Jason (cf. also *Reg. et imp. apoph.* 194D) seems to have further motivated him to become involved with Thessalian affairs. Westlake (1935:

89) suggests that Jason's long standing friendship with Thebes is probably due to the fact that his wife was very likely a Theban—she retired to Thebes after her husband's death, Xen. *Hell.* 6.4.37—and that his daughter was named Thebe. On Jason, see Berve, 1967, vol. 1: 285-290, Mandel, 1980: 47-77, Shrimpton, 1991: 76-77, 157-159, 161, Tuplin, 1993: 117-121.

28.8 καὶ γὰρ ⟨καὶ⟩ ἐγώ σε: it should read καὶ γὰρ ἐγὼ σέ, ("and", "but" I pity *you*!") according to the codices, with the emphatic σέ being in a pivotal position; a similar emphasis is given in the version of the anecdote preserved in *Reg. et imp. apoph.* 194D. Ziegler's added καί after γάρ provides the wrong emphasis ("and I too pity you"); see Denniston 108.

ἄδετος: "free", as opposed to Pelopidas who was imprisoned. The word-play is more colourful in the anecdote preserved in *Reg. et imp. apoph.* 194D.

28.9 μετὰ τῆς ἄλλης ἀσελγείας... πεποιημένου: for different versions of this story, cf. Xenophon (Xen. *Hell.* 6.4.37) and Valerius Maximus (9.13.3). Theopompos may have been P.'s source for his account on Thebe, on which see [**2c**] s.vv. "Theopompos and Xenophon".

28.10 συνεχῶς φοιτῶσα: "she was frequently visiting".

Ch. 29: Epameinondas frees Pelopidas (367)

29.1 Ἐπεὶ δ' οἱ στρατηγοὶ... ἐζημίωσεν: the first Theban expedition to liberate Pelopidas and Ismenias failed. Having suffered a considerable loss to Alexander's cavalry and having encountered treachery among their Thessalian allies and, also lacking supplies, the two boeotarchs, Kleomenes and Hypatos, were forced to make an ignominious retreat. Upon their return to Thebes they were punished with heavy fines (Diod. 15.71.4-7, Paus. 9.15.1-2). In his more informative account of the Theban rescue-campaign Diodoros tells us that when the Thebans were reduced to utter helplessness by Alexander's cavalry attack they appointed Epameinondas as commander-in-chief; also that it was only by his appointment as general and his successful maneuver that the army escaped serious disaster and the Thebans were allowed to retreat without further incident (cf. *Quaest. conv.* 680B and Teodorsson ad loc., vol. 2, p. 196, *An sen. resp. ger.* 797A-B, Nepos, *Epam.* 7.1-3). Pausanias, on the other hand, in his compressed narrative confuses Epameinondas' two operations and connects Pelopidas' release with the first rescue-mission (9.15.1-2); see Westlake, 1935: 142-143, Sordi, 1958: 215-216, Buckler, 1980a: 125-126.

Ἐπαμεινώνδαν... ἀπέστειλεν: Epameinondas was elected boeotarch for the following year (367) and undertook a campaign to gain the prisoners' release; cf. also *Reg. et imp. apoph.* 194E, Diod. 15.75.2.

29.4-8 P. lays out all the gruesome information about the tyrant's atrocities in a long series of principal clauses rapidly alternating them with subordinate clauses. His emphasis on Epameinondas' cautiousness and clever maneuvering in dealing with Alexander reinforces the contrast between the two characters, while at the same time it points unobtrusively to Epameinondas' ability to control his emotions and to avoid taking unnecessary risks, unlike his friend Pelopidas; see [**2e**], and note on 28.3.

29.4 οὐ μὴν ἀλλ': "nevertheless"; common in P.; for this conjunction of particles, see Denniston, 28-29.

Ἐπαμεινώνδας... τιθέμενος: his priority is to secure Pelopidas' safety, as in 385, at the Battle of Mantineia, on which see note on 4.7.

ὥσπερ θηρίον: a common simile in P.; cf. *Demetr.* 48.1, *Caes.* 66.10, *De Alex. magn. fort.* 344A.

29.5 ἐπῃωρεῖτο... τύραννον: Polyainos' description of Epameinondas' stratagem at 2.3.13 probably refers to this Thessalian campaign.

29.6 πυνθανόμενος... χρώμενος: on Alexander's ferocity, cf. Xen. *Hell.* 6.4.35, Ael. *V.H.* 14.40.

ὡς ζῶντας... κατώρυττεν: the live burial which is reported here as part of Alexander's atrocious conduct provides a link with an exceptional Roman ritual described in the paired life *Marcellus*, which required the live burial of Greek and Gallic couples (*Marc.* 3.5-6; cf. *De superst.* 171D).

29.7 Μελιβοίᾳ δὲ... ἀπέσφαξε: both Skotoussa and Meliboea were near the borders of Pherai and were probably suspected of intriguing against Alexander. Complaints about the tyrant's excesses in both cities must have reached Thebes before Epameinondas' arrival in

Thessaly. For this reason and also because of Alexander's record of treachery and perfidy (26.3, 27.6-7), Epameinondas was obliged to move cautiously against him (cf. 29.5) lest he murder Pelopidas and Ismenias; on the incidents in Skotoussa and Meliboea, cf. Diod. 15.75.1, Paus. 6.5.2-3; see further, Westlake, 1935: 145, Bengtson, 1975, No. 281. On the truce between the two towns and Pherai, see Sordi, 1958: 213-214n. 11.

Μελιβοίᾳ: a large village of Magnesia; cf. Paus. 6.5.2, Theopompos *FGrH* 115F372; see further, Stählin, *RE* s.v. "Meliboea, 2", 15.1, 1931: cols. 511-514.

Σκοτούσσῃ: a Thessalian town between Pherai and Pharsalos; see Steph. Byz. s.v. Σκοτοῦσσα, Stählin, *RE*, s.v. "Σκότουσσα, 1", 3A1, 1927: cols. 613-617.

ἡβηδὸν ἀπέσφαξε: Pausanias' expression ὅσον ἐν ἡλικίᾳ κατεφόνευσε (6.5.2-3) corresponds closely to P.'s. P. uses the same phrase regarding Marcellus' reported slaughter of the inhabitants of Leontini (14.3).

29.8 τὴν δὲ λόγχην... καταστέψας: the decking of the spear with garlands is reminiscent of the Dionysian thyrsos, which was wreathed in ivy and vine-leaves. The incident appears to be connected with the anecdote about Alexander leaving the theater in the middle of the performance of Euripides' *Troades*, which follows immediately (29.9-10). The murder by implication and its celebration take on a dramatic aspect and P. may be presenting them as a tragicomic event. For honoring a weapon as a god, see Hdt. 4.62, Aesch. *Sept.* 529-532 (T.G. Tucker, *The Seven against Thebes of Aeschylus* [Cambridge 1908] 106, G.O. Hutchinson, *Aeschylus: Septem Contra Thebas* [Oxford 1985] 127), Apoll. Rh. 1.466-470 (also *schol. Apoll. Rh.* 1.57-64), Akousilaos of Argos *FGrH* 2F22, Paus. 9.40.11-12 (cf. Hom. *Il.* 2.101-108; Frazer, vol. 5, 1898: 210-212), Verg. *Aen.* 10.773-775, 12.95-100; see further, Nilsson, 1967: 208-209.

Πολύφρονα... ἀπέκτεινε: Polyphron and Polydoros, Jason's brothers, succeeded their brother as ταγοί of Thessaly (Xen. *Hell.* 6.4.33-34). Polyphron killed Polydoros and was himself slain by Alexander. Diodoros in his account reverses the fate of the two brothers, and has Polydoros instead of Polyphron being killed by his nephew

Alexander (15.61.2); see note on 26.1, U. Köhler, "Attische Psephismen aus der ersten Hälfte des vierten Jahrhunderts", *Mittheilungen des Deutschen Archäologischen Institutes in Athen*, 2 (1877) 197-213, esp. 198n. 1, Tuplin, 1993: 120.

Τύχωνα: on Τύχων, see Nilsson (1967, vol. 1: 209 and n. 4), who notes that Alexander's offering to Tychon of the spear, with which he killed his uncle Polyphron, appears to belong to a late tradition.

29.9-10 τραγῳδὸν... δακρύων: the anecdote here quoted is essentially the same as in Ael. *V.H.* 14.40, but the play performed is different. P. concludes the description of Alexander's personality with an episode that reveals an altogether different aspect of his character. The tyrant's embarrassment and expression of compassion for the human sufferings enacted on the stage come as a complete surprise to the reader, especially since they immediately follow the list of his savage crimes. P. does not provide here any explanation for the peculiar co-existence of opposed traits in the tyrant's character (note also P.'s consternation at the contradiction between the political idealism reflected in the tragedies of the tyrant Dionysios I and his brutality, *De Alex. fort.* 338B-C). Elsewhere he is striving to justify the paradoxical combination of divergent traits and observes that even men who commit monstrous deeds feel guilty for their crimes (*De ser. num.* 556D). On P.'s conception of character, see note on 31.2-3.

29.11 ἔπτηξ' ἀλέκτωρ δοῦλος ὣς κλίνας πτερόν: this iambic trimeter by Phrynichos is also cited in *Alk.* 4.3 and *Amat.* 762E. The ἀλεκτρυόνων ἀγῶνες constituted a popular way of entertainment among the Greeks; cf. Plat. *Leg.* 789b, Aesch. *Eum.* 866, Aristoph. *Eq.* 494, 946, *Ach.* 166, *Pax* 502, Ael. *V.H.* 2.28, Luc. *Anach.* 37, Pl. *HN* 10.25, Colum. *De re rust.* 8.2, Varr. *De re rust.* 3; on cock-fighting, see Daremberg and Saglio, vol. 1, s.v. "Alektryonon Agones"; E. Csapo, "Deep Ambivalence: Notes on a Greek Cockfight," (Part I and Parts II-IV) *Phoenix* 47.1 (1993) 1-28, 115-124.

29.12 ὁ δὲ συνθέσθαι... ἀνεχώρησε: Buckler (1980a: 127) suggests that the treaty of peace and friendship for which Alexander's envoys offered to exchange the freedom of Pelopidas and Ismenias may have also required Theban recognition of Alexander's tageia in Thessaly (*ib.*, 300n. 31). Epameinondas, however, refused to agree to more than a thirty days' truce. Not wanting to risk a fight with Thebes under the leadership of its most capable general, Alexander agreed to Epameinondas' terms and returned the prisoners. The truce seems to have left Alexander with his power, such as it was, intact. Buckler also notes that Epameinondas' offer of the truce was the easiest means available of achieving his goals without committing Thebes to anything permanent.

Ch. 30: Pelopidas ambassador at Sousa (367)

30.1 Οἱ δὲ Θηβαῖοι... δόξαν αὐτοῦ: after an unsuccessful peace conference at Delphi in 368, which strengthened Sparta's military position (Xen. *Hell.* 7.1.27), the Thebans, probably realizing that the only way of maintaining their gains after Leuktra would be with the support of the Persian King, sent ambassadors together with their Peloponnesian allies to Sousa in 367 seeking a Common Peace. Xenophon relates that the Theban delegation was sent on the pretext of the Spartan embassy to Artaxerxes (*Hell.* 7.1.33; cf. Plut. *Artax.* 22.8-9, Nep. *Pel.* 4.3). This incident, in fact, is the first time that he mentions Pelopidas. The only Spartan envoy explicitly mentioned in our sources is Euthykles (Xen. *Hell.* 7.1.33), who was already in Persia when the Theban embassy arrived (Niese, *RE* 6, 1907: col. 1506, D.J. Mosley, "Euthycles: One or Two Spartan Envoys?", *CR,* n.s. 22.2 [1972] 167-169, Poralla, 1985, No. 301; he is probably the same Euthykles as the one listed separately in No. 302). Against the possibility of Antalkidas being one of the Spartan ambassadors at that time, see Buckler, 1977a: 139-145. The Theban Ismenias, who is not mentioned here, was also sent along with Pelopidas (*Artax.* 22.8). On the dating and duration of Pelopidas' embassy, see Roy, 1971: 569-599, esp. 591-592. On the Peace Conference at Sousa, see Bersanetti, 1949: 43-101, esp. 74-77, Ryder, 1965: 79-81, 136, Seager, 1974: 36-63, esp. 59-61, Buckler, 1980a: 151-157, id. 1982b: 180-204, esp. 186-187, 200-201, Wickersham, 1994: 113-115.

30.2-4 πρῶτον μὲν... βουλόμενος δοκεῖν: P. does not report Pelopidas' speech, but devotes a considerable amount of space to a eulogy of his achievements: he relates that Pelopidas' presence at Sousa caused a sensation at the Persian court, because of the fame he had gained at Leuktra and his leading role in the first invasion of the Peloponnese. It is only through Xenophon (*Hell.* 7.1.34) that we have an idea about Pelopidas' speech, which, in a rather unexpectedly neutral fashion, outlines the reasons why Persia should support Thebes. Xenophon gives a summary of Theban deeds, beginning with their support for the Persians at Plataia in 479—Theban medism was clearly not a matter to be proud of, even if Pelopidas used it to the advantage of the Thebans—continuing with the Thebans' opposition to Spartan policy after the Peloponnesian war, and ending with their ravaging of Spartan territory (*ib.* 7.1.34-35); see Buckler, 1980a: 153-154, id. 1982b: 180-204, esp. 200-201, Bernini, 1991: 74-111, esp. 80-84.

30.2 ἀλλ' ὡς πρῶτος... κατορθώματος: "when the account of the battle at Leuktra was the first to be spread abroad, it was ever increased by a new success"; πρῶτος functions here as predicative, on which, see Smyth, 1042.

30.3 ὑπὸ τὸ Ταΰγετον καὶ τὸν Εὐρώταν: between Taygetos and Eurotas; ὑπό + accus. in mathematics means "inscribed in" (LSJ, s.v. ὑπό: C.2.b). The gender of Ταΰγετος varies in P., neuter in *Lyk.* 15.10, *Agis.* 8.1, *Apoph. Lac.* 228B, masculine in *De ex.* 601D.

Σούσων καὶ Ἐκβατάνων: the capitals of Persia and Media. Sousa was the old capital of Elam, which had been the royal seat of the Persian empire under the Achaemenids (Str. 15.3.2). Ekbatana is the modern Hamadān, the ancient capital and treasury of the Medes (cf. Str. 11.13.1, Hdt. 1.98) and the summer residence of the Persian kings (Xen. *Kyr.* 8.6.22); see Weissbach, *RE* s.v. "Ekbatana, 1", 5.2, 1905: cols. 2155-2158, Olmstead, 1948: 162-171.

30.4 ταῦτ' οὖν ὁ Ἀρταξέρξης... τιμαῖς: Xenophon lays equal stress on the special honors which Artaxerxes conferred on Pelopidas (*Hell.* 7.1.34-35); he also informs us that Timagoras, the Athenian ambassador, confirmed the veracity of Pelopidas' words and, as a result, stood second in honor to him; on καί marking a result, see Smyth, 2288 and 2874.

Ἀρταξέρξης: the eldest son of Darius II and Parysatis; he began his reign in 405/4 and remained on the throne until 359/8 BC; see P. Briant, *Histoire de l'Empire perse de Cyrus à Alexandre*, ser. Achaemenid history 10, Nederlands instituut voor het Nabije Oosten, ch. 15 (Leiden 1996); also P.'s *Life of Artaxerxes*.

30.5 τοὺς λόγους... ἐκείνῳ: P.'s tendency not to give enough political or historical background for the narrative to be fully intelligible is quite apparent. His evaluation of Pelopidas' proposals as "firmer than those of the Athenians, and more honest than those of the Lacedaemonians" adds little to our understanding of the particular circumstances which made his arguments more credible to the King. Artaxerxes' preference for the Theban proposals is not surprising in view of the undesirable Spartan-Athenian alliance. He discerned that Thebes was the most advantageous power to support and that he could use it to counterbalance and weaken the other two city-states; see Buckler, 1980a: 154-155.

No oratorical ability on Pelopidas' part is here implied, since his inadequacy in this area is elsewhere attested (*Praec. ger. reip.* 819C) and is also implicitly confirmed at 30.13. For an analysis of the term ἁπλοῦς in this chapter, see Bernini, 1991: 74-111.

ἁπλουστέρους: "more honest". P. probably alludes to Spartan psychology, which was to think one thing and say another; cf. also *synkr. Lys./Su.* 41(3).2, *Apoph. Lac.* 223A-B. The duplicity of the Spartans was proverbial; cf. Hdt. 9.54, Eur. *Andr.* 451-452, Aristoph. *Lys.* 1233-1235; see further A.S. Bradford, "The Duplicitous Spartan", in A. Powell and S. Hodkinson, eds., *The Shadow of Sparta* (London, New York 1994) 59-85.

30.6 καίτοι δοκεῖ... Λακεδαιμόνιον: in *Artax.* 22.1 P. reports, on the authority of D(e)inon (*FGrH* 690F19), that Artaxerxes regarded the Spartans as very insolent, but was especially charmed by Antalkidas when he received him in audience at Sousa in 387; on the personal influence of Antalkidas over Artaxerxes, see Wilcken, 1941: 3-20, esp. 13, C.D. Hamilton, 1979: 306-308. On the controversy about Antalkidas' presence at Sousa in 367, see Buckler, 1977a: 139-145.

ᾧ τὸν στέφανον... ἀπέστειλε: the anecdote is also found in *Quaest. conv.* 713E, where the bad taste of the Persian King is described in pungent terms; see also Athen. 31e, Ael. *V.H.* 14.39. K. Hubert ("Zur indirekten Überlieferung der Tischgespräche Plutarchs", *Hermes* 73 [1938] 307-328, esp. 320) argues that Aelian and P. used a common source here, presumably a collection of anecdotes.

ᾧ: emendation by Coraes followed by Kronenberg; ὅτι codd.; ᾧγε ("to whom in fact") suggested by Ziegler to add emphasis, but not incorporated in the text; for a similar combination, cf. *Fab.* 14.6, *Su.* 10.2, 12.3, *Crass.* 5.2, *synkr. Dion/Brut.* 54(1).3/*Brut.*, *Adv. Col.* 1115E.

30.7 τὰς ἀξιώσεις... βασιλέως: Artaxerxes assented to Pelopidas' proposals, which included the recognition of Messenian independence—recorded also by Diodoros in a retrospective reference to Pelopidas' achievements, 15.81.3 (cf. 15.90.2)—and of the autonomy of all Greek states. Xenophon adds another term of the peace which required the Athenians to haul up their ships (*Hell.* 7.1.36); this provision was apparently an effective way to check Athenian seapower and was in the interest of both the Thebans and Artaxerxes; see Carrata Thomes, 1952: 23-25, Ryder, 1957: 199-205, id. 1965: 80-81, 136, Barbera, 1969-1970: 460-474, Seager, 1974: 36-63, esp. 58-61, Bengtson, 1975, No. 282, id. 1988: 173-174, Sealey, 1976: 426-427, Buckler, 1980a: 155-156, Cartledge, 1987: 200, Gray, 1989: 63-64, Jehne, 1994: 82-86. For a discussion of Pelopidas' use of the term autonomy as a means of promoting the Thebans' immediate political objectives, see Bosworth, 1992b: 122-152, esp. 140-142. On Thebes' earlier abortive attempt to win general recognition of the independence of Messenia during a peace conference held at Delphi in 368 (Xen. *Hell.* 7.1.27), see Ryder, *ib.*: 79, 134-135, Roy, 1971: 569-599, esp. 576, Buckler, *ib.*: 102-103.

αὐτονόμους μὲν εἶναι τοὺς Ἕλληνας: on Pelopidas' role as a panhellenic champion of liberty and autonomy (cf. also Polyb. 8.35.6), see [**2c.d**], Sordi, 1958: 196-198, Fuscagni, 1975: 31-55, and Georgiadou, in *Boeotia Antiqua*.

30.8 ταύτας ἔχων τὰς ἀποκρίσεις... διέβαλε: the offering of gifts to ambassadors was part of the diplomatic usage at the courts of the kings of Persia and Macedon, and the acceptance of these customary gifts was also a sign of proper diplomatic relations (see Perlman, 1976: 223-233, esp. 228). P. highlights the contrast between Pelopidas' incorruptibility and the venality of the other ambassadors, especially Timagoras of Athens. The Theban's words (30.5) are thus presented as entirely matching the morality of his deeds; for later evidence on the

Thebans' rejection both of direct attempts at bribery and their refusal to accept gifts, see Dem. 19.139 and Perlman, 1976: 223-233, esp. 227.

ἀνέζευξεν: "he returned home"; uncommon meaning in P. and in general; cf. Plut. *Cor.* 31 and Wyttenbach, s.v. ἀναζευγνύω.

30.9-12 Τιμαγόραν... τοὺς 'Αθηναίους: this is the best documented case of charges of taking bribes by an ambassador to the Persian court; see Perlman, 1976: 223-233, esp. 228-229. A more elaborate version of this anecdote is given in *Artax.* 22.9-12. Athenaios, on the authority of Phanias (Müller, *FHG* 2.296), has Timagoras [the Cretan?; text corrupt] or Entimos from Gortyna receive Artaxerxes' honors (*Deipn.* 48d), instead of Timagoras the Athenian; the latter, Athenaios explicitly says, never enjoyed the same honor as Entimos (48e). There is no indication that P. also derived his anecdote about Timagoras from Phanias, nor is it certain that Athenaios himself made direct use of Phanias—he may have fabricated the information about Entimos, and simply attributed it to Phanias in order to give his account more credibility; on Athenaios' manipulation of his sources, see Sansone, 1988: 311-318, esp. 311. Given, however, the striking parallels in phraseology in their anecdotes about Timagoras and the fact that they both mention Phanias as their source in their reference to the honours offered to Themistokles by the Persian King (*Deipn.* 29f, Plut. *Them.* 29.11), it is likely that they acquired it from the same tradition.

30.9 Τιμαγόραν... ἀπέκτειναν: depending on the legal form of the accusation, the punishment for bribe-taking could range from mere confiscation of the bribe to execution. In Timagoras' case the accusation seems to have been an *eisangelia*, for which death was the penalty. On the various procedures used for prosecuting bribery and embezzlement, see Hansen, 1975: 16-17, 63-64, Perlman, 1976: 223-233, esp. 226, 228-229, MacDowell, 1978: 172-174, Rhodes, 1981: 662.

εἰ μὲν ἐπὶ τῷ πλήθει... δικαίως: P. seems to know of another version, most likely Xenophon's, regarding the reasons for Timagoras' punishment by the Athenians, to which he makes a distressingly oblique reference later in his account (30.12-13). Xenophon (*Hell.* 7.1.38; cf. Dem. 19.191) tells us that Timagoras was convicted and sentenced to death by the Athenians on the indictment of Leon, who was Timagoras' fellow-envoy, for having agreed with the Theban proposals (cf. 7.1.35) and for his refusal to share quarters with him. P. chose to elaborate on the version which facilitated better the drawing of the contrast between Pelopidas' and Timagoras' attitude toward Artaxerxes (30.9-11). On Timagoras and Leon, see P. Cloché, "Les hommes politiques et la justice populaire dans l'Athènes du IVe siècle", *Historia* 9 (1960) 80-95, esp. 85, Mosley, 1968: 157-160, id. 1972: 12.

30.10 στρώτας θεράποντας: the entry in the *Suda*, Mosley notes ("Timagoras' bed-makers", *CR*, n.s. 22 [1972] 12), is an abbreviated form of that given by P. with each item presented in the same order. The difference between the two lists is that the *Suda* mentions στρατιώτας θεράποντας, while P. gives στρώτας θεράποντας. Mosley rightly suggests that the text of the *Suda* is in error, because Athenaios says in 48d that the Persians instituted στρώτας and quotes the account of Phanias (see earlier note) that Timagoras received a silver-footed bedstead, and coverings with an attendant to spread them, since the Greeks did not know how to make a bed (Müller, *FHG* 2.296).

30.12 'Επικράτους γοῦν... δῆμος: Athenaios (*Deipn.* 251a-b) cites Hegesander, a writer of 'Υπομνήματα, as his authority for the same anecdote (Müller, *FHG* 4.414).

'Επικράτους... τοῦ σακεσφόρου: σακεσφόρου, instead of σκευοφόρου, is a correction suggested by Palmer on the basis of the scholia on Aristophanes' *Ekkl.* 71. Plato, the comic poet, is mentioned in the scholia as the source of an anecdote regarding Epikrates—the scholiast notes that Epikrates was called σακεσφόρος because of his long beard (see J. Taillardat, *Les Images d'Aristophane: Études de la langue et de style* [Paris 1965] 63 and n. 1)—and it is probable that the anecdote about him in the *Pelopidas* derived from the comic poet's work *Πρέσβεις* (cf. fr. 119 [Kock, 1, 633]), as Wilamowitz suggests (*Kleine Schriften*, vol. 3 [Berlin 1969] 45n. 1). Epikrates seems to be the same as Epikrates of Kephisia mentioned by the Oxyrhynchus historian (Bruce II.2; Chambers 34-35; Paus. 3.9.8) as having received Persian bribes from Timokrates together with Kephalos of Kollytos; he was condemned to death for misconduct as ambassador (Dem. 19.277); see *Etym. Magn.* 361.3, Harpokr. and Suda s.v. 'Επικράτης, Bruce, 1966: 272-281, esp. 273-276, id. 1967: 56-57, Davies, *APF*, 1971: 4859, 181, Hansen, 1975: 69-72, 87-88, Cook, 1981: 186n. 12, Strauss, 1985: 67-74, esp. 68-69, Osborne and Byrne (1994, s.v. Ἐπικράτης, 70). Edmonds suggests (*The Fragments of Attic Comedy,* vol. 1 [Leiden 1957] 527) that the embassy of Epikrates to Artaxerxes to which Plato Comicus, P. and Athenaios (6.251a-b) refer was the occasion of Lysias' speech *against Epikrates*; he is refuted by Bruce, 1966: 272-281, esp. 276.

30.13 P.'s explanation of Pelopidas' success at Sousa is superficial. He attributes it solely to his reputation and his recent successes, omitting any reference to the support he received from the Athenian ambassador Timagoras (Xen. *Hell.* 7.1.35-38).

Chs. 31-32: Pelopidas' fatal expedition to Thessaly and the Battle of Kynoskephalai (364)

31.1 ἡ μὲν οὖν πρεσβεία... αὐτονομίαν: P.'s chief aim in reporting the embassy to Sousa again seems to have been to glorify Pelopidas by demonstrating the impressions his victories made on others (also 30.3). He does not discuss, however, the ephemeral outcome of Pelopidas' negotiations with Artaxerxes, but focuses his remarks on a eulogy of the panhellenic spirit of his proposals to the Persian King. We learn from Xenophon's account (*Hell.* 7.1.39-40) that Thebes' initiative to have the King's stipulations accepted by the Greek states met with a complete failure; see Bersanetti, 1949: 77-79, Ryder, 1965: 81, Roy, 1971: 569-599, esp. 595-596, id. 1994: 197, Buckler, 1980a: 157-160, Wickersham, 1994: 115.

αὐτονομίαν: for a discussion of the term as an instrument of propaganda to justify the hegemonic ambitions of the leading powers in the history of the fourth century, see Bosworth, 1992b: 122-152, esp. 140-143. On the meaning of αὐτονομία in the fifth century, see Lévy, 1983: 249-270.

31.2-3 Ἀλεξάνδρου δὲ... ἔσχεν: taking advantage of the Boiotian withdrawal from Thessaly since Epameinondas' campaign in 367, Alexander was striving to strengthen his position by waging war against the cities of the Thessalian Confederacy. He placed garrisons in Phthiotic Achaia and Magnesia threatening thereby to sever the Thessalian Confederacy in two. His opponents appealed to Thebes to send a relief force with Pelopidas as its commander. They succeeded in obtaining the active support of the Boiotians and Pelopidas was instructed to lead an army of 7,000 men into Thessaly. Just before Pelopidas' departure, an eclipse of the sun occurred (Diod. 15.80.1-2, Nep. *Pel.* 5.2); on the expedition, see Westlake, 1935: 147-148, Bersanetti, 1949: 43-101, esp. 79, Sordi, 1958: 217-218, Buckler, 1980a: 175, Roy, 1994: 202; for the dating of the solar eclipse to the 13th of July, see Boll, *RE*, s.v. "Finsternisse", 6.2, 1909: cols. 2329-2364, esp. 2356, F.K. Ginzel, *Handbuch der mathematischen und technischen Chronologie*, vol. 2 (Leipzig 1911) 527, Reincke, *RE*, s.v. "Pelopidas", 19.1, 1937: cols. 375-380, esp. 379; on P.'s treatment of lunar eclipses, see R. Flacelière, "Plutarque et les éclipses de la lune", *REA* 53 (1951) 203-221, Brenk, 1977: 38-48; for a review of important solar and lunar eclipses reported in Greek literature including the one mentioned here, see Préaux, 1973: 123-128, 156-201, Pritchett, 1979: 108-113; generally on eclipses, Th. Oppolzer, *Canon of Eclipses*, transl. by O. Gingerich (New York 1962), A. Demandt, *Verformungstendenzen in der Überlieferung antiker Sonnen-und Mondfinsternisse*, Mainz Verlag der Akademie der Wissenchaften und der Literatur, Abhandlungen der Geistes-und Socialwissenschaftlichen Klasse, no. 7 (Wiesbaden 1970).

Ἀλεξάνδρου δὲ... ἀναδραμόντος: having anticipated Alexander's unchangeability of character at 26.3—P. described him as "incurably brutal" (ἀνήκεστος καὶ θηριώδης)—he informs us here that the tyrant fell back into his old nature; he suggests thereby that his change of φύσις was in fact temporary, necessitated apparently by political expediency. P. several times speaks of human nature in terms of something changeable, responsive to habit, and closely allied with character (see Swain, 1989: 62-68, esp. 63). On P.'s conception of character-change, cf. *De ser. num.* 551E-552D; see also Russell, 1966 [1995] 139-154, esp. 144-147, Wardman, 1974: 132-140, Brenk, 1977: 176-181, id. 1983: 469-481, Gill, 1984: 469-481, Pelling, 1988a: 257-274, Stadter, 1989: 112.

31.4 ὁ δὲ Πελοπίδας... σημεῖον: the solar eclipse was interpreted as an ill omen by the soothsayers and induced Pelopidas' superstitious compatriots to rescind their consent (Diod. 15.80.3). Not deterred by the interdictions of the seers and the disapproval of his fellow-citizens Pelopidas set out for Thessaly taking with him three hundred mercenary cavalry. According to Diodoros' account (15.80.3) the eclipse of the sun foretold Pelopidas' death, but P. does not link this with his contempt for the eclipse, but blames it on his impetuosity (2.11, 32.9, *synkr. Pel./Marc.* 33(3).1-3). The eclipse and Pelopidas' failure in heeding the warnings of the soothsayers are parallelled by a similar incident in the *Marcellus* (28.8-10); cf. also Thuk. 7.50.

31.5 θερμότερος: for similar descriptions of anger in terms of heat, cf. ἀναφλεχθείς (32.9), παράθερμον (*synkr.* 33 (3).3), διεπυροῦτο τῷ θυμῷ, (*Phok.* 6.2); on the association of bodily heat and temperament, see A. Wardman, "Plutarch and Alexander", *CQ* 5 (1955) 96-107, esp. 102-104), Sansone, 1980: 63-74, esp. 66-68.

31.6 μάλιστα... κάλλος: in a contrived effort to make Pelopidas' act of rashness nobler than that of Marcellus P. attributes his desire to punish the tyrant primarily to the κάλλος of the achievement (cf. 4.4, *synkr.* 33(3).3-4) rather than to his elsewhere acknowledged uncontrollable anger (2.11, 25.4). This negative trait of his character is artfully balanced by his role as the liberator of the oppressed people in Greece (Θηβαίους μόνον... δυναστείας); see [**2d**], and Georgiadou, 1992a: 4222-4257, esp. 4241-4242, 4245-4248.

ἐν οἷς χρόνοις... ἔπεμπον: the Spartans were traditional allies of Syracuse and their record of aid to Dionysios I goes back to 404 (Diod. 14.10.2-4) soon after the beginning of his reign. They supplied him with mercenary troops for his war against the Carthaginians in 398/6 (Diod. 14.44.2) and sent thirty ships to relieve their siege of Syracuse (14.63.4, 70.1-3). We know, however, that in all the subsequent instances of cooperation between the Spartans and Dionysios I it was the Syracusans who sent aid to the Spartans (Xen. *Hell.* 6.2.33 [373/372], 7.1.20, 28, Diod. 15.70.1 [369/8]; cf. also Isocr. *Paneg.* 126, *Peace* 99), and not as P. reports here the other way around. Moreover, he makes no mention of the Athenian decree honoring the tyrant and his sons in 369/8 (*IG* II^2 103, Tod, 1962, No. 133) or, yet another Athenian decree, declaring Dionysios I and his descendants perpetual allies of the Athenians in 368/7 (*IG* II^2 105 and 523, *SEG* XIV 46, Tod, 1962, No. 136/Harding, 1985, No. 52); see further, Stroheker, 1958: 135-136, 139-146, Sanders, 1987: 10-11, Caven, 1990: 81-82, 115-116, 126, Usher, 1990: 184.

Ἀθηναῖοι δὲ... ὡς εὐεργέτην: no other source mentions a bronze statue of Alexander set up by the Athenians. We learn that they had made an alliance with Alexander in 368 (*Reg. et Imp. Apoph.* 193 D-E, Demosth. *contra Aristocr.* 120, Diod. 15.71.3, Bengtson, 1975: No. 276) and that they had a stele erected in his honor, as the inscription recording the Athenian alliance with the Thessalians in 361/360 testifies: cf. *IG* II[2] 116, ll. 39-40, E.L. Hicks, *A Manual of Greek Historical Inscriptions* (Oxford 1882) 97, Tod, 1962, No. 147, Harding, 1985, No. 59; see further, Westlake, 1935: 142, 154-155.

32 Our main sources for the battle of Kynoskephalai are P. and Diodoros (15.80.4-5), while Nepos refers to it only briefly (5.3-4). Both Diodoros and Nepos place the battle in Thessaly, but give no specific location; for a treatment of the battle, see Kromayer, vol. 2, 1907: 116-124, Bersanetti, 1949: 79-82, Pritchett, 1969: 118-119, Buckler, 1980a: 175-182.

32.1 Ὡς οὖν... Θετίδειον: Pelopidas first advanced to Pharsalos, where he augmented his forces by uniting with the army of the Thessalian League. No source gives the size of the Thessalian contingent, and it is likely that P.'s statement about Alexander having more than twice as many hoplites as the Thebans should not be taken literally; cf. Diodoros, who gives the figure of twenty thousand soldiers, 15.80.4. It is unlikely that after such an overwhelming victory it would have been necessary for the Thebans to send another army into Thessaly soon afterwards (35.1-3); furthermore, P. would not have missed the opportunity to stress the importance of a victory in which the Thebans were so significantly outnumbered by Alexander's army; note, for instance, his emphasis on Pelopidas' achievement at the Battle of Tegyra, 17.11-12. He is, in fact, rather restrained in his praise of the exploit (cf. 33.4); see also Westlake, 1935: 149n. 5, Buckler, 1980a: 181.

Pelopidas proceeded from Pharsalos to the valley of the Enipeus from the west, while Alexander set off from Pherai and entering the same valley from the east advanced as far as the temple of Thetis.

Θετίδειον: Strabo (9.5.6) locates the Thetideion in the neighborhood of the old and the new Pharsalos (9.5.6; cf. Polyb. 18.20.6). Three different locations have been proposed as suitable candidates for the location of the site, Palaiokastro-Derengli (or Ambelia), Pnakakia and Agios Athanasios, on which see the discussion by Pritchett, 1969: 114-117.

32.2 εἰπόντος... νικήσομεν: for a similar type of ironic response, see *Pel.* 2.4 (cf. *Reg. et imp. apoph.* 183D), 17.2 (cf. *Reg. et imp. apoph.* 194D).

32.3-4 ἀντιτεινόντων... ἔπρασσον: proceeding from opposite directions—Pelopidas in a westerly, Alexander in an easterly—they dispatched their infantry to occupy the ridge which lay between them. Pelopidas also sent his cavalry against Alexander's horsemen and chased them out into the plain. The latter's infantry succeeded in seizing the ridge first thus gaining a more advantageous position which would enable them to withstand cavalry attacks more easily. When the Thessalian hoplites attempted to storm the hills they were beaten off by Alexander's troops.

ἀντιτεινόντων: Bryan's emendation to ἀνατεινόντων ("stretching out") finds support in Polybios' description of the ridge of Kynoskephalai (18.22.9): λόφοι καλοῦνται μὲν Κυνὸς Κεφαλαί, τραχεῖς δ' εἰσὶ καὶ περικεκλασμένοι καὶ πρὸς ὕψος ἱκανὸν ἀνατείνοντες; cf. also Hdt. 7.176.

τὰς καλουμένας Κυνὸς Κεφαλὰς: the ancient toponym Kynoskephalai ("Dog Heads") is identified with the ridge of Sitochoron by Pritchett (1969: 118), the ridge of Bekidhes by Buckler (1980a: 176), and Kendani by Munn (1987: 106-138, esp. 113-115).

32.5-7 κατιδὼν οὖν... ἀναχώρησιν: when Pelopidas saw his hoplites' plight, he recalled his horsemen from their pursuit of the enemy cavalry on the plain and ordered them to charge Alexander's infantry on the hill. At the same time, he advanced to the head of his infantry encouraging his men to make a frontal assault. This, combined with the flanking movement of his cavalry, forced Alexander's line gradually to retreat. On Pelopidas' tactic of using his hoplites to pin the enemy line and his cavalry to assault the unprotected left flank of Alexander's infantry, see Bersanetti, 1949: 43-101, esp. 80-81, Westlake, 1935: 150, Pritchett, 1969: 119, Buckler, 1980a: 179-181; more generally, on the idea of coordinating cavalry and infantry in Theban warfare in the fourth century, see L.J. Worley, *Hippeis, The Cavalry of Ancient Greece*, (Boulder 1994) 141-152.

ἐπὶ σκέλος ποιούμενοι τὴν ἀναχώρησιν: a military phrase denoting a step by step retreat facing the enemy; cf. Eur. *Phoen.* 1400-1401, Aristoph. *Av.* 383.

32.8-11 When Pelopidas noticed that Alexander's troops had still not been put to flight and that Alexander, standing on the right wing, was encouraging and putting his mercenaries in array, he surrendered to the passion for revenge and rushed at the tyrant with challenging cries. The latter, P. tells us, did not await the assault, but retired among his guards, while Pelopidas was at once mortally wounded by the tyrant's mercenaries. The Thessalians ran down from the hills to Pelopidas' aid, but it was already too late; together with the cavalry they completed the rout of the phalanx, of which more than three thousand were slain (cf. Diod. 15.80.5, Nep. *Pel.* 5.4).

32.9 οὐ κατέσχε... τύραννον: Diodoros dwells solely on the heroism of Pelopidas' act (15.80.5). P. on the other hand leaves this aspect unmentioned here—he deals with it at 34.7 and in *synkr.* 32(2).5—and in a vivid and emotional narrative, lays the emphasis on Pelopidas' inability to subject his anger to his judgment (see also *De cohib. ira* 458D-E); for a parallel, cf. Cyrus the Younger's attack on his brother, in Xen. *Anab.* 1.8.26-27. On the Theban's inability to check his passion, which provides a major point of contact with Marcellus, see [**2d, e**], and Georgiadou, 1992a, passim. For a discussion of P.'s attitude toward ὀργή, see R. Laurenti, "Lo stoicismo romano e Plutarco di fronte al tema dell'ira", in I. Gallo, ed., *Aspetti dello stoicismo e dell'epicureismo in Plutarco, Quaderni del Giornale Ferrarese* 9 (1988) 33-56. For Kallisthenes as P.'s possible source in this section, see [**2c**] s.v. "Polybios", and Sordi, 1958: 196-198.

Chs. 33-34: Funeral rites in honor of Pelopidas

P. gives an elaborate account of the Thessalians' bestowal of honors on the dead Pelopidas, which he exploits in order to reintroduce the hero's significant role as a Panhellenic champion of freedom. After a poignant contrast between the dignified and funeral rites offered to Pelopidas and the extravagance exhibited at the burial of Dionysios I and Alexander the Great P. swiftly recapitulates, by way of conclusion, the glorified career of Pelopidas.

33.1 πατέρα... ἐκεῖνον: Pelopidas is here represented, Fuscagni notes (1975: 47), as the true model of εὐεργέτης, who maintained the roles of διαλλακτής, δικαστής, σύμμαχος and βοηθὸς τοῦ δοκεῖσθαι ἀδικεῖσθαι that the Thessalians had earlier assigned to him (26.4) without displaying any hegemonic aspirations.

33.2 οἱ δὲ Θεσσαλοὶ... χάριν: some trace of these honorary decrees survives in a metrical inscription from a statue of Pelopidas sculpted by Lysippos and erected at Delphi by the Thessalians (*SEG* XXII 460, Harding, 1985, No. 49). It is uncertain, however, whether the statue had been commissioned by the Thessalians posthumously or after his first expedition to Thessaly. P. makes no mention of the statue, although he must have known about it (cf. Nep. *Pel.* 5.4: *omnes Thessaliae civitates interfectum Pelopidam... statuis aeneis... donarunt*); for his silence on these matters, see [**2c**] s.v. "Autopsy"; on the inscription, see Bousquet, 1939: 125-132, id. 1963: 188-208, esp. 206-208, A. Wilhelm, "Zu Ehren des Pelopidas", *Wiener Jahreshefte* 33 (1941) 35-45, esp. 38, Sjöqvist, 1973: 19-33, esp. 10-11, Hornblower, 1991: 227.

33.3 μήτε θώρακα... δήσασθαι: on this cumulative type of negation in P., see Ambrosini, 1991: 19-34, esp. 20.

κεῖραι δ' ἵππους: the clipping of the horses as a form of mourning is attested as a Persian practice by Herodotos (9.24); cf. also Plut. *Arist.* 14.8 and note on 34.2.

33.4 ὥσπερ οὐ νενικηκότων... καταδεδουλωμένων: the same point is also made by Diodoros (15.81.1) in a more thorough manner: Οἱ δὲ Θηβαῖοι περιβόητον νίκην ἀπενηνεγμένοι, πρὸς ἅπαντας ἔφασαν ἑαυτοὺς ἡττῆσθαι διὰ την Πελοπίδου τελευτήν· ἀξιόλογον γὰρ ἀπολωλεκότες ἄνδρα, κατὰ λόγον ἔκρινον τὴν νίκην ἥττονα ὑπάρχειν τῆς Πελοπίδου ζωῆς. It is tempting to assume that P. is quoting here directly from Ephoros.

33.5 τρόπαια... ἐπιφέροντες: Nepos concludes the *Pelopidas* with a description of the honors bestowed posthumously upon Pelopidas and on his children (5.4). Only he mentions statues and the gift of land to Pelopidas' children, on which see Bradley, 1968: 165-166.

33.8 δόξομεν ὑμῖν οὐκ ἀπιστεῖν: Coraes suggests ὑμᾶς instead of ὑμῖν, Madvig ὑμᾶς ἡμῖν, and Reiske ἄπιστα εἰπεῖν instead of οὐκ ἀπιστεῖν. Kronenberg (1937: 303-314, esp. 308-309) sees in the reading δόξομεν ὑμῖν οὐκ ἀμνηστεῖν the desired wording. Both the reading of the manuscript (ὑμῖν with δοκῶ + dative) and Coraes' emendation (ὑμᾶς with accusative + infinitive) are acceptable. The former, accepted with some reservations by Ziegler, may do some injustice to the motives of the Thessalians in offering to bury Pelopidas themselves; it implies that they desire to do so simply in order to make a good impression on their Theban allies. Coraes' reading avoids creating this impression: "we will be sure that you will not disbelieve us".

33.10 πῶς γὰρ <ἂν> ἔτι τολμήσαιμεν: the future tense preserved in codd. CL^2 (τολμήσομεν) fits aptly with the preceding δόξομεν (33.8); furthermore, in terms of the rhetoric of the Thessalian speech, the future seems more appropriate to the exigencies of their threatening situation than does the more indefinite ⟨ἂν⟩ τολμήσαιμεν. The use of the future also seems to anticipate more forcefully yet another request to the Thebans for assistance against Alexander; cf. 35.1, where such a request is probably implied.

34.1-3 P. likes to draw attention to his heroes' burial, even when they meet with a natural death (see Van der Valk, 1982: 302-337 and n. 90). He reflects on the vulgar exhibition of luxury and pomp at the funerals of Dionysios I and Hephaestion in order to conclude that the dignified funeral rites accorded to Pelopidas by the Thessalians far exceeded those of the other two. P. elicits, in a series of blunt contrasts between freedom and compulsion, manifestation of gratitude and envy, dignity and vain splendour, the main aspects of Pelopidas' character described previously. For P.'s views on the modesty of burials, cf. *Consol. ad uxor.* 608F.

34.1 τοῖς τὸ λαμπρὸν... γενομένην: P. acknowledges Philistos (*FGrH* 556F40) as the source of his comment on the lavish funeral of Dionysios; this must have been included in the concluding book of his Περὶ Διονυσίου; on the extravagance of the tyrant's funeral, see also Diod. 15.74.5, Theon, *Progymn.* 2.68 (Spengel). The remark that the pomposity of the funeral marked a fitting conclusion to the tragedy of Dionysios' tyranny may be part of P.'s comment, as it fits in with his opinion about Dionysios' tyranny; it may have also been inspired by Timaios, who compared Dionysios to a tragic actor (τῶν τραγικῶν παθῶν... ἀγωνιστήν, *Quaest. Conv.* 717C [*FGrH* 566F105]); see also Sanders, 1987: 83-84.

οὐκ ἐν ἐλέφαντι καὶ χρυσῷ: for a similar description of solemn funeral rites, see *Aem.* 39.7.

ἐξόδιον θεατρικὸν: the final, retiring song of the chorus in Greek drama. P. often uses such theatrical terms to heighten the emotion of his narrative, and often to express censure; note, however, the use of the term προαγών at 16.1 for exactly the opposite effect. P. often links the tragic and theatrical with the mythical and false, in contrast to the historical and true (P. de Lacy, "Biography and Tragedy in Plutarch", *AJP* 73 [1952] 159-171). On P.'s use of theatrical terms and concepts, see Tagliasacchi, 1960: 24-142, O'Donnell, 1975, Digregorio, 1976: 151-174, F. Albini, *Plutarco. Non Posse Suaviter Vivi Secundum Epicurum*, Università di Genova, Facoltà di Lettere (Genova 1993) 43-44. The simile (οἷον... ἐξόδιον θεατρικόν) is found also in *Alex.* 75.5, *Crass.* 33.7 and is particularly appropriate here because Dionysios had written tragedies; on his poetic activity, cf. *Tim.* 15.7, *De Alex. fort.* 338B-C, Tzetz. *Chil.* 5.178-181; see further, Stroheker, 1958: 97-98, W. Bühler, "Tzetzes über die Ἕκτορος Λύτρα des Dionysios", *ZPE* 11 (1973) 69-79, Sanders, 1975: 1-20, Caven, 1990: 222-223.

34.2 Ἀλέξανδρος... ἀναλαμβάνουσαι: only P. mentions the clipping of the horses by Alexander; cf. *Alex.* 72.3 and J.R. Hamilton's note on it (1969: 200). Arrian (7.14.4) reports, without naming his sources, that Alexander cut his hair. He also points out that these accounts vary greatly according to the writers' attitude to Hephaestion or Alexander (7.14.2). Aelian (*V.H.* 7.8) regards the clipping of Alexander's hair as an act in imitation of Achilles (cf. *Il.* 23.135-136), but considers the dismantling of the battlements of the acropolis at Ekbatana as a violent and barbaric manifestation of grief. On Alexander's extravagant mourning for Hephaestion's death, cf. Arr. 7.14.8, Diod. 17.114.4. There are scanty fragments of a monograph by Ephippos on the deaths of Alexander and Hephaestion (*FGrHist* 126). On clipping one's hair when somebody dies, see Vernant, 1991: 66.

34.4 ἀνὴρ δὲ δημοτικός: not to be translated "commoner" as Perrin (Loeb, vol. 5, 1968: 428) or "issu du peuple" as Flacelière and Chambry (Budé, vol. 4, 1966: 176), but rather with the meaning of popular and democratic man, since we know that Pelopidas came from a noble family (cf. 3.1); on P.'s varied use of δημοτικός, see Stadter, 1989: 94-95.

οὐ γυναικός... οὐκ ἀναγκάζοντος: the accumulation of negative cola with οὐ rather than οὔτε is common in P.; cf. 13.7, *Per.* 12.3, 39.3 (also Holden, 1894, notes ad locc.), *Sol.* 7.1, *Lyk.* 9.3, *Fab.* 17.2.

εἰκότως... εὐδαιμονισμόν: "he rightly seemed to attain fully supreme happiness" (ἀπέχω + accus.: have, receive in full, LSJ, IV.1). P. turns away from his initial condemnation of rashness and excessive boldness in battle (chs. 1-2) and raises the discussion to a more general level. In his quest for the absolute condition of εὐδαιμονισμός, he embraces the notion that the death of men in the hour of their triumph is the ultimate human accomplishment. Clearly determined not to exclude Pelopidas from this privileged group of people he disregards the special circumstances that caused his death. For other instances of people who died at the high point of their life, cf. Kleobis and Biton (Hdt. 1.30-31), the legend of Polykrete (Parthen. 9.8, Hercher; cf. Plut. *De mul. virt.* 254C-E), of Drakon (Suid. s.v. Δράκων).

34.5-6 οὐ γάρ... ἀναβήσῃ: the use of *chreiai* to introduce and conclude the *Pelopidas* suggests ring composition in P.'s narrative method. In the proem emphasis is laid on the utilitarian purposes of staying alive in battle, here on the glory of an honorable death.

34.5 ὡς Αἴσωπος... ἀπολιπών: Perry notes (*Aesopica* [Urbana 1952] 249-250) that this *chreia* is attributed to Anacharsis (*Gnom. Vat.* 21), Diogenes the Cynic (Dio Chrys. *Orat.* 6.46 [Stob. 49.27]), and Plato (Anton. Mel. 1.58), probably not the well-known philospher. Diogenes Laertios (6.1.5) assigns a similar saying to Antisthenes, and Stobaios (4.41.2) to Dionysios' tragedy Λήδα. The lines quoted by Stobaios encapsulate P.'s saying in a more compact form:

θνατῶν δὲ μηδεὶς μηδέν' ὄλβιόν ποτε
κρίνῃ, πρὶν αὐτὸν εὖ τελευτήσαντ' ἴδῃ·
ἐν ἀσφαλεῖ γὰρ τὸν θανόντ' ἐπαινέσαι

The essence of the *chreia* reminds one of Herodotos' story about the Athenian Tellos (1.30; cf. 1.32; Plut. *Sol.* 27.6) and may owe its existence to the long tradition of a more general theme: the impossibility of judging someone happy before his death. This is often found in tragedy (Aesch. *Ag.* 928, Soph. *Oed. Rex* 1528-1530, *Trach.* 1-3, Eur. *Androm.* 100-102, *Tr.* 510, *Heracl.* 865-866, *I.A.* 161-162; cf. also Plat. *Apol.* 29A, where Sokrates says that nobody knows whether death is not really the greatest of blessings; yet men fear it as if they knew it to be the greatest of evils) and in consolatory literature (Cic. *Tusc.* 1.46.3, Sen. *ad Polyb.* 9.9); on this Aesopic *chreia*, see Gallo, 1980, vol. 2: 437, 443-444. For the belief that death safeguards the virtuous acts of heroes against deterioration, see Vernant, 1991: 50-74. On P.'s praise of Aesopic wisdom, see M.J. Luzzatto, "Plutarco, Socrate e l'Esopo di Delfi", *ICS* 13.2 (1988) 427-445, esp. 429, 441.

καὶ <τῇ> τύχῃ μεταβάλλεσθαι ⟨μὴ⟩ ἀπολιπών: "not allowing a change by means of fortune" (cf. "non consente successivi cambiamenti di sorte", transl. by Gallo, 1980: 443); Ziegler's addition of the article is unnecessary; cf. also *Agis/Kleom.* 52.9 (τύχῃ σφαλέντες), *synkr. Per./Fabius Maximus* 2.3.

34.6 διὸ βέλτιστον ὁ Λάκων... ἀναβήσῃ: P. may have Pindar in mind when he mentions the example of Diagoras to whom the seventh Olympian Ode is dedicated (note also P.'s reference to the Olympian and Pythian victories below). Diagoras of Rhodes, himself a victor at all four Panhellenic festivals, became the father of three—Damagetos, Dorieus, and Akousilaos— and grandfather of two Olympic victors—Eukles and Peisirhodos—; cf. also Paus. 6.7.1-2, Cic. *Tusc.* I.46.110-111.

34.7 τὰς δ' 'Ολυμπιακὰς... ἀπέθανεν: Westlake suggests (1939: 11-22, esp. 14) that this sentence seems to contain the substance of an honorary inscription or proclamation, which was perhaps reproduced verbatim in the historical source P. is following.

ἐν τῇ τρισκαιδεκάτῃ βοιωταρχίᾳ: while P. affirms that Pelopidas was in his thirteenth year as a boeotarch at the time of his death, Diodoros records (15.81.4) that he was elected boeotarch annually without interruption a total of fifteen times, after the liberation of Thebes (379). P.'s report is the more plausible, since we know of at least one year (371, *Pel.* 20.1, Paus. 9.13.6-7) during which Pelopidas did not hold the office. There has been much debate and disagreement as to the second year. A Boiotian decree honoring the Carthaginian ambassador

Nobas, dated with reservations to the year 368—a year before Dionysios I waged a new war against Carthage—contains the list of seven boeotarchs (*IG* VII 2407); Pelopidas does not figure in the list. If the decree can be assigned with any certainty to 368, then so can the mysterious second year; see Köhler, 1889: 636-643, esp. 637-639, Bersanetti, 1949, 43-101, esp. 89-93 and Fuscagni, 1972: 415-433, Roesch, 1984: 45-60, esp. 46-48, 51.

Ch. 35: The Thebans' second expedition to Thessaly (364) and Alexander's death (358)

P. often continues the narrative beyond the protagonist's death (cf. *Arist.* 27.2-7, *Dem.* 31, *Sert.* 27, *Num.* 22.3-12, *Ant.* 71-87) and occasionally calls attention, not without some delight, to the punishment or downfall of the enemies or murderers of his heroes; cf. *Dem.* 31.4, *Phok.* 38.2, *Arat.* 54.4, *Crass.* 33.7; see further C.P. Jones, 1971: 74, 102, Wardman, 1974: 18, Van Der Valk, 1982: 301-337, esp. 320.

35.1 'Ο δὲ θάνατος... ὠφέλησε: P.'s description of the Thessalians as being saddened by Pelopidas' death, but also benefiting from it, is probably an ironic reference to the double loss claimed earlier by the Thessalians (33.9).

35.2-3 Upon learning of Pelopidas' death the Thebans sent immediately an expeditionary force of 7,000 hoplites and 700 cavalry to Thessaly. Having suffered heavy losses in the battle at Kynoskephalai (32.11) Alexander was defeated in the second encounter and forced to withdraw from the Thessalian cities he had reduced and to surrender Magnesia and Phthiotic Achaia to the Boiotians (cf. Diod. 15.80.6). He became their subject-ally and was obliged to supply troops for Theban campaigns along with his Thessalian opponents (cf. Xen. *Hell.* 7.5.4); see Westake, 1935: 150-152, Sordi, 1958: 219-220, Buckler, 1980a: 181-182, Roy, 1994: 202-203. On the date of the expedition, see note below.

ἡγουμένων Μαληκίδου καὶ Διογείτονος: Lindskog's emendation to a plural participle is unnecessary. It is not uncommon to have a singular verb with a series of singular subjects (cf. *Phok.* 25.5, Thuk. 1.29.2, 1.45.2, Theokr. 7.3); see Dover, 1960 [1988]: 61-77, esp. 6363, Kühner-Gerth, vol. 2.1, p. 79 (ch. 370.2). Both names, Μαληκίδας and Διογίτων, are listed as boeotarchs in an inscription (*IG* VII 2408; Roesch, 1984: 45-60, esp. 47-48) dated to 364/3 (Köhler

[1889] 636-643, esp. 637-638), while Pelopidas and Epameinondas are not. Based on the proposed date of the inscription the second expedition has been assigned to the spring of 363 (Beloch, 1923: 244). It has been pointed out, however, that the dating is not secure, since these two generals may well have been colleagues more than once (M. Cary, "Notes on the ἀριστεία of Thebes", *JHS* 42 [1922] 184-191, esp. 190-191, followed by Westlake, 1935: 151n. 1, Bersanetti, 1949: 82n. 1, Sordi, 1958: 219 and Buckler, 1980a: 181). P. insists that the second expedition set out for Thessaly immediately, in which case it must have taken place in the early autumn of 364, soon after the Battle of Kynoskephalai.

Μαληκίδου: emendation by Köhler (1889: 636-643, esp. 638) based on the reading of *IG* VII 2408; Μαλκίτου codd.; on the various readings of this name, see Κουμανούδης, 1979, no. 1282.

Διογείτονος: emended to Διογίτονος following *IG* VII 2408; see further Κουμανούδης, 1979, nos. 515 and 516.

35.4-12 P.'s account of Alexander's murder may derive from a digression on Alexander in Theopompos' *Philippika*, on which see [**2c**] s.vv. "Theopompos and Xenophon", and Westlake, 1939: 14. The account of his murder appears to have excited the interest of several writers; cf. Xen. *Hell.* 6.4.35-37, Diod. 16.14.1, Cic. *De off.* 2.7.25, Val. Max. 9.13, ext. 3. On Xenophon's account of Alexander's death and the similarities it bears with Herodotos' story of the murder of Kandaules, see Gray, 1989: 70-72.

35.4 Θηβαῖοι... ἠρκέσθησαν: the main purpose of the second expedition to Thessaly was achieved. Even though the Thebans had not been able to force Alexander to give up his tyranny, they succeeded in restricting his authority to Pherai; see Westlake, 1935: 152, Buckler, 1980a: 182.

ὀλίγον ὕστερον: Alexander's murder seems to have taken place in 358, according to Diodoros' calculation of his reign (15.61.2) as eleven years (369-358); this date is accepted by Westlake, 1935: 156, Hammond, 1937: 66-67n. 91, Sordi, 1958: 230, 233; cf. also Diod. 16.14.1, where he tells the story under 357-356.

ἣν δ' ὀλίγον... διηγήσομαι: P. seems to think that Alexander's humiliating defeat by the Thebans in 364 (35.1-3) was not sufficient punishment for the tyrant but that some act of divine intervention was called for (cf. also Valerius Maximus, who portrays the tyrant as subject to divine anger because he could not control his fear and his passion for Thebe, 9.13.3). Divine retribution for Pelopidas' death comes also partly as a fulfillment of the latter's wish expressed during his imprisonment by the tyrant (28.4): "may you perish sooner by becoming more hateful to the gods than now". Although P. himself probably agreed with the Platonic view that the divine by its very nature is not subject to anger, he seems to have dwelt with pleasure on the punishment of evildoers. For an extensive discussion of the theme of divine retribution in P.'s *Lives*, see Brenk, 1977: 256-275; more generally, on the inevitability of punishment as a notion going back to Hesiod, see McGlew, 1993: 59-60, 90-91.

35.5 Θήβην... φυλακῶν οὖσαν: on Pelopidas' role in inculcating in Thebe hatred for Alexander, see 28.9.

τῆς τυραννίδος: metonymically for τύραννος; cf. 17.3 ἵππον.

ἐντὸς τῶν ὅπλων καὶ τῶν φυλακῶν: Coraes' emendation of φυγάδων (codd.) to φυλάκων renders aptly the notion of ὡπλισμένων φυλάκων (armed guards) conveyed here by the hendiadys; cf. parallel examples: ἄσκευον... ἀσπίδων τε καὶ στρατοῦ (Soph. *El.* 36), "sine armis ac multitudine" (Cic. *Caec.* 93). Ziegler prefers Schaefer's φυλακῶν, which is apparently motivated by the φυλακαὶ τῶν παρανυκτερευόντων in 35.7. For other examples of hendiadys in P., see Stadter, 1989: 190 and note on 26.3; generally, D. Sansone, "On Hendiadys in Greek", *Glotta* 62 (1984) 16-25.

35.6 ἔπειτα δὲ... ὠμότητα: both Cicero (*De off.* 2.7.25) and Valerius Maximus (9.13, ext. 3)—the latter probably deriving his account from Cicero or from the same source as he—ascribe Alexander's murder solely to Thebe's suspicions of his infidelity. Xenophon's account of Thebe's motivations for murdering Alexander is in general agreement with P.'s; he also adds that because Alexander had no children from her, he was trying to win as his wife the widow of Jason (*Hell.* 6.4.37). On Alexander's debaucheries and brutality, cf. 26.3, 27.6-7, 28.9, 29.6-8, Xen. *ib.* 6.4.35.

συνθεμένη... τρόπον: Teisiphonos, Peitholaos (or Pytholaos) and Lykophron together with Thebe were Jason's children (stepchildren according to Konon, *FGrH* 26F1[L]). Diodoros (16.14.1) omits Peitholaos from his account, while he emphatically attributes the act of murder to both Thebe and her brothers. P. and Xenophon (*Hell.* 6.4.35) make it clear that she was only the moral instigator of that murder, while the former mentions in *Amat.* 768F that it was Pytholaos who actually killed Alexander. In Xenophon's account (6.4.36) Thebe succeeded in getting her brothers' assistance for her husband's murder after she reported to them that Alexander was plotting against them.

35.7-8 παρανυκτερευόντων: attested only in P.; cf. *Eum.* 8.11, *Gracch.* 35.5, *Arat.* 8.5, *De soll. anim.* 969F.

ὁ δὲ θάλαμος... ἐκεῖνον: P. explicitly states that at night she used to go in alone to their bed-chamber which was guarded only by a dog; In Cicero's version (*De off.* 2.7.25) Alexander, apparently fearing that Thebe might be conspiring against him, would visit her in her chamber preceded always by a Thracian slave with a drawn sword; some bodyguards had already searched her garment chest for concealed weapons (cf. Val. Max. 9.13.7, ext. 3).

35.9 εἶθ' οὕτως... ἔδειξεν: Xenophon gives a slightly different version of the incident (6.4.36): Thebe received Alexander in a drunken state and, after she put him to bed, carried his sword out of the chamber.

35.11 ὁ δὲ τρίτος... διεχρήσατο: in *Amatorius* 768F P. informs us that the person who committed the actual act of murder was Πυθόλαος, who, he suggests, had been his lover; see also note on 35.6.

35.12 P. expresses momentary disappointment at the swiftness of Alexander's death, thinking that his crimes were not sufficiently punished. He finally concedes that Alexander may, after all, have suffered what his lawless deeds deserved, a belief which is in line with the idea expounded in *De ser. num.* 560B that god allots us our deserts. Moreover, P.'s emphasis on the speediness of Alexander's death should be seen in the light of the Platonic view expressed in the same treatise that the divine does not expedite punishment for all alike, but removes from life and amputates at once what is incurable (τὸ μὲν ἀνήκεστον, 551E; cf. Plat. *Leg.* 862e-863a), as was the case with Alexander (ὡς δ' ἦν ἀνήκεστος, 26.3).

τῷ δὲ μόνον... ἀπολέσθαι: in the *Moralia* P. mentions some other instances of women who killed their husbands: Semiramis, who killed Ninos (*Amat.* 753E; cf. Diod. 2.20.3-5) and Aretaphila of Cyrene (*De mul. virt.* 255E-258C), who, emulating the glorious and far-famed daring of Thebe, had her tyrant husband killed. He occasionally transmits the names of persons who first perpetrated or invented a thing (cf. *Rom.* 22.5, *synkr. Thes./Rom.* 35(6).4, *synkr. Lyk./Num.* 25(3).12-13; see further R. Flacelière, "Sur quelques passages des Vies de Plutarque", *REG* 61 [1948] 67-103, esp. 102-103). Ephoros' book Περὶ εὑρημάτων (70 *FGrH* 104-106) seems to have dealt with this topic; for a general discussion, see A. Kleingünther, "ΠΡΩΤΟΣ ΕΥΡΕΤΗΣ, Untersuchungen zur Geschichteeiner Fragestellung", *Philologus Suppl.* 26.1 (1933) 1-65. In the *De Herod. malign.* 856A P. condemns unnamed authorities for claiming that the assassination of the tyrant Alexander by Thebe was not prompted by a noble spirit, but was an act of jealousy and womanly passion. Elsewhere he praises her for her far-famed daring (*Mulier. virt.* 256A).

ὑπὸ γυναικὸς ἰδίας: "by his own wife". Although Thebe did not actually kill Alexander, she is described here as killing him in every way but making the final stroke.

καὶ τῇ μετὰ θάνατον... Φεραίων: P. often records gruesome details about the physical disfigurement inflicted upon some of his characters after their death, cf. *Cic.* 48-49, *Crass.* 32.1-2, *Galba* 27.1-5.

ἄξια πεπονθέναι... παρανομημάτων: P. concludes the *Life of Pelopidas* with a didactic message, thereby reminding his readers of the moral function of his biographies. Alexander seems to have received the just reward for his misbehavior, just as the Spartans had for seizing the Theban acropolis (cf. Xen. *Hell.* 5.4.1). Although the *Life* has ended, P.'s book has not; it continues immediately with the *Life of Marcellus*, and concludes with the *synkrisis*, on which see [**2d**]; on endings of *Lives* in P., see Pelling, "Is Death the End? Closure in Plutarch's *Lives*", forthcoming.

Select Bibliography

G.J.D. Aalders, *Political Thought in Hellenistic Times* (Amsterdam 1975).

— — —, *Plutarch's Political Thought*, Verhandelingen der Koninklijke Nederlandse Akademie van Wetenschappen, Afd. Letterkunde, Nieuwe Reeks, Deel 116 (Amsterdam, Oxford, New York 1982).

G.J.D. Aalders, H. Wzn and L. De Blois, "Plutarch und die politische Philosophie der Griechen", *ANRW* II.36.5 (1992) 3384-3404.

S. Accame, "L'Attentato di Pelopida contro i Polemarchi", *RFIC* 14 (1936) 337-355.

— — —, *La lega ateniense del IV secolo a.C.*, Studi pubblicati dall'istituto italiano per la storia antica, vol. 2 (Rome 1941).

F.E. Adcock, *The Greek and Macedonian Art of War* (Berkeley, Los Angeles 1957).

S.E. Alcock, "Greece. A Landscape of Resistance?", in D.J. Mattingly, ed., *Dialogues in Roman Imperialism. Power, Discourse and Discrepant Experience in the Roman Empire*, *JRS Suppl. Series* 23 (Portsmouth 1997) 103-115.

A. Aloni, "Ricerche sulla forma letteraria del *De Genio Socratis* di Plutarco", *Acme* 33.1-2 (1980) 45-112.

— — — and G. Guidorizzi, *De genio Socratis - I ritardi della punizione divina* (Milan 1982).

J.M. Alonso-Núnez, "The Emergence of Universal Historiography from the 4th to the 2nd Centuries B.C.", in H. Verdin, G. Schepens and E. De Keyser, eds., *Purposes of History: Studies in Greek Historiography from the 4th to the 2nd Centuries B.C.*, *Studia Hellenistica* 30 (Leuven 1990) 173-192.

— — —, Plutarco, *Il demone di Socrate* (Milan 1982).

R. Ambrosini, "Funzione espressiva della sintassi nelle lingua di Plutarco", in G. d'Ippolito and I. Gallo, eds., *Strutture Formali dei Moralia di Plutarco*, Atti del III Convegno Plutarcheo, Palermo 3-5 maggio 1989 (Naples 1991) 19-34.

J.K. Anderson, *Military Theory and Practice in the Age of Xenophon* (Berkeley, Los Angeles 1970).

A. Andrewes, "Diodoros and Ephoros. One Source of Misunderstanding", in J.W. Eadie and J. Ober, eds., *The Craft of the Ancient Historian: Essays in Honor of Chester G. Starr* (Lanham 1985) 189-195.

D. Asheri, "La diaspora e il ritorno dei Messeni", in E. Gabba, ed., *Tria corda. Scritti in onore di A. Momigliano*, Biblioteca di Athenaeum 1 (Como 1983) 27-42.

A.E. Astin, *Cato the Censor* (Oxford 1978).

A. Aymard, "Philippe de Macédoine otage à Thèbes", *REA* 56 (1954) 15-36; repr. in A. Aymard, *Études d'Histoire ancienne* (Paris 1967) 418-435.

D. Babut, *Plutarque et le Stoicisme* (Paris 1969).

— — —, "Le dialogue de Plutarque sur le démon de Socrate. Essai d'interprétation", *BAGB* 1 (1984) 51-76; repr. in *Parerga, Choix d'articles de Daniel Babut (1974-1994)*, Collection de la Maison de l'Orient Méditerranéen No 24, Sér. Littér. et Philos. 6 (Lyon 1994) 51-76.

— — —, "La part du rationalisme dans la religion de Plutarque. l'exemple du *De genio Socratis*", *ICS* 13.2 (1988) 383-408; repr. in *Parerga, Choix d'articles de Daniel Babut (1974-1994)*, Collection de la Maison de l'Orient Méditerranéen No 24, Sér. Littér. et Philos. 6 (Lyon 1994) 77-102.

G.L. Barber, *The Historian Ephorus* (Cambridge 1935; repr. New York 1979).

L. Barbera, "Il problema interpretativo del trattato conchiuso a Tebe nel 366/5 a.C.", *Helicon* 9-10 (1969-1970) 460-474.

N.I. Barbu, *Les procédés de la peinture des caractères et la vérité historique dans les biographies de Plutarque* (Paris 1933).

A. Barigazzi, "Una nuova interpretazione del *De genio Socratis*", *ICS* 13.2 (1988) 409-425; repr. in *Studi su Plutarco*, Università degli Studi di Firenze, Dipartimento di Scienze dell'Antichità (Florence 1994) 213-234.

— — —, "Una introduzione alle *Vite*", in *Studi su Plutarco*, Università degli Studi di Firenze, Dipartimento di Scienze dell'Antichità (Florence 1994) 287-301.

R.H. Barrow, *Plutarch and his Times* (Bloomington, London 1969).

J. Barthelmess, "Recent Work on the Moralia", in *Miscellanea Plutarchea, Quaderni del Giornale Filologico Ferrarese* 8 (1986) 61-81.

H. Beister, *Untersuchungen zu der Zeit der thebanischen Hegemonie* (Munich 1970).

K.J. Beloch, *Griechische Geschichte*², vols. 3:1 and 3:2 (Berlin, Leipzig 1922/1923).

H. Bengtson, ed. *Die Staatsverträge des Altertums*, vol. II²: *Die Verträge der griechisch-römischen Welt von 700 bis 338 v. Chr.* (Munich, Berlin 1975).

— — —, *Griechische Staatsmänner des 5. und 4. Jahrhunderts v. Chr.* (Munich 1983).

— — —, *History of Greece from the Beginnings to the Byzantine Era*, transl. and updated by E.F. Bloedow (Ottawa 1988).

U. Bernini, "Ἁπλοῦς e Ἁπλότης. In margine a Plut., Pelopidas, 30-31", *Studi Italiani di Filologia Classica,* ser. 3, 9 (1991) 74-111.

G.M. Bersanetti, "Pelopida, con due Appendici sulle Beotarchie di Pelopida e sul Processo di Epaminonda e Pelopida", *Athenaeum,* n.s. 27 (1949) 43-101.

H. Berve, *Die Tyrannis bei den Griechen*, 2 vols. (Munich 1967).

L. Bizard, "Fouilles du Ptoïon (1903): II. Inscriptions", *BCH* 44 (1920) 227-262.

J. Boardman et al., *Lexicon Iconographicum Mythologiae Classicae*, 2.1 (Zurich, Munich 1984).

F. Bock, *Untersuchungen zu Plutarchs Schrift Περὶ τοῦ Σωκράτους δαιμονίου,* Diss. Munich 1910.

G. Bockisch, "Ἁρμοσταί (431-387)", *Klio* 46 (1965) 129-239.

H. Bolkestein, *Wohltätigkeit und Armenpflege im vorchristlichen Altertum* (Utrecht 1939; repr. New York 1979).

R.J. Bonner and G. Smith, *The Administration of Justice from Homer to Aristotle*, vol. 1 (Chicago 1970).

— — — , "Administration of Justice in Boeotia", *CPh* 40 (1945) 11-23.

L.V. Borrelli, "Ptoion" in *Enciclopedia dell'arte antica classica e orientale*, vol. 6 (Rome 1965) 535-538.

E.K. Borthwick, "The Scene on the Panagjurischte Amphora. A New Solution", *JHS* 96 (1976) 148-151.

A. B. Bosworth, "History and Artifice in Plutarch's *Eumenes*", in P.A. Stadter, ed., *Plutarch and the Historical Tradition* (London, New York 1992a) 56-89.

— — — , "*Autonomia*. The Use and Abuse of Political Terminology", *SIFC*, ser. 3, 10 (1992b) 122-152.

J. Bousquet, "Une statue de Pélopidas à Delphes signée de Lysippe", *Révue Archéologique* 14 (1939) 125-132.

— — — , "Inscriptions de Delphes", *BCH* 87 (1963) 188-208.

J.R. Bradley, *The Sources of Cornelius Nepos. Selected Lives*, Ph.D. diss. Harvard University 1968.

L. Breitenbach, *Xenophons Hellenika*, 3 vols. (Berlin 1873-1876).

J.M. Bremmer, "Plutarch and the Naming of Greek Women", *AJP* 102 (1981) 425-426.

F. Brenk, "The Dreams of Plutarch's *Lives*", *Latomus* 34 (1975) 336-349.

— — — , *In Mist Apparelled. Religious Themes in Plutarch's 'Moralia' and 'Lives'*, *Mnemosyne Supplement* 48 (1977).

— — — , "The Religious Spirit of Plutarch", *ANRW* 2.36.1 (1987) 248-349.

— — — , "Plutarch's Life 'Markos Antonios'", *ANRW* 2.33.6 (1992) 4347-4469.

— — — , "Memoria sociale e non tanto sociale. La liberazione di Tebe nel *Daimonion di Socrate* di Plutarco", in P. Volpe Cacciatore, ed., *Studi in Omaggio di Italo Gallo* (Naples 1998); forthcoming.

P. Briant, *Antigone le Borgne. Les débuts de sa carrière et les problèmes de l'Assemblée Macédonienne* (Paris 1973).

I.A.F. Bruce, "Internal Politics and the Outbreak of the Corinthian War", *Emerita* 28 (1960) 76-80.

— — — , "Athenian Embassies in the Early Fourth Century BC", *Historia* 15 (1966) 272-281.

— — — , *An Historical Commentary on the 'Hellenica Oxyrhynchia'* (Cambridge 1967).

B. Bucher-Isler, *Norm und Individualität in den Biographien Plutarchs* (Bern, Stuttgart 1972).

R.J. Buck, *A History of Boeotia* (Edmonton 1979).

— — — , "Boeotia, its Development Of Institutions and Oligarchic and Democratic Theory in The Fifth and Fourth Centuries B.C.", in P. Roesch, ed., *Colloques internationaux du CNRS "La Béotie Antique"*, Éditions du CNRS, Lyon-Saint Étienne 16-20 Mai 1983 (Paris 1985) 291-295.

— — — , "The Athenians at Thebes in 379/8 B.C.", *The Ancient History Bulletin* 6.3 (1992) 103-109.

— — — , *Boiotia and the Boiotian League, 423-371 B.C.* (Edmonton 1994).

J. Buckler, "Dating the Peace of 375/4 B.C.", *GRBS* 12 (1971) 353-361.

— — — , "A Second Look at the Monument of Chabrias", *Hesperia* 41 (1972) 466-474.

— — — , "Plutarch and the Fate of Antalkidas", *GRBS* 18 (1977a) 139-145.

— — — , "The Thespians at Leuctra", *WS* 90 (1977b) 76-79.

— — — , "Plutarch on the Trials of Pelopidas and Epameinondas (369 BC)", *CP* 73 (1978a) 36-42.

— — — , "The Alleged Achaian Arbitration after Leuktra", *SymbOslo* 53 (1978b) 85-96.

— — — , "The Re-establishment of the Boiotarchia (378 BC)", *AJAH* 4 (1979) 50-64.

— — — , *The Theban Hegemony, Harvard Historical Studies 98* (Cambridge, Mass. 1980a).

— — — , "Plutarch on Leuktra", *SymbOslo* 55 (1980b) 75-93.

— — — , "The Alleged Theban-Spartan Alliance of 386 B.C.", *Eranos* 78 (1980c) 179-185.

——— , "Χωρὶς ἱππεῖς. A Note on the Battle of Chaironeia in 338 B.C.", *Teiresias Supplement* 3 (1990) 75-80.

——— , "Plutarch and Autopsy", *ANRW* 2.33.6 (1992) 4788-4830.

——— , "Epameinondas and Pythagoreanism", *Historia* 42.1 (1993) 104-108.

——— , "Some Thoughts on Ploutarkhos' Approach to History", *Boeotia Antiqua* 3 (1993) 69-77.

——— , "The Battle of Tegyra, 375 BC", *Boeotia Antiqua* 5 (1995) 43-58.

——— , "Helikon and Klio", in A. Hurst and A. Schachter, eds., *La montagne des Muses, Fondation Hardt pour l'Étude de l'antiquité classique, Recherches et Entretiens* 7 (Geneva 1996) 127-139.

W. Burkert, "Greek Tragedy and Sacrificial Ritual", *GRBS* 7 (1966) 87-121.

——— , *Lore and Science in Ancient Pythagoreanism* (Cambridge, Mass. 1972).

——— , *Structure and History in Greek Mythology and Ritual*, Sather Classical Lectures, vol. 47 (Berkeley, Los Angeles 1979).

——— , *Homo Necans* (Berkeley, Los Angeles 1983).

——— , *Greek Religion* (Cambridge, Mass. 1985).

A.P. Burnett, "Thebes and the Expansion of the Second Athenian Confederacy: *IG* II^2 40 and *IG* II^2 43", *Historia* 11 (1962) 1-17.

G. Busolt, "Der zweite athenische Bund und die auf der Autonomie beruhende hellenische Politik von der Schlacht bei Knidos bis zum Frieden des Eubulos", *Jahrbücher für classische Philologie Suppl.* 7 (1873-1875) 679-684.

G. Busolt - H. Swoboda, *Griechische Staatskunde*, 2 vols. (Munich 1920/1926).

A. Caiazza, *Plutarco, Precetti Politici*, Dipartimento di Filosofia e Politica dall'Istituto Universitario Orientale di Napoli (Naples 1993).

J. Cargill, *The Second Athenian League. Empire or Free Alliance?* (Berkeley, Los Angeles 1981).

P. Carrara, "Plutarco ed Euripide. alcune considerazioni sulle citazioni euripidee in Plutarco (*De aud. poet.*)", *ICS* 13.2 (1988) 447-455.

F. Carrata Thomes, *Egemonia beotica e potenza marittima nella politica di Epaminonda,* Pubblicazioni della Facoltà di Lettere e Filosofia, vol. 4, fasc. 4 (Turin 1952).

P. Cartledge, "Hoplites and Heroes. Sparta's Contribution to the Technique of Ancient Warefare", *JHS* 97 (1977) 11-27.

— — —, *Sparta and Lakonia* (London 1979).

— — —, *Agesilaos and the Crisis of Sparta* (Baltimore 1987).

M. Cary, "The Trial of Epaminondas ", *CQ* 18 (1924) 182-184.

J. Casabona, *Recherches sur le vocabulaire des sacrifices en Grec des origines à la fin de l'époque classique*, Publications des Annales de la Faculté des Lettres, Aix-en-Provence, n.s. 56 (Aix-en-Provence 1996).

B. Caven, *Dionysius I. War-Lord of Sicily* (New Haven 1990).

G.L. Cawkwell, "Notes on the Peace of 375/4", *Historia* 12 (1963) 84-95.

— — —, "Epaminondas and Thebes", *CQ* 66 (1972) 254-278.

— — —, "The Foundation of the Second Athenian Confederacy", *CQ* 67 (1973) 47-60.

— — —, "Agesilaus and Sparta", *CQ* 70 (1976a) 62-84.

— — —, "The Imperialism of Thrasybulus", *CQ* 70 (1976b) 270-277.

— — —, *Philip of Macedon* (London, Boston 1978).

M. Chambers, ed., *Hellenica Oxyrhynchia* (Stuttgart, Leipzig 1993).

K.M.T. Chrimes, *Ancient Sparta. A Re-examination of the Evidence* (New York 1952).

W. Christ, *Plutarchs Dialog vom Daimonion des Sokrates*, Sitzungsberichte der philosophisch-philolog. und der histor. Klasse der Akad. der Wissenschaften zu München (Munich 1901).

K. Clinton, *The Sacred Officials of the Eleusinian Mysteries, Transactions of the American Philosophical Society*, n.s., vol. 64.3 (Philadelphia 1974).

P. Cloché, *La restauration democratique à Athènes en 403 av. J.-C.* (Paris 1915).

— — —, "La politique Thébaine de 404 à 396 av. J.-C.", *REG* 31 (1918) 315-343.

— — —, *La politique étrangère d' Athènes de 404 à 338 avant Jésus-Christ* (Paris 1934).

— — —, *La démocratie athénienne* (Paris 1951).

— — —, *Thèbes de Béotie* (Namur 1952).

W.R. Connor, "History without Heroes. Theopompus' Treatment of Philip of Macedon", *GRBS* 8 (1967) 133-154.

— — —, *The New Politicians of Fifth-Century Athens* (Princeton 1971).

M.L. Cook, *Boeotia in the Corinthian War. Foreign Policy and Domestic Politics*, Ph.D. diss. University of Washington 1981.

A. Corlu, *Le démon de Socrate* (Paris 1970).

S. Costanza, "La Synkrisis nello schema biografico di Plutarco", *Messana*, Studi diretti da Michele Catalano, 4 (1955) 127-156.

F. Croissant and F. Salviat, "Aphrodite gardienne des magistrats. Gynéconomes de Thasos et polémarques de Thèbes", *BCH* 90 (1966) 460-471.

P.J. Cuff, "The Trials of Epaminondas - A Note", *Athenaeum,* n.s. 32 (1954) 259-264.

C. Daremberg-E. Saglio, *Dictionnaire des antiquités grecques et romaines*, vol. 1 (Paris 1877).

J.K. Davies, *Athenian Propertied Families 600-300 BC* (Oxford 1971).

L. De Blois, "The Perception of Politics in Plutarch's Roman 'Lives'", *ANRW* 2.33.6 (1992) 4568-4623.

H. Delbrück, *Geschichte der Kriegskunst in Rahmen der politischen Geschichte*, vol. 1: *Das Altertum*, (Berlin 1908), transl. by W.J. Renfroe, *Warfare in Antiquity*, *History of the Art of War*, vol. 1 (Lincoln, London 1975).

J.D. Denniston, *The Greek Particles* (Oxford 1981).

R. Develin, *Athenian Officials 648-321 BC* (Cambridge 1989).

C.J. De Vogel, *Pythagoras and Early Pythagoreanism* (Assen 1966).

J.G. De Voto, *Agesilaos II and the Politics of Sparta, 404-377 B.C.*, Ph.D. diss. Loyola University of Chicago 1982.

— — —, "Agesilaos in Boiotia in 378 and 377 B.C.", *AHB* 1 (1987) 75-82.

——— , "The Liberation of Thebes in 379/8 B.C.", in R. Sutton, ed., *Daidalikon: Studies in Memory of R.V. Schoder* (Wauconda 1989) 101-116.

——— , "The Theban Sacred Band", in *Warfare in Antiquity* 23.2 (1992) 3-19.

L. Digregorio, "Plutarco e la tragedia greca", *Prometheus* 2 (1976) 151-174.

——— , "Lettura diretta e utilizzazione di fonti nelle citazioni plutarchee dei tre grandi tragici I", *Aevum* 53 (1979) 11-50; Id., "Lettura diretta e utilizzazione di fonti nelle citazioni plutarchee dei tre grandi tragici II", *Aevum* 54 (1980) 46-79.

J. Dillon, *The Middle Platonists (80 B.C. to A.D. 220)* (Ithaca 1977).

G. Dipersia, "La nuova popolazione di Messene al tempo di Epaminonda", *Contributi dall'Istituto di Storia Antica,* Università Cattolica del Sacro Cuore Milano, vol. 2 (1974) 54-61.

P.L. Donini, "Plutarco, Ammonio e l'Academia", in *Miscellanea Plutarchea, Quaderni del Giornale Filologico Ferrarese* 8 (1986a) 97-110.

——— , "Lo scetticismo academico, Aristotele e l'unità della tradizione platonica secondo Plutarco", in G. Cambiano, ed., *Storiografia e Dossografia nella Filosofia Antica* (Turin 1986b) 203-226.

K.J. Dover, "ΔΕΚΑΤΟΣ ΑΥΤΟΣ", *JHS* 80 (1960) 61-77; repr. in K.J. Dover, *The Greeks and their Legacy* (Oxford 1988) 159-180.

K. Dowden, *Death and the Maiden. Girls' Initiation Rites in Greek Mythology* (London, New York 1989).

R. Drews, "Diodorus and his Sources", *AJP* 83 (1962) 383-392.

——— , "Ephorus and History Written κατὰ γένος", *AJP* 84 (1963) 244-255.

M. Dubuisson, *OI AMFI TINA-OI PERI TINA, L'Évolution des Sens et des Emplois*, Mémoire: Université de Liège (Liège 1976-1977).

J. Ducat, *Les Kouroi du Ptoïon* (Paris 1971).

S. Dusanic, "L'Académie de Platon et la Paix Commune de 371 av. J.-C.", *REG* 92 (1979) 319-347.

——— , "Le médisme d' Isménias et les relations Gréco-Perses dans la politique de l'Académie Platonicienne (383-378 av. J.-C.", in P.

Roesch, ed., *Colloques internationaux du CNRS "La Béotie antique"*, Lyon-Saint Étienne 16-20 Mai 1983 (Paris 1985) 227-235.

H. Erbse, "Plutarchs Schrift *Peri deisidaimonias*", *Hermes* 70 (1952) 295-314.

— — —, "Die Bedeutung der Synkrisis in den Parallelbiographien Plutarchs", *Hermes* 84 (1956) 398-424.

W. Fellmann, *Antigonos Gonatas, König der Makedonen, und die Griechischen Staaten*, Diss. Würzburg 1930.

M.R. Fernández, "Los silencios de Jenofonte en el *Agesilao* de Plutarco", *Habis* 15 (1984) 41-70.

C. Ferone, "A proposito di Plutarco, *Pelopidas* 7, 1-2 (*PHeid.* 209)", in Mario Capasso, ed., *Papiri Letterari Greci e Latin*, Dipartimento di Filologia Classica e Medioevale, Papyrologica Lupiensia 1 (Università degli Studi de Lecce 1992) 275-279.

A. Ferrill, *The Origins of War* (London 1985).

R. Flacelière, É. Chambry, M. Juneaux, *Plutarque. Vies, Thésée/Romulus — Lycurgue/Numa*, vol. 1 (Paris 1957).

— — —, *Plutarque.Vies, Timoléon/Paul-Émile, Pélopidas/Marcellus*, vol. 4 (Paris 1966).

R. Flacelière and J. Irigoin, "Plutarque dans ses *Oeuvres Morales*", in *Plutarque: Oeuvres Morales*, vol. 1.1 (Paris 1987) vii-ccxxvi.

R. Flacelière, "Plutarque et la Pythie", *REG* (1943) 72-111.

— — —, "Plutarque et l'épicurisme", in *Epicurea in memoriam Hectoris Bignone*, Pubblicazioni dall'Istituto di Filologia Classica (Genova 1959) 197-215.

M.A Flower, *Theopompus of Chios. History and Rhetoric in the Fourth Century BC* (Oxford 1994).

J. Fontenrose, *Python. A Study of Delphic Myth and its Origins* (Berkeley, Los Angeles 1959).

— — —, *The Delphic Oracle. Its Responses and Operations with a Catalogue of Responses.* (Berkeley, Los Angeles 1978).

W.G. Forrest, *A History of Sparta 950-152 BC* (New York 1968).

M. Fortina, *Epaminonda* (Turin 1958).

J.-C. Fraisse, *Philia. La notion d' amitié dans la philosophie antique* (Paris 1974).

G. François, *Le polythéisme et l'emploi au singulier des mots θεός, δαίμων dans la littérature grecque d' Homère à Platon* (Paris 1957).

J.G. Frazer, *Pausanias's Description of Greece*, vol. 5 (London 1898).

F. Frazier, "Remarques à propos de l'usage des citations en matière de chronologie dans les *Vies*", *ICS* 13.2 (1988a) 297-309.

— — — , "À propos de la 'philotimia' dans les 'Vies': quelques jalons dans l'histoire d'une notion", *RPh* 62 (1988b) 109-127.

C. Froidefond, "Plutarque et le platonisme", *ANRW* 2.36.1 (1987) 184-233.

F.J. Frost *Plutarch's Themistocles. A Historical Commentary* (Princeton 1980).

S. Fuscagni, "Callistene di Olinto e la 'Vita di Pelopida' di Plutarco", in M. Sordi, ed., *Contributi dall'Istituto di Storia Antica, Storiografia e Propaganda*, Pubblicazioni della Università Cattolica di Milano, 3 (1975) 31-55.

— — — , "Le Beotarchie di Pelopida e il numero dei Beotarchi dopo la liberazione della Cadmea del 379", *RIL* 106 (1972) 415-433.

I. Gallo, *Frammenti biografici da papiri*, 2 vols. (Rome 1975/1980).

— — — , ed., *Aspetti dello Stoicismo e dell'Epicureismo in Plutarco, Quaderni del Giornale Filologico Ferrarese* 9 (Ferrara 1988).

H.J. Gehrke, *Stasis. Untersuchungen zu den inneren Kriegen in den griechischen Staaten des 5. und 4. Jahrhunderts v. Chr.*, in *Vestigia. Beiträge zur Alten Geschichte*, vol. 35 (Munich 1985).

J. Geiger, "Zum Bilde Julius Caesars in der Römischen Kaiserzeit", *Historia* 24 (1975) 444-453.

— — — , "Plutarch's Parallel *Lives*: The Choice of the Heroes", *Hermes* 109 (1981) 85-104; repr. in B. Scardigli, ed., *Essays on Plutarch's Lives*, (Oxford 1995) 165-190.

— — — , *Cornelius Nepos and Ancient Political Biography, Historia Einzelschriften* 47 (Stuttgart 1985).

— — — , "Nepos and Plutarch. From Latin to Greek Political Biography", *ICS* 13.2 (1988) 245-256.

A. Georgiadou, "The *Lives* of the Caesars and Plutarch's Other *Lives*", *ICS* 13.2 (1988) 349-356.

— — —, "A Note on Claviger's emendation in Plutarch's Text (*Pel.* 18.7)", *Teiresias Supplement* 3 (1990) 81-82.

— — —, "Bias and Character-portrayal in Plutarch's *Lives of Pelopidas and Marcellus*", *ANRW* 2.33.6 (1992a) 4222-4257.

— — —, "The Corruption of Geometry and the Problem of Two Mean Proportionals", in I Gallo, ed., *Plutarco e le Scienze* (Genova 1992b) 147-164.

— — —, "Vita Activa and Vita Contemplativa. Plutarch's *De Genio* and Euripides' *Antiope*", in I. Gallo and B. Scardigli, eds., *Teoria e Prassi politica nelle opere di Plutarco* (Naples 1995) 187-200.

— — —, "Epameinondas and the Socratic Paradigm in the *De Genio Socratis*", in *Plutarchea Lovaniensia, Studia Hellenistica* 32 (1996) 113-122.

— — —, "Pro-Boiotian Traditions in the Fourth Century BC. Kallisthenes and Ephoros as Plutarch's Sources in the *Pelopidas*", *Boeotia Antiqua* 6 (1996); forthcoming.

C. Gill, "The Question of Character-Development. Plutarch and Tacitus", *CQ* 33 (1983) 469-481.

S. Stern-Gillet, *Aristotle's Philosophy of Friendship*, Suny Series in Ancient Greek Philosophy (Albany, New York 1995).

A.W. Gomme, *A Historical Commentary on Thucydides*, vol. 2 (Oxford 1956; repr. 1962).

J.P. González, "El surgimento de una facción democrática Tebana (382-379 AC), *Faventia* 8.2 (1986) 69-83.

— — —, "La politica Tebana de 379 a 371 a.C.", *Studia Zamorensia* 7 (1986) 383-409.

— — —, "Les facciones politicas Tebanas en el periodo de la formacion de la hegemonia (379-371 a.C.)", *Polis* 3 (1991a) 121-135.

— — —, "La contribución de Plutarco al conocimiento de la topografía de Beocia en el siglo IV a.C.", in J.G. López and E.C. Dorda, eds., *Estudios sobre Plutarco: paisaje y naturaleza*, Actas del II Simposio Espanol sobre Plutarco, Murcia 1990 (Madrid 1991b) 89-100.

A.J. Gossage, "Plutarch", in T.A. Dorey, ed., *Latin Biography* (London 1967) 45-77.

V.J. Gray, "The Value of Diodorus Siculus for the Years 411-386 B.C.", *Hermes* 115. 1 (1987) 72-89.

— — —, *The Character of Xenophon's Hellenica* (London 1989).

G.T. Griffith, "Athens in the Fourth Century", in P.D.A. Garnsey and C.R. Whittaker, eds., *Imperialism in the Ancient World* (Cambridge 1978) 127-144.

P. Guillon, *Les trépieds du Ptoion* (Paris 1943).

H.M. Hack, *The Rise of Thebes. A Study of Theban Politics and Diplomacy, 386-371 B.C.*, Ph.D. diss. Yale University 1975.

— — —, "Thebes and the Spartan Hegemony, 386-382 B.C.", *AJP* 99 (1978) 210-227.

C.D. Hamilton, *Politics and Diplomacy in the Corinthian War*, Ph.D. diss. Cornell University 1968.

— — —, *Sparta's Bitter Victories, Politics and Diplomacy in the Corinthian War* (Ithaca 1979).

— — —, "The Generalship of King Agesilaus of Sparta", *AW* 8 (1983) 119-127.

— — —, *Agesilaus and the Failure of Spartan Hegemony* (Ithaca 1991).

— — —, "Plutarch's *Life of Agesilaus*", *ANRW* 2.33.6 (1992) 4201-4221.

— — —, "Plutarch and Xenophon on Agesilaus", *The Ancient World* 25 (1994) 205-212.

J.R. Hamilton, *Plutarch. Alexander, A Commentary* (Oxford 1969).

N.G.L. Hammond, "Diodorus' Narrative of the Sacred War", *JHS* 57 (1937) 44-78.

— — —, "The Two Battles of Chaeronea (338 B.C. and 86 B.C.)", *Klio* 31 (1938) 186-218.

— — —, *A History of Greece to 322 B.C.*, 3rd ed. (Oxford 1986).

— — —, *The Macedonian State. Origins, Institutions and History* (Oxford 1989).

N.G.L. Hammond and G.T. Griffith, *A History of Macedonia*, vol. 2 (Oxford 1979).

J. Hani, *Plutarque, Oeuvres Morales*, vol. 8 (Paris 1980).

M.H. Hansen, *Eisangelia: the Sovereignty of the People's Court in Athens in the Fourth Century B.C. and the Impeachment of Generals and Politicians*, ser. Odense University Classical Studies, vol. 6 (Odense 1975).

V. Hanson, "Epameinondas, the Battle of Leuktra (371 B.C.), and the 'Revolution' in Greek Battle Tactics", *Classical Antiquity* 7 (1988) 190-207.

P.R. Hardie, "Plutarch and the Interpretation of Myth", *ANRW* 2.33.6 (1992) 4743-4787.

— — —, "Sign Language in *On the Sign of Socrates*", *Studia Hellenistica* 32 (1996) 123-136.

P. Harding, *From the End of the Peloponnesian War to the Battle of Ipsus* (Cambridge 1985).

A.R.W. Harrison, *The Law of Athens: Procedure*, vol. 2 (Oxford 1971).

M.B. Hatzopoulos, *'Η ὁμηρεία τοῦ Φιλίππου τοῦ 'Αμύντα στὶς Θῆβες, Μνήμη Εὐγενίας Χατζηδάκη* (Athens 1983).

— — —, "La Béotie et la Macédoine à l'époque de l'hégémonie Thébaine: Le point de vue Macédoine", in *Colloques internationaux du CNRS "La Béotie Antique"*, Éditions du CNRS, Lyon-Saint Étienne 16-20 Mai 1983 (Paris 1985) 247-257.

M. Haug, *Die Quellen Plutarchs in den Lebensbeschreibungen der Griechen neu untersucht* (Tübingen 1854).

B.V. Head, *On the Chronological Sequence of the Coins of Boeotia* (Chicago 1974).

W.C. Helmbold and E.N. O' Neil, *Plutarch's Quotations*, The American Philological Association, vol. 19 (Baltimore 1959).

A. Henrichs, "Human Sacrifice in Greek Religion. Three Case Studies", in *Le sacrifice dans l'antiquité, Fondation Hardt pour l' étude de l'antiquité classique, Entretiens* 27 (1980) 195-235.

K. Herbert, *Ephorus in Plutarch's Lives: A Source Problem*, Ph.D. diss. Harvard University 1954.

— — —, "The Identity of Plutarch's Lost Scipio", *AJP* 78.1 (1957) 83-88.

G. Herman, *Ritualised Friendship and the Greek City* (Cambridge, New York 1987).

J.P. Hershbell, "Plutarch's Portrait of Socrates", *ICS* 13.2 (1988) 365-381.

— — —, "Plutarch and Stoicism", *ANRW* 2.36.5 (1992a) 3336-3352.

— — —, "Plutarch and Epicureanism", *ANRW* 2.36.5 (1992b) 3353-3383.

W.E. Higgins, *Xenophon the Athenian* (Albany 1977).

C.A. Hignett, *History of the Athenian Constitution* (Oxford 1952).

R. Hirzel, *Der Dialog*, 2 vols. (Leipzig 1895).

R.F. Hock and E.N. O'Neil, *The Ancient Chreia in Ancient Rhetoric*, vol. 1: *The Progymnasmata*, Graeco-Roman Religion Ser. 9 (Atlanta 1986).

H.A. Holden, *Plutarch's Life of Pericles* (London 1894); repr. with no date.

S. Hornblower,, *The Greek World 479-323 BC* (London, New York 1991).

D. Hughes, *Human Sacrifice in Ancient Greece* (London, New York 1991).

J. Irigoin, "Histoire du texte des *Oeuvres Morales* de Plutarque", in *Plutarque: Oeuvres Morales*, vol. 1.1 (Paris 1987) ccxxvii-cccx.

M. Jehne, *Koine Eirene. Untersuchungen zu den Befriedungs- und Stabilisierungsbemühungen in der griechischen Poliswelt des 4. Jahrhunderts v. Chr.*, *Hermes Einzelschriften* 63 (Stuttgart 1994).

C.P. Jones, "Towards a Chronology of Plutarch's Works", *JRS* 56 (1966a) 61-74; repr. in B. Scardigli, ed., *Essays on Plutarch's Lives* (Oxford 1995) 95-123.

— — —, "The Teacher of Plutarch", *HSCP* 71 (1966b) 205-213.

— — —, *Plutarch and Rome* (Oxford 1971).

R.M. Jones, *The Platonism of Plutarch* (Menasha, Wisconsin 1916); repr. in L. Taran, ed., ser. Ancient Philosophy (New York, London 1980).

W. Judeich, "Athen und Theben vom Königsfrieden bis zur Schlacht bei Leuktra", *RhM* 76 (1927) 171-197.

L. Kallet, "Iphikrates, Timotheos, and Athens, 371-360 B.C.", *GRBS* 24 (1983) 239-252.

R.M. Kallet-Marx, "Athens, Thebes, and the Foundation of the Second Athenian League", *Classical Antiquity* 4.1 (1985) 127-151.

E. Kearns, *The Heroes of Attica, BICS Supplement* 57, University of London, 1989.

A.G. Keen, "Were the Boiotian Poleis Autonomoi?", in M.H. Hansen and K. Raaflaub, eds., *More Studies in the Ancient Greek Polis*, Papers from the Copenhagen Polis Centre 3, *Historia Einzelschriften* 108 (1996) 113-125.

N.M. Kennell, *The Gymnasium of Virtue, Education and Culture in Ancient Sparta* (Chapel Hill, London 1995).

A. Keramopoulos, "Θηβαϊκά", *Deltion* 3 (1917) 1-503.

C. Kerenyi, *Eleusis: Archetypal Image of Mother and Daughter*, transl. by R. Manheim, Bollingen Series LXV, vol. 4 (New York 1967).

J. Knauss, *Die Melioration des Kopaisbeckens durch die Minyer im 2. Jt. v. Chr., Kopais 2* (Munich 1987).

U. Köhler, "Über Boiotische Inschriften aus der thebanischen Zeit", *Hermes* 24 (1889) 636-643.

D. Konstan, *Friendship in the Classical World*, ser. Key Themes in Ancient History (Cambridge 1997).

Σ.Ν. Κουμανούδης, *Θηβαϊκὴ Προσωπογραφία, Βιβλιοθήκη τῆς ἐν Ἀθήναις* Ἀρχαιολογικῆς Ἑταιρείας, vol. 90 (Athens 1979).

C. Kraay, *Archaic and Classical Greek Coins* (Berkeley, Los Angeles 1976).

P. Krentz, *The Thirty at Athens* (Ithaca, London 1982).

— — —, "The Nature of Hoplite Battle", *California Studies in Classical Antiquity* 16.1 (1985) 50-61.

J. Kromayer, *Antike Schlachtfelder in Griechenland*, vol. 1 [1903], vol. 2 [1907], vol. 4 [1924-1931] (Berlin).

J. Kromayer and G. Veith, *Heerwesen und Kriegführung der Griechen und Römer*, vol. 4, 3.2 (Munich 1963).

A.J. Kronenberg, "Ad Plutarchi Vitas", *Mnemosyne* 5 (1937) 303-314.

R. Kühner and B. Gerth, *Ausführliche Grammatik der griechischen Sprache*[3] (Hannover 1890-1904).

L. Lacroix, "Le bouclier, emblème des Béotiens", *Révue Belge de Philologie et d'Histoire* 36 (1958) 5-30.

E. Lanzilotta, "La politica spartana dopo la pace di Antalcida", *Miscellanea Greca e Romana*, Studi pubblicati dall'istituto italiano per la storia antica, fasc. 31, 7 (1980) 129-178.

G. Larfeld, *Sylloge Inscriptionum Boeticarum* (Berlin 1883).

D.H.J. Larmour, "Plutarch's Compositional Methods in the *Theseus* and *Romulus*", *TAPA* 118 (1988) 361-375.

— — —, "Making Parallels: *Synkrisis* and Plutarch's "Themistocles and Camillus', *ANRW* 2.33.6 (1992) 4154-4200.

J.A.O. Larsen, *Representative Government in Greek and Roman History* (Berkeley, Los Angeles 1966; repr. 1976).

— — —, *Greek Federal States* (Oxford 1968).

T. Larsen, *Papyri Graecae Haunienses Instituti Papyrologiae Graecae Universitatis Hauniensis*, vol. 1, *Literarische Texte und Ptolemäische Urkunden* (Copenhagen 1942).

J. Larson, *Greek Heroine Cults* (Madison 1995).

G.M. Lattanzi, *Il 'De Genio Socratis' di Plutarco* (Rome 1933).

B. Latzarus, *Les idées religieuses de Plutarque* (Paris 1920); repr. in L. Tarán, ed., vol. 30, ser. Greek and Roman Philosophy (New York, London 1987).

S. Lauffer, *Kopais 1* (Frankfurt a.M. 1986).

R. Laurenti, *Aristotele. I Frammenti dei Dialoghi*, 2 vols. (Naples 1987).

J.F. Lazenby, *The Spartan Army* (Chicago 1985).

R.P. Legon, "Phliasian Politics and Policy in the Fourth Century BC", *Historia* 16 (1967) 324-337.

P. Lévêque and P. Vidal-Naquet, "Epaminondas Pythagoricien ou le Problème tactique de la Droite et de la Gauche", *Historia* 9 (1960) 294-308.

M.A. Levi, *Plutarco e il V secolo*, Biblioteca storica universitaria, ser. 2, Monografie vol. 5 (Milan 1955).

S. Levin, "The Old Greek Oracles in Decline", *ANRW* 2.18.2 (1989) 1599-1649

E. Lévy, "Autonomia et éleuthéria au Ve siècle", *RPh* 57 (1983) 249-270.

— — —, "L'art de la déformation historique dans les *Hélleniques* de Xénophon", in H. Verdin, G. Schepens and E. De Keyser, eds., *Purposes of History. Studies in Greek Historiography from the 4th to the 2nd centuries B.C., Studia Hellenistica* 30 (Louvain 1990) 125-157.

R. Lonis, *Guerre et religion en Grèce à l'époque classique: Recherches sur les rites, les dieux, l'idéologie de la victoire* (Paris 1979).

— — —, "La guerre en Grèce, quinze années de recherche: 1968-1983", *REG* 98 (1985) 321-379.

N. Loraux, "HBH et ANDREIA: Deux versions de la mort du combattant athénien", *Ancient Society* 6 (1975) 1-31.

— — —, "La belle mort Spartiate", *Ktèma* 2 (1977) 105-120.

— — —, *The Invention of Athens. The Funeral Oration in the Classical City,* transl. by A. Sheridan (Cambridge, Mass., London 1986).

— — —, *Tragic Ways of Killing a Woman* (Cambridge, Mass. 1991).

D. Lührmann, "*De Invidia et Otio* (*Moralia* 536 E-538 E)", in H.D. Betz, ed., *Plutarch's Ethical Writings and Early Christian Literature*, ser. Studia ad Corpus Hellenisticum Novi Testamenti, vol. 4 (Leiden 1978) 363-366.

A. MacDonald, "A Note on the Raid of Sphodrias", *Historia* 21 (1972) 38-44.

D.M. MacDowell, *The Law in Classical Athens* (Ithaca, New York 1978).

— — —, "The *Oikos* in Athenian Law", *CQ* 39 (1989) 10-21.

P.R. Mckechnie and S.J. Kern, *Hellenica Oxyrhynchia* (Warminster, Wiltshire 1988).

J. Mandel, "Jason. The Tyrant of Pherae, Tagus of Thessaly as Reflected in Ancient Sources and Modern Literature. The Image of the New Tyrant", *RSA* 10 (1980) 47-77.

H.I. Marrou, "Les classes d'age de la jeunesse Spartiate", *REA* 48 (1946) 216-230.

— — —, *Histoire de l'éducation dans l'antiquité* (Paris 1965).

H. Martin, "The Concept of Πρᾳότης in Plutarch's *Lives*", *GRBS* 3 (1960) 65-73.

— — — , "The Concept of Philanthropia in Plutarch's *Lives*", *AJP* 82 (1961) 164-175.

V. Martin, "Le traitement de l'histoire diplomatique dans la tradition littéraire du IVème siècle avant J.-C.", *Museum Helveticum* 1 (1944) 13-30.

A. Mastrocinque, "La liberazione di Tebe (379 a.C.) e le origini della storiografia tragica", in A. Mastrocinque, ed., *Omaggio a Piero Treves*, Università di Venezia: Faccoltà di Lettere e Filosofia (Padova 1983) 237-247.

J.F. McGlew, *Tyranny and Political Culture in Ancient Greece* (Ithaca 1993).

P. McKechnie, *Outsiders in the Greek Cities in the Fourth Century BC* (London, New York 1989).

K. Meister, *Historische Kritik bei Polybios, Palingenesia* 9 (Wiesbaden 1975).

J. Mewaldt, "Selbstcitate in den Biographien Plutarchs", *Hermes* 42 (1907) 564-578.

T. Michaelis, *De Ordine Vitarum parallelarum Plutarchi*, Diss. Berlin 1875.

H. Michell, *Sparta* (Cambridge 1964).

C.S.J. Mitchell, *An Analysis of Plutarch's Quotations from Euripides*, Ph.D. diss. University of South California 1968.

J.C. Mitchell, *Epaminondas and Philip II. A Comparative Study of Military Reorganization*, Ph.D. diss. University of Kentucky 1980.

J. Moles, "Nepos and Biography", rev. in *CR* 109.2 (1989) 229-233.

A.D. Momigliano, "Le fonti della Storia greca e macedone nel libro xvi di Diodoro", *RIL* 65 (1932) 523-543.

— — — , *Filippo il Macedone. Saggio sulla storia greca del IV secolo a.C.* (Florence 1934).

— — — , "L'egemonia tebana in Senofonte e Eforo", *Atene e Roma* 38 (1935a) 101-117.

— — — , "La storia di Eforo e le *Elleniche* di Teopompo", *RFIC*, n.s. 13 (1935b) 180-204.

D.J. Mosley, "The Athenian Embassy to Sparta in 371 B.C.", *PCPS*, n.s. 8 (1962) 41-46.

— — — , "Leon and Timagoras: Co-envoys for Four Years?", *GRBS* 9 (1968) 157-160.

— — — , "Diplomacy and Disunion in Ancient Greece", *Phoenix* 25 (1971) 319-330.

— — — , "Theban Diplomacy in 371 B.C.", *REG* 85 (1972a) 312-318.

— — — , "Timagoras' Bed-makers", *CR*, n.s. 22 (1972b) 12.

— — — , *Envoys and Diplomacy in Ancient Greece* (Wiesbaden 1973).

— — — , "Xenophon and Iphicrates", *RSA* 4 (1974) 65-68.

M.H. Munn, "Agesilaos' Boiotian Campaigns and the Theban Stockade of 378-377 B.C.", *Classical Antiquity* 6 (1987) 106-138.

— — — , *The Defense of Attica. The Dema Wall and the Boiotian War of 378-5 B.C.* (Berkeley, Los Angeles 1993).

A.G. Nicolaides, "Is Plutarch Fair to Nicias?", *ICS* 13.2 (1988) 319-333.

M.P. Nilsson, *Griechische Feste von religiöser Bedeutung, mit Ausschluß der attischen* (Leipzig 1906; repr. Stuttgart 1957).

— — — , *Geschichte der griechischen Religion*, 3rd ed., 2 vols. (Munich 1967-1974).

K. Nipperdey, *Cornelius Nepos* (Dublin, Zurich 1967).

J. Ober, *Athenian Reactions to Military Pressure and the Defense of Attica, 404-322 B.C.*, Ph.D. diss. University of Michigan 1980.

E.A.M.E. O'Connor-Visser, *Aspects of Human Sacrifice in the Tragedies of Euripides* (Amsterdam 1987).

E.A. O'Donnell, *The Transferred Use of Theatre Terms as a Feature of Plutarch's Style*, Ph.D. diss. University of Pennsylvania, Philadelphia 1975.

P. Oliva, *Sparta and her Social Problems* (Prague 1971).

A.T Olmstead, *History of the Persian Empire* (Chicago 1948).

E.N. O'Neil, "De Cupiditate Divitiarum (Moralia 523 C-528 B)", in H.D. Betz, ed., *Plutarch's Ethical Writings and Early Christian Literature*, ser. Studi ad Corpus Hellenisticum Novi Testamenti, vol. 4 (Leiden 1978) 289-362.

J. Opsomer, "Divination and Academic 'Scepticism' according to Plutarch", *Studia Hellenistica* 32 (1996) 165-194.

M.J. Osborne and S.G. Byrne, eds., *A Lexicon of Greek Personal Names* (P.M. Fraser and E. Matthews, gen. eds.); vol. 2: *Attica*, (Oxford 1994).

H.W. Parke, "Herippidas, Harmost at Thebes", *CQ* 21 (1927) 159-165.

— — — , "The Development of the Second Spartan Empire", *JHS* 50 (1930) 37-79.

— — — , "The Evidence for Harmosts in Laconia", *Hermathena* 46 (1931) 31-38.

— — — , *Greek Merecenary Soldiers from the Earliest Times to the Battle of Ipsus* (Oxford 1933).

R. Parker, *Miasma: Pollution and Purification in Early Greek Religion* (Oxford 1983).

P. Pédech, "La date de la bataille de Leuctres (371 av. J.C)", *RSA* 2 (1972) 1-6.

— — — , *Historiens compagnons d' Alexandre: Callisthène, Onésicrite, Néarque, Ptolémée, Aristobule* (Paris 1984).

C.B.R. Pelling, "Plutarch's Method of Work in the Roman *Lives*", *JHS* 99 (1979) 74-96; repr. in B. Scardigli, ed., *Essays on Plutarch's Lives* (Oxford 1995) 265-318.

— — — , "Plutarch's Adaptation of his Source Material", *JHS* 100 (1980) 127-139; repr. in B. Scardigli, ed., *Essays on Plutarch's Lives* (Oxford 1995) 125-154.

— — — , "Synkrisis in Plutarch's Lives", *Giornale Filologico Ferrarese* 8 (1986) 83-96.

— — — , "Aspects of Plutarch's Characterisation", *ICS* 13.2 (1988a) 257-274.

— — — , *Plutarch. Life of Antony* (Cambridge 1988b).

— — — , "Plutarch: Roman Heroes and Greek Culture", in M. Griffin and J. Barnes, eds., *Philosophia Togata. Essays on Philosophy and Roman Society* (Oxford 1989) 199-232.

— — — , "Truth and Fiction in Plutarch's *Lives*", in D.A. Russell, ed., *Antonine Literature* (Oxford 1990) 19-52.

— — — , "The Moralism of Plutarch's *Lives*", in D. Innes, H. Hine and Ch. Pelling, eds., *Ethics and Rhetoric. Classical Essays for*

Donald Russell on his Seventy-Fifth Birthday (Oxford 1995) 205-220.

— — — , "Is Death the End? Closure in Plutarch's *Lives*", in F. Dunn, D. Fowler, and D. Roberts, eds., *Classical Closure. Endings in Ancient Literature* (Princeton) forthcoming.

L. Peper, *De Plutarchi Epaminonda*, Diss. Jena, 1912.

S. Perlman, "The Politicians in the Athenian Democracy of the Fourth Century B.C.", *Athenaeum* 41 (1963) 327-355.

— — — , "On Bribing Athenian Ambassadors", *GRBS* 17 (1976) 223-233.

A. Podlecki, "Plutarch and Athens", *ICS* 13.2 (1988) 231-243.

— — — and S. Duane, "A Survey of Work on Plutarch's Greek *Lives*, 1951-1988", *ANRW* 2.33.6 (1992) 4053-4127.

P. Poralla, *Prosopographie der Lakedaimonier*, (Breslau 1913); repr. 1985, with introduction and addenda by A.S. Bradford.

J. Pouilloux, *Fouilles de Delphes*, vol. 3, Épigraphie, fasc. 4^4, École Française d'Athènes (Paris 1976).

L. Prandi, *Callistene di Olinto* (Milan 1985).

A.W. Price, *Love and Friendship in Plato and Aristotle* (Oxford 1989).

W.K. Pritchett, "Observations on Chaeronea", *AJA* 61 (1958) 307-311.

— — — , *Studies in Ancient Greek Topography*, part. I, University of California Publications: Classical Studies 1 (Berkeley, Los Angeles 1965).

— — — , *Studies in Ancient Greek Topography*, part II: *Battlefields*, University of California Publications: Classical Studies 4 (Berkeley, Los Angeles 1969).

— — — , *Studies in Ancient Greek Topography*, part IV: *Passes*, University of California Publications: Classical Studies 28 (Berkeley, Los Angeles 1982).

— — — , *The Greek State at War*, part II (Berkeley, Los Angeles 1974).

— — — , *The Greek State at War*, part III: *Religion* (Berkeley, Los Angeles 1979).

— — — , *The Greek State at War*, part IV (Berkeley, Los Angeles 1985).

G. Proietti. *Xenophon's Sparta. An Introduction, Mnemosyne Supplement* 98 (Leiden 1987).

B. Puech, "Prosopographie des amis de Plutarque", *ANRW* 2.33.6 (1992) 4831-4893.

G. Queck, *De Fontibus Plutarchi in 'Vita Pelopidae'*, Diss. Dramburg 1876.

P.A. Rahe, "The Annihilation of the Sacred Band at Chaeroneia", *AJA* 84 (1980) 84-87.

C. Rehdantz, *Vitae Iphicratis, Chabriae, Timothei Atheniensium* (Berlin 1844).

C.I. Reid, *Diodorus and his Sources*, Ph.D. diss. Harvard University 1969.

A. Rescigno, *L'Eclissi degli Oracoli* (Naples 1995).

P.J. Rhodes, "On Labelling 4th-Century Politicians", *LCM* 3 (1978) 207-211.

— — — , *A Commentary on the Aristotelian Athenaion Politeia* (Oxford 1981).

D.G. Rice, *Why Sparta Failed: A Study of Politics and Policy from the Peace of Antalcidas to the Battle of Leuctra, 386-371 B.C.*, Ph.D. diss. Yale University 1971.

— — — , "Agesilaus, Agesipolis, and Spartan Politics, 386-379 B.C.", *Historia* 23.2 (1974) 164-182.

— — — , "Xenophon, Diodorus and the Year 379/378 B.C. Reconstruction and Reappraisal", *YCS* 24 (1975) 95-130.

J.-C. Riedinger, *Étude sur les Hélleniques, Xénophon et l'histoire* (Paris 1991).

M. Riley, "The Purpose and the Unity of Plutarch's *De Genio Socratis*", *GRBS* 18 (1977) 257-273.

R.W. Roberts, *The Ancient Boeotians: Their Character and Culture, and their Reputation* (Cambridge 1895).

R.W. Roberts and B.V. Head, *The Ancient Boeotians and the Coinage of Boeotia* (Chicago 1974). Reprint of R.W. Roberts, *The Ancient Boeotians: Their Character and Culture and Their Reputation* (Cambridge 1895) and B.V. Head, *The Coins of Ancient Boeotia. A Chronological Sequence* (London 1891).

C.A. Roebuck, *A History of Messenia from 369 to 146 B.C.* (Chicago 1941).

P. Roesch, *Thespies et la confédération béotienne* (Paris 1965).

— — —, "Les lois fédérales Béotiennes", *Teiresias Supplement* 1 (1972) 61-70.

— — —, *Études Béotiennes* (Paris 1982).

— — —, "Un décret inédit de la Ligue Thébaine et la flotte d' Épaminondas", *REG* 97 (1984) 45-60.

J. Roisman, "Kallikratidas–A Greek Patriot?", *CJ* 83.1 (1987) 21-33.

J. de Romilly, *La douceur dans la pensée grecque* (Paris 1979).

H.J. Rose, "La notion du divin depuis Homère jusqu' à Platon", in *Fondation Hardt, Entretiens sur l'antiquité classique* 1 (1952) 3-43.

V. Rose, *Aristotelis qui ferebantur librorum fragmenta* (Stuttgart 1968).

T.G. Rosenmeyer, "Beginnings in Plutarch's *Lives*", *YCS* 29 (1992) 205-230.

W.D. Ross, *Aristotelis Fragmenta Selecta* (Oxford 1955).

J. Roy, "Arcadia and Boeotia in Peloponnesian Affairs, 370-362 B.C.", *Historia* 20 (1971) 569-599.

— — —, "Thebes in the 360s B.C.", in *CAH*, 2nd. ed., vol. 6 (1994) 187-208.

S. Tromp de Ruiter, "De vocis quae est φιλανθρωπία significatione atque usu", *Mnemosyne* 59 (1932) 271-306.

D.A. Russell, "On Reading Plutarch's *Lives*", *G&R* 13 (1966) 139-154; repr. in B. Scardigli, ed., *Essays on Plutarch's Lives* (Oxford 1995) 75-94.

— — —, *Plutarch* (New York 1973).

T.T.B. Ryder, "The Supposed Common Peace of 366/5 B.C.", *CQ* 51 (1957) 199-205.

— — —, "Athenian Foreign Policy and the Peace Conference at Sparta in 371 B.C.", *CQ* 56 (1963) 237-241.

— — —, *Koine Eirene* (London, New York, Toronto 1965).

K.S. Sacks, *Diodorus Siculus and the First Century* (Princeton 1990).

P. Salmon, "L'armée fédérale des Béotiens", *Antiquité Classique* 22 (1953) 347-360.

— — — , *Étude sur la confédération béotienne (447/6-386)*, Académie Royale de Belgique, Mémoires de la Classe des Lettres, vol. 63, fasc. 3 (Brussels 1978).

— — — , "Le rôle des Béotarques dans la confédération béotienne", in Éd. Frezous and A. Jacquemin, eds., *Les Rélations Internationales*, Actes du Colloque de Strasbourg 15-17 juin 1993, Université des Sciences Humaines de Strasbourg, Travaux du Centre de Recherche sur le Proche-Orient et la Grèce Antique 13 (Leiden 1995) 365-383.

A.E. Samuel, *Greek and Roman Chronology, Calendars and Years in Classical Antiquity* (Munich 1972).

F.H. Sandbach, *Plutarch. Moralia, Fragments*, vol. 15, The Loeb Classical Library (London, Cambridge, Mass. 1969).

— — — , "Plutarch and Aristotle", *ICS* 7 (1982) 207-232.

L.J. Sanders, *Dionysius I of Syracuse and Greek Tyranny* (London, New York, Sydney 1987).

D. Sansone, "Plutarch, Alexander and the Discovery of Naphtha", *GRBS* 21 (1980) 63-74.

— — — , "Notes on Plutarch: Pericles and Fabius", *ICS* 13.2 (1988) 311-318.

— — — , *Plutarch. The Lives of Aristeides and Cato* (Warminster, Wiltshire 1989).

H. Gallet de Santerre, *Délos primitive et archaïque* (Paris 1958).

B. Scardigli, "Scritti Recenti sulle *Vite* di Plutarco (1974-1986)", *Quaderni del Giornale Filologico Ferrarese* 8, *Miscellanea Plutarchea* (Ferrara 1986) 7-59.

— — — , "Introduction", in B. Scardigli, ed., *Essays on Plutarch's Lives* (Oxford 1995) 1-46.

A. Schachter, *Cults of Boiotia*, vol. 1: "Acheloos to Hera", *BICS Supplement* 38.1, University of London 1981.

— — — , vol. 2: "Herakles to Poseidon", *BICS Supplement* 38.2, University of London 1986.

— — — , vol. 3: "Potnia to Zeus", *BICS Supplement* 38.3, University of London 1994.

A. Schaefer, *Die Berichte Xenophons, Plutarchs und Diodors über die Besetzung und Befreiung Thebens (382, 379 v. Chr.)*, Diss. Munich 1930.

G. Schneeweiss, "Gegenstand und Absicht in den Biographien Plutarchs", *Festschrift für Franz Egermann zu seinem 80. Geburstag am 13. Februar 1985* (Munich 1985) 147-162.

S. Schröder, *Plutarchs Schrift De Pythiae Oraculis. Text, Einleitung und Kommentar* (Stuttgart 1990).

P.M. Schuhl, *Aristote. Fragments et Témoignages*, Bibl. de la Fac. des Lettres et Sciences Humaines de Paris, Sorbonne, Sér. Textes et Documents 17 (Paris 1968).

E. Schwartz, "Kallisthenes' *Hellenika*", *Hermes* 35 (1900) 106-130.

E. Schwyzer, *Griechische Grammatik auf den Grundlage von Karl Brugmanns Griechischer Grammatik*, vol. 2 (Munich 1975).

R. Seager, "Thrasybulus, Conon and Athenian Imperialism, 396-386 B.C.", *JHS* 87 (1967) 95-115.

— — —, "The King's Peace and the Balance of Power in Greece, 386-362 B.C.", *Athenaeum*, n.s. 52 (1974) 36-63.

— — —, "The King's Peace and the Second Athenian Confederacy", in *CAH*, 2nd ed., vol. 6 (1994) 156-186.

R. Sealey, "Callistratos of Aphidna and his Contemporaries", *Historia* 5 (1956) 178-203.

— — —, *A History of the Greek City States* (Berkeley, Los Angeles 1976).

J. Seibert, *Die politischen Flüchtlinge und Verbannten in der griechischen Geschichte*, 2 vols., Impulse der Forschung 30 (Darmstadt 1979).

C. Seltman, *Greek Coins*^{2} (London 1955).

B. Sergent, "La date de la bataille de Leuctres et celle de la fête des gymnopaidiai", *Rivista Storica dell'Antichità* 21 (1991) 137-143.

G. Shrimpton, *The Epaminondas Tradition*, Ph.D. diss. Stanford University 1970.

— — —, "Plutarch's Life of Epaminondas", *Pacific Coast Philology* 6 (1971) 55-59.

— — — , "The Theban Supremacy in Fourth-Century Literature", *Phoenix* 25 (1971) 31-318.

— — — , *Theopompus the Historian* (Montréal 1991).

E. Siegmann, *Literarische Griechische Texte der Heidelberger Papyrussamlung* (Heidelberg 1956).

R.K. Sinclair, "Diodorus and the Writing of History", *The Proceedings of the African Classical Associations* 6 (1963) 36-45.

E. Sjôqvist, "Lysippus" in *Lectures in Memory of Louise Taft Semple*, 2nd. ser., 1966-1970 (Norman 1973) 19-33.

R.E. Smith, "The Opposition to Agesilaus' Foreign Policy", *Historia* 2 (1953-54) 274-288.

M. Sordi, "La pace di Atene del 371/370", *RFIC* 79 (1951) 34-64.

— — — , *La lega tessala*, Studi pubblicati dall'istituto italiano per la storia antica (Rome 1958).

— — — , *Diodori Siculi Bibliothecae Liber Sextus Decimus* (Florence 1969).

— — — , "La restaurazione della lega Beotica nel 379-8 A.C.", *Athenaeum*, n.s. 51 fasc. I-II (1973) 79-91.

— — — , "Il soggiorno di Filippo a Tebe nella propaganda storiografica", *Contributi dall'Istituto di storia antica*, Università Cattolica del Sacro Cuore di Milano 3 (1975) 56-64.

— — — , "Pelopida da Tegira a Leuttra" in H. Beister - J. Buckler, eds., *BOIOTIKA. Vorträge vom 5. Internationalen Böotien-Kolloquium zu Ehren von Professor Dr. Siegfried Lauffer* (Munich 1989) 123-130.

— — — , "Tendenze storiographiche e realtà storica nella liberazione della Cadmea in Plut. *Pel.* 5-13", in I. Gallo and B. Sacrdigli, eds., *Teoria e Prassi politica nelle opere di Plutarco* (Naples 1995) 415-423.

G. Soury, "Plutarque, prêtre de Delphes. l'inspiration prophétique" *REG* 54 (1941) 50-69.

— — — , *La démonologie de Plutarque* (Paris 1942).

P.A. Stadter, *Plutarch's Historical Methods. An Analysis of the Mulierum Virtutes* (Cambridge, Mass. 1965).

— — — , "Plutarch's Comparison of Pericles and Fabius Maximus", *GRBS* 16 (1975) 77-85; repr. in B. Scardigli, ed., *Essays on Plutarch's Lives* (Oxford 1995) 155-164.

— — — , "The Rhetoric of Plutarch's *Pericles*", *Ancient Society* 18 (1987) 251-269.

— — — , "The Proems of Plutarch's *Lives*", *ICS* 13.2 (1988) 275-295.

— — — , *A Commentary on Plutarch's Pericles* (Chapel Hill, London 1989).

— — — , "Subject to the Erotic: Male Sexual Behaviour in Plutarch", in D. Innes, H. Hines and C.B.R. Pelling, eds., *Ethics and Rhetoric: Classical Essays for Donald Russell on his Seventy-fifth Birthday*, (Oxford 1995) 221-236.

C. Starr, *Political Intelligence in Classical Greece*, *Mnemosyne Supplement* 31 (Leiden 1974).

E. v. Stern, *Geschichte der spartanischen und thebanischen Hegemonie vom Königsfrieden bis zur Schlacht bei Mantineia*, Diss. Dorpat 1884.

— — — , *Xenophons Hellenika und die böotische Geschichtsüberlieferung* (Dorpat 1887).

C. Stoltz, *Zur relativen Chronologie der Parallelbiographien Plutarchs*, Diss. Lund 1929.

B.S. Strauss, "The Cultural Significance of Bribery and Embezzlement in Athenian Politics: the Evidence of the Period 403-386 B.C.", *The Ancient World* 11 (1985) 67-74.

— — — , *Athens after the Peloponnesian War. Class, Faction and policy 403-386 BC* (Ithaca 1987).

— — — , "Oikos/Polis: Towards a Theory of Athenian Paternal Ideology 450-399 B.C.", in W.R. Connor, M.H. Hansen, K.A. Raaflaub, B.S. Strauss, eds., *Aspects of Athenian Democracy*, *Classica et Mediaevalia Dissertationes* 11 (Copenhagen 1990).

K.F. Stroheker, *Dionysios I* (Wiesbaden 1958).

R. Syme, "Biographers of the Caesars", *Museum Helveticum* 37 (1980) 104-128.

S. Symeonoglou, *The Topography of Thebes from the Bronze Age to Modern Times* (Princeton 1985).

S. Swain, *Plutarch and Rome: Three Studies* , D. Phil. thesis Oxford 1987.

— — — , "Plutarch's *Philopoemen* and *Flamininus*", *ICS* 13.2 (1988) 335-347.

— — — , "Character Change in Plutarch", *Phoenix* 43.1 (1989) 62-68.

— — — , "Hellenic Culture and Roman Heroes of Plutarch", *JHS* 110 (1990) 126-145; repr. in B. Scardigli, ed., *Essays on Plutarch's Lives* (Oxford 1995) 228-264.

H. Swoboda, "Zur Geschichte des Epameinondas", *RhM* NS 55 (1900) 460-475.

A.M. Tagliasacchi, "Plutarco e la tragedia greca", *Dioniso* 34 (1960) 24-142.

W. Tarn, W. *Antigonos Gonatas* (Chicago 1969).

— — — , "The battles of Cos and Andros", *JHS* 29 (1909) 264-285.

S.T. Teodorsson, A *Commentary on Plutarch's Table Talks, vol. 2 (Books 4-6)*, Studia Graeca et Latina Gothoburgensia 53 (Göteborg 1990).

C. Theander, *Plutarch und die Geschichte,* Bulletin de la Societé Royale des Lettres de Lund, 1950-51, fasc. 1 (Lund 1951) 1-86.

— — — , "Zur Zeitfolge der Biographien Plutarchs", *Eranos* 56 (1958) 12-20.

W.E. Thompson, "Isocrates on the Peace Treaties" *CQ* 33 (1983) 75-80.

— — — , "Plutarchs Forschungen in Rom", *Eranos* 57 (1959) 99-131.

E.N. Tigerstedt, *The Legend of Sparta in Classical Antiquity*, 3 vols. (Uppsala 1965-1978).

F. Titchener, "Critical Trends in Plutarch's Roman *Lives*, 1975-1990", *ANRW* 2.33.6 (1992) 4128-4153.

M.N. Tod, *A Selection of Greek Historical Inscriptions*, vol. 2 (Oxford 1948; repr. 1950, 1962).

A. Toynbee, *Some Problems of Greek History* (London, New York, Toronto 1969).

T. Tracy, "Perfect Friendship in Aristotle's *Nicomachean Ethics*", *ICS* 4 (1979) 65-75.

J.S. Traill, *Persons of Ancient Athens*, vols. 1-3 (1994-1995).

C.J. Tuplin, "The Athenian Embassy to Sparta 372/1", *LCM* 2 (1977) 51-56.

— — —, "Pausanias and Plutarch's *Epameinondas*", *CQ*, n.s. 34.2 (1984) 346-358.

— — —, "Military Engagements in Xenophon's *Hellenica*", in I.S. Moxon, J.D. Smart, A.J. Woodman, eds., *Past Perspectives. Studies in Greek and Roman Historical Writing* (Cambridge 1986) 37-66.

— — —, "The Leuctra Campaign. Some Outstanding Problems", *Klio* 69 (1987) 72-107.

— — —, *The Failings of Empire, Historia Einzelschriften* 76 (Stuttgart 1993)

H.N. Ulrichs, *Reisen und Forschungen in Griechenland*, 2 vols. (Bremen 1840-1863).

R. Urban, *Der Königsfrieden von 387/86 v. Chr.*, in *Historia Einzelschriften* 68 (Stuttgart 1991).

S. Usher, *Isocrates' Panegyricus and to Nicocles* (Warminster 1990).

M. Garcia Valdés, ed., *Estudios sobre Plutarco: ideas religiosas*, Actas del III Simposio Internacional Sobre Plutarco, 30 April-2 May 1992 (Madrid 1994).

E. Valgiglio, *Divinità e religione in Plutarco* (Genova 1988).

— — —, "Dagli 'Ethica" ai 'Bioi' in Plutarco", *ANRW* 2.33.6 (1992) 3963-4051.

M. Van der Valk, "Notes on the Composition and Arrangement of the Biographies of Plutarch", *Studi in Onore di Aristide Colonna*, Istituto di Filologia Classica (Università degli Studi di Perugia 1982) 301-337.

F. Vater, "Leben des Pelopidas", *Neue Jahrbücher für Philologie und Pädagogik*, ser. 2, Suppl. 8 (1842) 325-382.

G. Verbeke, "Plutarch and the Development of Aristotle", in I. Düring and G.L. Owen, eds., *Aristotle and Plato in the Mid-Fourth Century* (Göteborg 1960) 236-247.

J.-P. Vernant, *Mortals and Immortals, Collected Essays* (Princeton 1991).

H.S. Versnel, "Self-Sacrifice, Compensation and the Anonymous Gods", in *Le sacrifice dans l'antiquité, Fondation Hardt pour l'étude de l'antiquité classique, Entretiens* 27 (1980) 135-185.

— — — , "The Festival for Bona Dea and the Thesmophoria", *G&R* 39.1 (1992) 31-55.

F. Vian, *La guerre des Géants* (Paris 1952).

— — — , *Les origines de Thèbes* (Paris 1963).

— — — , "Epameinondas" in *Kleine Schriften* (Leipzig 1877) 272-307.

P. Walcot, *Envy and the Greeks. A Study of Human Behaviour* (Warminster, Wiltshire 1978).

A. Wardman, "Plutarch's Methods in the *Lives*", *CQ* 21 (1971) 254-261.

— — — , *Plutarch's Lives* (Berkeley, Los Angeles 1974).

H.D. Westlake, *Thessaly in the Fourth Century* (London 1935).

— — — , "The Sources of Plutarch's *Pelopidas*", *CQ* 33 (1939) 11-22.

— — — , "Xenophon and Epaminondas", *GRBS* 16.1 (1975) 23-40.

J. Wickersham and G. Verbrugghe, *Greek Historical Documents. The Fourth Century* (Toronto 1973).

J. Wickersham, *Hegemony and Greek Historians* (Maryland, London 1994).

U.v. Wilamowitz-Moellendorf, "Plutarch als Biograph", in *Reden und Vorträge*, vol. 2 (Berlin 1926) 247-279; repr. in one vol. (Zurich 1967); transl. as "Plutarch as Biographer", in B. Scardigli, ed., *Essays on Plutarch's Lives* (Oxford 1995) 47-74.

U. Wilcken, "Über Entstehung und Zweck des Königsfriedens", *Abhandlung der Preussischen Akademie, Phil.-hist. Klasse*, no. 15 (Berlin 1941) 3-20.

J. Wilkins, "The State and the Individual: Euripides' Plays of Voluntary Self-sacrifice", in A. Powell, ed., *Euripides, Women, and Sexuality* (London, New York 1990).

J. Wiseman, "Epaminondas and the Theban Invasions", *Klio* 51 (1969) 177-199.

J. Wolter, "Die Schlacht bei Leuktra", in J. Kromayer, vol. 4, *Antike Schlachtfelder in Griechenland* (Berlin 1924-1931) 290-316.

I. Worthington, *A Historical Commentary on Dinarchus. Rhetoric and Conspiracy in Late Fourth-Century Athens* (Ann Arbor 1992).

D.W. Wyttenbach, *Lexicon Plutarcheum*, 2 vols. (Oxford 1830; repr. Hildesheim 1962).

K. Ziegler, "Plutarchstudien", *RhM*, n.s. 83 (1934) 211-250.

— — —, *Plutarchos von Chaironeia* (Stuttgart 1964); repr. from *RE* 21 (1949).

— — —, *Plutarch, Grosse Griechen und Römer*, vol. 3 (Zurich, Stuttgart 1955).